Semigroups for Delay Equations

Research Notes in Mathematics

Volume 10

Semigroups for Delay Equations

András Bátkai
Loránd Eötvös University Budapest

Susanna Piazzera
University of Tübingen

CRC Press
Taylor & Francis Group
Boca Raton London New York

CRC Press is an imprint of the
Taylor & Francis Group, an **informa** business

AN A K PETERS BOOK

First published 2005 by A K Peters, Ltd.

Published 2019 by CRC Press
Taylor & Francis Group
6000 Broken Sound Parkway NW, Suite 300
Boca Raton, FL 33487-2742

First issued in paperback 2019

No claim to original U.S. Government works

ISBN 13: 978-0-367-45416-6 (pbk)
ISBN 13: 978-1-56881-243-4 (hbk)

Visit the Taylor & Francis Web site at
http://www.taylorandfrancis.com

and the CRC Press Web site at
http://www.crcpress.com

Library of Congress Cataloging-in-Publication Data

Bátkai, András, 1972-
    Semigroups for delay equations / András Bátkai, Susanna Piazzera.
        p. cm.-- (Research notes in mathematics ; v. 10)
    Includes bibliographical references and index.
    ISBN 1-56881-243-4
    1. Semigroups. 2. Delay differential equations. I. Piazzera, S. (Susanna). II. Title. III. Research notes in mathematics (Boston, Mass.) ; 10.

QA182.B38 2005
512'.27--dc22

2005048729

Contents

Preface

Partial differential equations with delay arise in many areas of applied mathematics. In most instances physical, chemical, biological, or economical phenomena naturally depend not only on the present state but also on some past occurrences.

Let us show you some of the typical examples of such equations. Consider an age-dependent population equation with delay

$$\begin{cases} \partial_t u(t,a) + \partial_a u(t,a) = -\mu(a)\,u(t,a) + \nu(a)\,u(t-r,a), \\ \qquad\qquad\qquad\qquad\qquad\qquad t \geq 0, \ a \in \mathbb{R}_+, \\ u(t,0) = \int_0^{+\infty} \beta(a)\,u(t,a)\,da, \qquad t \geq 0, \end{cases}$$

for a positive $r > 0$.

As a further example, you can consider a diffusion process with delayed feedback depending on the flux

$$\partial_t u(t,s) = \Delta u(t,s) + \sum_{i=1}^{N} c_i \partial_i u(t-r_i,s), \quad t \geq 0, \ s \in \Omega,$$

with $c_i \in \mathbb{R}$ and $r_i > 0$, or a diffusion process with distributed delay

$$\partial_t u(t,s) = \Delta u(t,s) + \int_{-r}^{0} d\eta(\sigma)u(t+\sigma,s), \quad t \geq 0, \ s \in \Omega,$$

with η being a function of bounded variation, or a wave equation with delayed damping

$$\partial_t^2 u(t,s) = \Delta u(t,s) + c\,\partial_t u(t-r,s), \quad t \geq 0, \ s \in \Omega,$$

or a parabolic equation with delays in the highest order derivatives

$$\partial_t u(t,s) = \Delta u(t,s) + c\Delta u(t-r,s), \quad t \geq 0, \ s \in \Omega,$$

with $r > 0$ and appropriate initial and boundary conditions where $\Omega \subset \mathbb{R}^N$ is a bounded smooth domain.

In this book, we develop a general and systematic theory of these equations with emphasis on the qualitative behavior and asymptotic properties. The theory of strongly continuous operator semigroups provides a most adequate tool to achieve this goal.

The systematic investigation of delay equations started at the beginning of the twentieth century with the work of V. Volterra [211]. However, the asymptotic behavior of the solutions was not well understood at that time.

The idea that the solutions of delay equations give rise to an operator semigroup on an appropriate function space, called *history* or *phase space*, goes back to N. Krasovskii [124]. The general theory was first formulated by J. Hale [100] and S. N. Shimanov [190] and became established with the publication of Hale's book in 1971 [101]. Subsequently, he, together with S. Verduyn Lunel [103], described the asymptotic properties of the solution in the finite-dimensional case using semigroup theory. Later, G. Webb [212] modified this procedure by first constructing a semigroup and then showing that this semigroup yields the solutions of the delay equation. This is the idea we will systematically employ throughout this book.

To this end, we proceed as follows.

1. We associate to a given delay equation a phase (or history) space $\mathcal{E}$ and an (unbounded) operator $(\mathcal{A}, D(\mathcal{A}))$.

2. We show that the solutions of the abstract Cauchy problem associated to $(\mathcal{A}, D(\mathcal{A}))$ in $\mathcal{E}$ correspond to those of the delay equation in a natural way.

3. We show that $(\mathcal{A}, D(\mathcal{A}))$ in many cases generates a strongly continuous semigroup $(\mathcal{T}(t))_{t \geq 0}$ of linear operators on $\mathcal{E}$. This implies that the delay equation is "well-posed."

4. We study the qualitative and asymptotic behavior of the semigroup $(\mathcal{T}(t))_{t \geq 0}$ and in this way obtain detailed information about the solutions of the delay equation.

We point out that at the very beginning of this approach, one has to choose an appropriate phase space $\mathcal{E}$. This choice is not unique. J. Hale [101] made his investigations in spaces of continuous functions. The same spaces were chosen by O. Diekmann et al. [62], J. Hale, S. Verduyn Lunel [103], J. Wu [222], and many other authors.

Although this choice certainly has its advantages, we follow G. Webb [213] and work in L^p instead. This is more natural and useful for several reasons. First, it allows us to work in a Hilbert space. This is important for possible applications to (optimal) control theory, and also allows us to apply stability criteria such as Gearhart's theorem. In Chapter 9, the discussion of wave equations with delay is based on this choice and contains many new results. Further, the choice of L^p-phase space also allows us to consider equations with discontinuous initial functions.

Structure of the Book

In Part I, we give a short introduction to the theory of strongly continuous semigroups of linear operators on Banach spaces. In particular, we concentrate on the results that we will need in later chapters.

In Part II, we introduce abstract delay equations on a Banach space X and discuss several examples. Then we show the equivalence of the abstract delay equation to an abstract Cauchy problem on the product space $\mathcal{E} := X \times L^p([-1,0], X)$, and prove well-posedness for a large class of delay problems. Moreover, we show how to construct the solutions of the delay equation from the semigroup associated to the equivalent abstract Cauchy problem. Spectral properties are also discussed.

In Part III, we systematically study the qualitative behavior of the solutions (more precisely, of the solution semigroup). In particular, we investigate regularity, stability, positivity, and other properties such as asymptotic almost periodicity. For this purpose, we apply some very recent results from semigroup theory.

In Part IV, we extend the theory to further problems. First, we turn our attention to second-order Cauchy problems with delay. This topic has not been investigated in depth until now. Just as before, we show how to reduce such an equation to an abstract Cauchy problem and then apply very recent tools to investigate the asymptotic behavior of the solutions. Then we consider parabolic problems with delays in the highest order derivatives. We use similar ideas as in Part II and III and investigate the asymptotic behavior of the solutions, which also shows that these methods can be applied to a wide range of problems.

Acknowledgments

The idea of this book arose when the authors did their PhD theses under the supervision of Rainer Nagel. We are indebted to him for his continuous interest and encouragement during the preparation of the manuscript.

The results presented here are mainly based on work by the authors and their co-authors, P. Binding, V. Casarino, A. Dijksma, K.-J. Engel, B. Farkas, E. Fašanga, R. Hryniv, H. Langer, R. Nagel, R. Schnaubelt, and R. Shvidkoy, [11], [12], [13], [14], [17], [19], [18], [29], [20], [153], [166].

We also present related work by S. Boulite, S. Brendle, L. Maniar, A. Rhandi, M. Stein, J. Voigt, and H. Vogt [25, 26, 139, 197] and discuss many new examples. Some of the results appear here for the first time.

Sarah McAllister, Klaus-Jochen Engel, Bálint Farkas, Heinz Langer, Rainer Nagel, Abdelaziz Rhandi, Ulf Schlotterbeck, and Roland Schnaubelt read (parts of) the manuscript, and their suggestions contributed enormously to the improvement of our presentation.

During the years of the preparation of the manuscript, the authors were supported by the Konrad-Adenauer-Stiftung; the OTKA grants Nr. F034840, F049624; the FKFP grant Nr. 0049/2001; the Research Training Networks HPRN-CT-2000-00116 and HPRN-CT-2002-00281 of the European Union; the János Bolyai Research Fellowship of the Hungarian Academy of Sciences; and the Margarete von Wrangell Habilitationstipendium of the Wissenschaftsministerium Baden-Württemberg.

Budapest, Rome, and Tübingen   András Bátkai
April 2005   Susanna Piazzera

Part I

Preliminary Results in Semigroup Theory

Part I
Preliminary Results in Semigroup Theory

Chapter 1

Semigroup Theory

For the convenience of the reader, we present in this and the following chapter a short introduction of those parts of the theory of strongly continuous semigroups that will be needed in the main text. We generally refer to [72] for the proofs.

In Section 1.1 we introduce strongly continuous semigroups, their generator, resolvent and state basic properties as well as the Hille-Yosida generation theorem. In Section 1.2 we show how semigroups naturally occur as solutions of Banach space valued differential equations. In Section 1.3 we introduce special classes of semigroups having more regularity properties than only strong continuity. In Section 1.4 we present a perturbation theory for strongly continuous semigroups. We give proofs only for those results that are not included in some monographs. Experts in semigroup theory can immediately jump to Part II.

1.1 Strongly Continuous Semigroups

Let us fix some notations. From now on, we take X to be a complex Banach space with norm $\| \cdot \|$. We denote by $\mathcal{L}(X)$ the Banach algebra of all bounded linear operators on X endowed with the operator norm, which again is denoted by $\| \cdot \|$. The identity operator on X is denoted by $\mathrm{Id} \in \mathcal{L}(X)$, and $\mathbb{R}_+$ denotes the interval $[0, +\infty)$.

Definition 1.1. A family $(T(t))_{t \geq 0}$ of bounded linear operators on a Banach space X is called a *strongly continuous semigroup* (or C_0-*semigroup*) if the following properties hold:

(i) $T(0) = \mathrm{Id}$.

(ii) $T(t + s) = T(t)T(s)$ for all $t, s \geq 0$.

3

(iii) The *orbit maps* $t \mapsto T(t)x$ are continuous from $\mathbb{R}_+$ into X for every $x \in X$.

From now on, by a *semigroup* we will always mean a strongly continuous semigroup. Sometimes C_0-semigroups are also called linear semidynamical systems.

We will see in Section 1.2 that the orbit maps of semigroups occur as solutions of differential equations in Banach spaces. The key definition for this fact is the following.

Definition 1.2. Let $(T(t))_{t \geq 0}$ be a strongly continuous semigroup on a Banach space X and let $D(A)$ be the subspace of X defined as

$$D(A) := \{x \in X \ : \ \lim_{h \searrow 0} \frac{1}{h}\big(T(h)x - x\big) \ \text{ exists}\}.$$

For every $x \in D(A)$, we define

$$Ax := \lim_{h \searrow 0} \frac{1}{h}\big(T(h)x - x\big).$$

The operator $A : D(A) \subseteq X \to X$ is called the *generator* of the semigroup $(T(t))_{t \geq 0}$.

In the following, we will denote the operator A with domain $D(A)$ by the pair $(A, D(A))$.

Lemma 1.3. *For the generator $(A, D(A))$ of a strongly continuous semigroup $(T(t))_{t \geq 0}$, the following properties hold:*

(i) $A : D(A) \subseteq X \to X$ is a linear operator.

(ii) If $x \in D(A)$, then $T(t)x \in D(A)$ and

$$\frac{d}{dt}T(t)x = T(t)Ax = AT(t)x \quad \text{for all } t \geq 0.$$

As a consequence the orbit map is continuously differentiable.

(iii) For every $t \geq 0$ and $x \in X$, one has

$$\int_0^t T(s)x \, ds \in D(A).$$

(iv) For every $t \geq 0$ the identities

$$T(t)x - x = A \int_0^t T(s)x \, ds \quad \text{if } x \in X,$$

$$= \int_0^t T(s)Ax \, ds \quad \text{if } x \in D(A),$$

hold.

With the help of this lemma, one can show the following theorem.

Theorem 1.4. *Let $(T(t))_{t \geq 0}$ be a strongly continuous semigroup on a Banach space X with generator $(A, D(A))$. Then $(A, D(A))$ is a closed operator and the domain $D(A)$ is dense in X. Moreover, if $(S(t))_{t \geq 0}$ is another strongly continuous semigroup with the same generator $(A, D(A))$, then $S(t) = T(t)$ for all $t \geq 0$.*

As a consequence of the above theorem, there is a one-to-one correspondence between strongly continuous semigroups and their generators. Therefore, we will say that an operator $(A, D(A))$ generates a strongly continuous semigroup $(T(t))_{t \geq 0}$ on a Banach space X if $(A, D(A))$ is the generator of the semigroup $(T(t))_{t \geq 0}$. Since $(A, D(A))$ is a closed operator, $D(A)$ is a Banach space with the graph norm $\| \cdot \|_A$. We will denote this Banach space by X_1, i.e., $X_1 := (D(A), \| \cdot \|_A)$.

Example 1.5. (Uniformly Continuous Semigroups.) Let X be a Banach space and $A \in \mathcal{L}(X)$ a linear bounded operator. Define

$$T(t) := e^{tA} := \sum_{n=0}^{\infty} \frac{(tA)^n}{n!} \quad \text{for } t \geq 0.$$

Then $(T(t))_{t \geq 0}$ is a strongly continuous semigroup with generator (A, X). Actually, the semigroup is uniformly continuous, i.e., the map $t \mapsto T(t)$ is continuous from $\mathbb{R}_+$ to $\mathcal{L}(X)$. Moreover, one can prove that a semigroup is uniformly continuous if and only if its generator is a bounded linear operator.

Example 1.6. (Multiplication Semigroups.) Let Ω be a locally compact metric space, $q : \Omega \to \mathbb{C}$ a continuous function with real part bounded above, that is, $\sup_{\omega \in \Omega} \Re q(\omega) < \infty$. On the Banach space $X := C_0(\Omega)$ of continuous functions that vanish at infinity, define the multiplication operators

$$T(t)f := e^{tq} \cdot f, \quad f \in X \text{ and } t \geq 0.$$

Then the family $(T(t))_{t\geq 0}$ is a strongly continuous semigroup on X, called the multiplication semigroup, and its generator is given by the multiplication operator

$$Af = q \cdot f,$$

with domain $D(A) = \{f \in X \ : \ qf \in X\}$.

An analogous result holds on the spaces $L^p(\Omega, \mu)$, $1 \leq p < \infty$, (Ω, μ) σ-finite measure space, with $q : \Omega \to \mathbb{C}$ a measurable function with real part essentially bounded above.

Example 1.7. (Shift Semigroups.) Let X be one of the following Banach spaces:

- $C_{ub}(\mathbb{R})$ of all bounded, uniformly continuous functions on $\mathbb{R}$ endowed with the supremum norm $\|\cdot\|_\infty$.

- $C_0(\mathbb{R})$ of all continuous functions on $\mathbb{R}$ vanishing at infinity endowed with the supremum norm $\|\cdot\|_\infty$.

- $L^p(\mathbb{R})$, $1 \leq p < \infty$, of all p-integrable functions on $\mathbb{R}$ endowed with the corresponding p-norm $\|\cdot\|_p$.

For $f \in X$ and $t \geq 0$, we call

$$(T_l(t)f)(s) := f(s+t), \quad s \in \mathbb{R},$$

the left shift or translation (of f by t), while

$$(T_r(t)f)(s) := f(s-t), \quad s \in \mathbb{R},$$

is the right shift or translation (of f by t).

The families $(T_l(t))_{t\geq 0}$ and $(T_r(t))_{t\geq 0}$ are strongly continuous semigroups on X with generators

$$A_l f = f' \qquad \text{and} \qquad A_r f = -f',$$

respectively, and domains

$$D(A_l) = D(A_r) = \{f \in X \ : \ f \text{ is differentiable and } f' \in X\},$$

if $X = C_{ub}(\mathbb{R})$ or $C_0(\mathbb{R})$, and

$$D(A_l) = D(A_r) = W^{1,p}(\mathbb{R}),$$

if $X = L^p(\mathbb{R})$.

Up to now we have seen that every strongly continuous semigroup has a generator determining the semigroup uniquely (see Theorem 1.4). We have also seen that generators are closed, densely defined operators. In the following we will show some more properties of strongly continuous semigroups. This should help characterize all closed, densely defined operators that generate semigroups.

One of the important features is the following proposition.

Proposition 1.8. *For every strongly continuous semigroup* $(T(t))_{t\geq 0}$, *there exist constants* $w \in \mathbb{R}$ *and* $M \geq 1$ *such that*

$$\|T(t)\| \leq Me^{wt}$$

for all $t \geq 0$.

By means of Proposition 1.8 we can define a very important constant.

Definition 1.9. Let $(A, D(A))$ be the generator of a strongly continuous semigroup $(T(t))_{t\geq 0}$. We call

$$\omega_0(A) := \inf \left\{ \omega \in \mathbb{R} : \exists M > 0 \text{ such that } \|T(t)\| \leq Me^{\omega t} \ \forall t \geq 0 \right\} \quad (1.1)$$

the semigroup's *growth bound*.

Example 1.10. For the uniformly continuous semigroup of Example 1.5, we have $\omega_0(A) \leq \|A\|$.

For the multiplication semigroups of Example 1.6, we have $\omega_0(A) = \sup_{\omega \in \Omega} \Re q(\omega)$ if $X = C_0(\Omega)$, and $\omega_0(A) = \text{ess-sup}_{\omega \in \Omega} \Re q(\omega)$ if $X = L^p(\Omega, \mu)$, $1 \leq p < \infty$.

For the shift semigroup of Example 1.7, we have $\omega_0(A) = 0$.

We have seen in Theorem 1.4 that the generator $(A, D(A))$ of a strongly continuous semigroup is always a closed operator, therefore by the closed graph theorem, if $(A, D(A))$ is bijective, its inverse becomes a bounded operator on X. This motivates the following definition.

Definition 1.11. Let $(A, D(A))$ be a closed operator on a Banach space X. We call the sets

$$\rho(A) := \{\lambda \in \mathbb{C} : \lambda - A \text{ is bijective}\}$$

the *resolvent set* of A, and

$$\sigma(A) := \mathbb{C} \setminus \rho(A)$$

the *spectrum* of A, respectively. For $\lambda \in \rho(A)$, we call $R(\lambda, A) := (\lambda - A)^{-1}$ the *resolvent* of A at λ.

The following result states that the resolvent of the generator is given by the Laplace transform of the semigroup (at least in a right half plane).

Theorem 1.12. *Let $(A, D(A))$ be the generator of a strongly continuous semigroup $(T(t))_{t \geq 0}$ on a Banach space X. Then the following properties hold:*

(i) *For every $\lambda \in \mathbb{C}$ with $\Re\lambda > \omega_0(A)$, we have $\lambda \in \rho(A)$ and*

$$R(\lambda, A)x = \int_0^\infty e^{-\lambda s} T(s) x \, ds$$

for all $x \in X$.

(ii) *If $\lambda \in \mathbb{C}$ such that $R(\lambda)x := \int_0^\infty e^{-\lambda s} T(s) x \, ds$ exists for all $x \in X$, then $\lambda \in \rho(A)$ and $R(\lambda, A) = R(\lambda)$.*

Now we can state the basic theorem in semigroup theory, which characterizes generators of strongly continuous semigroups by means of their resolvents only.

Theorem 1.13. (Hille, Yosida, Feller, Miyadera, Phillips.) *Let $(A, D(A))$ be a linear operator on a Banach space X and let $w \in \mathbb{R}$, $M \geq 1$ be constants. Then the following properties are equivalent.*

(i) *$(A, D(A))$ generates a strongly continuous semigroup $(T(t))_{t \geq 0}$ satisfying*
$$\|T(t)\| \leq M e^{wt} \qquad \text{for } t \geq 0.$$

(ii) *$(A, D(A))$ is closed, densely defined, and for every $\lambda > w$ one has $\lambda \in \rho(A)$ and*

$$\|[(\lambda - w)R(\lambda, A)]^n\| \leq M \qquad \text{for all } n \in \mathbb{N}.$$

(iii) *$(A, D(A))$ is closed, densely defined, and for every $\lambda \in \mathbb{C}$ with $\Re\lambda > w$ one has $\lambda \in \rho(A)$ and*

$$\|R(\lambda, A)^n\| \leq \frac{M}{(\Re\lambda - w)^n} \qquad \text{for all } n \in \mathbb{N}.$$

The previous result has an important form concerning dissipative operators. We formulate the results only in the Hilbert space setting.

Definition 1.14. Let H be a Hilbert space. The operator $(A, D(A))$ is called *dissipative* if
$$\Re\langle Ax, x\rangle \leq 0$$
for all $x \in D(A)$.

Dissipative operators have an important property needed later in Part IV.

Lemma 1.15. *Let A be a dissipative operator in a Hilbert space H. Then A is closable.*

Semigroup generation via dissipative operators is characterized in the following result.

Theorem 1.16. (Lumer-Phillips.) *Let $(A, D(A))$ be a dissipative operator in the Hilbert space H. The closure $\bar{A}$ of A generates a strongly continuous semigroup if and only if $\mathrm{rg}\,(\lambda - A)$ is dense in H for one $\lambda > 0$.*

The symbol rg denotes the *range* of an operator.

1.2 Abstract Cauchy Problems

The aim of this section is to show how to solve abstract (i.e., Banach space-valued) initial value problems using operator semigroups.

Definition 1.17. Let X be a Banach space, $A : D(A) \subseteq X \to X$ a linear operator, and $x \in X$.

(i) The initial value problem
$$\begin{cases} u'(t) = Au(t) & \text{for } t \geq 0, \\ u(0) = x \end{cases} \qquad \text{(ACP)}$$

is called the *abstract Cauchy problem* associated to $(A, D(A))$ with initial value x.

(ii) A function $u : \mathbb{R}_+ \to X$ is called a (*classical*) *solution* of (ACP) if u is continuously differentiable, $u(t) \in D(A)$ for all $t \geq 0$, and (ACP) holds.

The following proposition follows from Lemma 1.3.

Proposition 1.18. *Let $(A, D(A))$ be the generator of a strongly continuous semigroup $(T(t))_{t \geq 0}$. Then, for every $x \in D(A)$, the function*

$$u : t \mapsto u(t) := T(t)x$$

is the unique classical solution of (ACP) with initial value x.

Actually, there is no hope to have a classical solution of (ACP) if the initial value x is not in $D(A)$. This suggests that more general concepts of "solutions" might be useful.

Definition 1.19. A continuous function $u : \mathbb{R}_+ \to X$ is called a *mild solution* of (ACP) if $\int_0^t u(s) \, ds \in D(A)$ for all $t \geq 0$ and

$$u(t) = x + A \int_0^t u(s) \, ds \qquad \text{for all } t \geq 0.$$

We can now generalize Proposition 1.18 to mild solutions.

Proposition 1.20. *Let $(A, D(A))$ be the generator of a strongly continuous semigroup $(T(t))_{t \geq 0}$. Then, for every $x \in X$, the function*

$$u : t \mapsto u(t) := T(t)x$$

is the unique mild solution of (ACP) with initial value x.

So far, if $(A, D(A))$ generates a semigroup, we have existence and uniqueness of solutions of (ACP). In addition, we can characterize generators of semigroups by means of the associated abstract Cauchy problem.

Definition 1.21. The abstract Cauchy problem

$$\begin{cases} u'(t) = Au(t) & \text{for } t \geq 0, \\ u(0) = x \end{cases} \qquad \text{(ACP)}$$

associated to an operator $A : D(A) \subseteq X \to X$ is called *well-posed* if the domain $D(A)$ is dense in X, for every $x \in D(A)$ there exists a unique classical solution u_x of (ACP), and for every sequence $(x_n)_{n \in \mathbb{N}}$ in $D(A)$ satisfying $\lim_{n \to \infty} x_n = 0$ one has $\lim_{n \to \infty} u_{x_n}(t) = 0$ uniformly for all t in compact intervals $[0, T]$.

We are now ready to state the main result of this section.

Theorem 1.22. *For a closed operator $A : D(A) \subseteq X \to X$, the associated abstract Cauchy problem (ACP) is well-posed if and only if $(A, D(A))$ generates a strongly continuous semigroup on X.*

Therefore, to solve an abstract Cauchy problem means to show that the operator $(A, D(A))$ generates a strongly continuous semigroup. We will pursue this strategy systematically in Part II for delay equations.

1.3   Special Classes of Semigroups

We have seen in Section 1.2 that semigroups are solutions of differential equations. Therefore, to know properties of semigroups means to know properties of the solutions. In this section, we concentrate on regularity properties such as analyticity, compactness, and norm continuity.

1.3.1   Analytic Semigroups, Fractional Powers, and Maximal Regularity

In the following, let X be a Banach space, and for $\theta \in (0, \pi]$ let $\Sigma_\theta \subseteq \mathbb{C}$ be the sector

$$\Sigma_\theta := \{\lambda \in \mathbb{C} \ : \ |\arg\lambda| < \theta\} \setminus \{0\}.$$

Definition 1.23. A family of operators $(T(z))_{z \in \Sigma_\theta \cup \{0\}} \subset X$ is called an *analytic semigroup* (of angle $\theta \in (0, \frac{\pi}{2}]$) if

(i) $T(0) = I$ and $T(z_1 + z_2) = T(z_1)T(z_2)$ for all $z_1, z_2 \in \Sigma_\theta$,

(ii) the map $z \mapsto T(z)$ is analytic on Σ_θ, and

(iii) $\lim_{\Sigma_{\theta'} \ni z \to 0} T(z)x = x$ for all $x \in X$ and $0 < \theta' < \theta$.

If, in addition,

(iv) $\sup_{z \in \Sigma_{\theta'}} \|T(z)\| < \infty$ for every $0 < \theta' < \theta$,

we call $(T(z))_{z \in \Sigma_\theta \cup \{0\}}$ a *bounded analytic semigroup*.

Let $(A, D(A))$ be the generator of an analytic semigroup $(T(t))_{t \geq 0}$. It follows immediately from the differentiability of the semigroup that $T(t)x \in D(A)$ for all $t > 0$ and all $x \in X$. In particular, the orbit maps of the semigroup are continuously differentiable with derivative

$$\frac{d}{dt}T(t)x = AT(t)x \qquad \text{for } t > 0 \text{ and } x \in X,$$

and the operators $AT(t)$ are bounded for $t > 0$.

Generators of analytic semigroups can be characterized by means of their resolvent.

Theorem 1.24. *Let $\theta \in (0, \frac{\pi}{2}]$. For an operator $(A, D(A))$ on a Banach space X the following statements are equivalent:*

(i) $(A, D(A))$ generates a bounded analytic semigroup of angle θ.

(ii) $(A, D(A))$ is sectorial, i.e., closed, densely defined, $\Sigma_{\theta + \frac{\pi}{2}} \subseteq \rho(A)$, and

$$\sup_{\lambda \in \Sigma_{\theta + \frac{\pi}{2} - \varepsilon}} \|\lambda R(\lambda, A)\| < \infty \quad \text{for all } \varepsilon > 0.$$

A very important property of strongly continuous semigroups is that it is possible to define fractional powers of their generators in the following way.

Let $(A, D(A))$ be the generator of a strongly continuous semigroup $(T(t))_{t \geq 0}$ on a Banach space X and let $\delta > \omega_0(A)$. Then $(A - \delta, D(A))$ generates the bounded semigroup $(e^{-\delta t} T(t))_{t \geq 0}$ and $i\mathbb{R} \subset \rho(A - \delta)$. Consequently, for $0 < \theta < 1$ the integral

$$(-A + \delta)^{-\theta} := \frac{\sin \pi \theta}{\pi} \int_0^\infty t^{-\theta} R(t, A - \delta) \, dt \tag{1.2}$$

exists and defines a linear bounded operator $(-A + \delta)^{-\theta} \in \mathcal{L}(X)$.

Lemma 1.25. *For every $0 < \theta < 1$, the operator $(-A + \delta)^{-\theta}$ defined by Equation (1.2) is injective.*

This lemma motivates the following definition.

Definition 1.26. Let $(A, D(A))$ be the generator of a strongly continuous semigroup on a Banach space X and let $\delta > \omega_0(A)$. We define

$$(-A + \delta)^{\theta} := \begin{cases} [(-A + \delta)^{-\theta}]^{-1} & \text{if } 0 < \theta < 1, \\ \text{Id} & \text{if } \theta = 0, \end{cases}$$

with domain

$$D((-A+\delta)^\theta) := \begin{cases} \text{rg}\,((-A + \delta)^{-\theta}) = \text{ the range of } (-A + \delta)^{-\theta} \\ \hspace{5cm} \text{if } 0 < \theta < 1, \\ X \hspace{4.2cm} \text{if } \theta = 0. \end{cases}$$

The operator $(-A + \delta)^{\theta}$ is called a *fractional power*.

Fractional powers have many properties and are very useful in applications of analytic semigroups to partial differential equations. However, we will only need the following properties.

Theorem 1.27. *Let $(A, D(A))$ be the generator of an analytic semigroup and let $\delta > \omega_0(A)$. Then the following properties hold:*

(i) *$D((-A + \delta)^\theta) = D((-A + \nu)^\theta)$ for all $\nu > \omega_0(A)$ and $0 < \theta < 1$.*

(ii) *$T(t) : X \to D((-A + \delta)^\theta)$ for all $t > 0$, and $0 < \theta < 1$.*

(iii) *For every $x \in D((-A + \delta)^\theta)$, we have $T(t)(-A + \delta)^\theta x = (-A + \delta)^\theta T(t)x$.*

(iv) *For every $t > 0$, the operator $(-A + \delta)^\theta T(t)$ is bounded and*

$$\|(-A + \delta)^\theta T(t)\| \leq M_\theta t^{-\theta}$$

for a constant $M_\theta \geq 0$ depending on θ.

The following proposition will be useful in the next parts of this book.

Proposition 1.28. *Let $(A, D(A))$ be the generator of an analytic semigroup and B a closed operator satisfying $D(B) \supseteq D(A^\theta)$, $0 < \theta < 1$. Then there exist constants $C, C_1 \geq 0$ such that*

$$\|Bx\| \leq C\|(-A + \delta)^\theta x\| \qquad \text{for every } x \in D((-A + \delta)^\theta),$$

and

$$\|Bx\| \leq C_1(\rho^\theta\|x\| + \rho^{\theta-1}\|Ax\|) \qquad \text{for every } \rho > 0 \text{ and } x \in D(A).$$

Important examples of analytic semigroups are selfadjoint semigroups in Hilbert spaces, as a consequence of Theorem 2.9 and Equation (2.1). We recall it in a form usually referred to as von Neumann's theorem.

Theorem 1.29. (von Neumann.) *Let H be a separable Hilbert space and let C be a closed, densely defined operator. Then $A := -C^*C$ is a negative semidefinite selfadjoint operator generating an analytic semigroup. One has $D(C) = D\left((-A)^{\frac{1}{2}}\right)$ and $\|Cx\| = \left\|A^{\frac{1}{2}}x\right\|$ for all $x \in D(C)$.*

An important subclass among operators generating analytic semigroups is the one having maximal regularity. The operator A has *maximal regularity of type L^q* if, for some $q \in (1, \infty)$ and $a > 0$ and every $f \in L^q([0, a], X)$, the mild solution $u(t) = \int_0^t T(t - s)f(s)\, ds$ of the inhomogeneous problem

$$u'(t) = Au(t) + f(t), \quad t \geq 0, \qquad u(0) = 0, \tag{1.3}$$

actually belongs to $L^q([0,a], X_1) \cap W^{1,q}([0,a], X)$ and solves Equation (1.3) for a.e. $t \in [0,a]$. If A has this property, then it holds in fact for all $q \in (1,\infty)$ and $a > 0$. It can be seen that sectoriality of angle $\phi > \pi/2$ is a necessary condition for maximal L^q-regularity and that it is also sufficient if X is a Hilbert space. On large classes of Banach spaces (containing $L^s(\Omega)$, $s \neq 2, \infty$, and $C_0(\Omega)$) there do exist generators of analytic C_0-semigroups that do not have maximal L^q-regularity. However, semigroups on $X = L^s(\Omega)$, $1 < s < \infty$, arising from partial differential equations typically possess maximal L^q-regularity. Recently, Weis obtained a breakthrough in this subject [217]. He characterized those operators having maximal regularity of type L^q by a "randomized" version of the notion of sectoriality if X is a UMD-space (e.g., $X = L^s(\Omega)$, $1 < s < \infty$).

Furthermore, we need the real interpolation space $Y = (X, X_1)_{1-1/p,p}$ for some fixed $p \in (1,\infty)$ associated to a sectorial operator A. If X is a Hilbert space and $p = 2$, then Y is again (isomorphic to) a Hilbert space (see Equation (1.4)). It is well known that $X_1 \hookrightarrow Y \hookrightarrow X$ and that

$$x \in Y \iff T(\cdot)x \in L^p_{loc}(\mathbb{R}_+, X_1), \tag{1.4}$$
$$\|x\|_Y \leq c \, \|T(\cdot)x\|_{L^p([0,a], X_1)} \leq c' \, \|x\|_Y ,$$

where the constants may depend on a and p. Since real interpolation spaces appear only in the last chapter in this book and in the context of operator semigroup theory, we consider Equation (1.4) as the definition of Y. Moreover, the parts of A in X_1 and Y (also denoted by A) are again sectorial with the same d and ϕ. The analytic C_0-semigroups generated by the parts of A are just the restrictions of $T(t)$ to Y and X_1 (thus also denoted by $T(t)$). See [2, Theorem V.2.1.3] for these properties. Moreover, the parts of A have the same spectrum as A by [72, Proposition IV.2.17], and hence the spectra of $T(t)$ and its restrictions also coincide due to the Spectral Mapping Theorem. We also need the continuous embedding

$$W^{1,p}([0,a], X) \cap L^p([0,a], X_1) \hookrightarrow C([0,a], Y), \qquad a > 0, \tag{1.5}$$

proved in [2, Theorem III.4.10.2].

1.3.2   Norm Continuous and Compact Semigroups

Besides analyticity, there are other important regularity properties of strongly continuous semigroups. We consider here norm continuity and compactness.

Definition 1.30. Let $(A, D(A))$ be the generator of a strongly continuous semigroup $(T(t))_{t \geq 0}$ on a Banach space X.

(i) We say that $(T(t))_{t\geq 0}$ is *eventually norm continuous*, or *norm continuous for $t > t_0$* if there exists $t_0 \geq 0$ such that the function

$$t \mapsto T(t)$$

is norm continuous from (t_0, ∞) to $\mathcal{L}(X)$. If t_0 can be chosen to be 0, we say that $(T(t))_{t\geq 0}$ is *immediately norm continuous*.

(ii) We say that $(T(t))_{t\geq 0}$ is *eventually compact*, or *compact for $t \geq t_0$*, if there exists $t_0 > 0$ such that the operator $T(t_0)$ is compact. We say that $(T(t))_{t\geq 0}$ is *immediately compact* if $T(t)$ is compact for all $t > 0$.

The class of eventually/immediately compact semigroups is contained in the class of eventually/immediately norm continuous semigroups as shown in the following lemma.

Lemma 1.31. *Let $(T(t))_{t\geq 0}$ be a strongly continuous semigroup with generator $(A, D(A))$. Then the semigroup is compact for $t \geq t_0$ if and only if it is norm continuous for $t \geq t_0$ and the operator $R(\lambda, A)T(t_0)$ is compact for some (and hence all) $\lambda \in \rho(A)$.*

Corollary 1.32. *Let $(T(t))_{t\geq 0}$ be an eventually norm continuous semigroup with generator $(A, D(A))$ having compact resolvent. Then the semigroup $(T(t))_{t\geq 0}$ is eventually compact.*

For immediately compact semigroups, we have the following characterization.

Theorem 1.33. *For a strongly continuous semigroup $(T(t))_{t\geq 0}$, the following properties are equivalent:*

(i) $(T(t))_{t\geq 0}$ is immediately compact.

(ii) $(T(t))_{t\geq 0}$ is immediately norm continuous and its generator has compact resolvent.

We give here a simple, but useful example.

Example 1.34. Let X be a Banach space and $1 \leq p < \infty$. On the Banach space $L^p([-1, 0], X)$ of all Bochner p-integrable functions we consider the left shift semigroup $(T_0(t))_{t\geq 0}$ defined by

$$(T_0(t)f)(\sigma) := \begin{cases} f(\sigma + t) & \text{if } \sigma + t \leq 0, \\ 0 & \text{if } \sigma + t > 0. \end{cases} \tag{1.6}$$

Then $(T_0(t))_{t\geq 0}$ is a strongly continuous semigroup on the Banach space $L^p([-1,0],X)$ with generator

$$A_0 f = \frac{d}{d\sigma} f$$

and domain

$$D(A_0) = \{f \in W^{1,p}([-1,0],X) \ : \ f(0) = 0\},$$

where $\frac{d}{d\sigma}$ denotes the weak derivative on $L^p([-1,0],X)$.

Moreover, the semigroup $(T_0(t))_{t\geq 0}$ is *nilpotent*, i.e., $T_0(t) = 0$ for $t \geq 1$. In particular, it is norm continuous and compact for $t \geq 1$, but its resolvent is compact if and only if the space X is finite-dimensional. This is because the embedding $W^{1,p}([-1,0],X) \hookrightarrow L^p([-1,0],X)$ is compact if and only if X is finite-dimensional.

1.4   Perturbation Theory

Although the generation theorem of Hille-Yosida (see Theorem 1.13) characterizes all generators of strongly continuous semigroups, it is very difficult to verify the estimate of all powers of the resolvent in concrete applications. Thus, one tries to build up the operator (and therefore the semigroup) from simpler ones using perturbation techniques.

In this section, we present three perturbation theorems that will be needed later on in this book.

1.4.1   Bounded Perturbation

We start with bounded perturbations.

Theorem 1.35. *Let $(A, D(A))$ be the generator of a strongly continuous semigroup $(T(t))_{t\geq 0}$ on a Banach space X satisfying*

$$\|T(t)\| \leq Me^{wt} \quad \text{for all } t \geq 0$$

and some $w \in \mathbb{R}$, $M \geq 1$. If $C \in \mathcal{L}(X)$, then

$$D := A + C \qquad \text{with} \qquad D(D) := D(A)$$

generates a strongly continuous semigroup $(U(t))_{t\geq 0}$ satisfying

$$\|U(t)\| \leq Me^{(w+M\|C\|)t} \quad \text{for all } t \geq 0.$$

Moreover, the perturbed semigroup $(U(t))_{t\geq 0}$ satisfies the Variation of Parameters Formulas

$$U(t)x = T(t)x + \int_0^t T(t-s)CU(s)x\,ds$$

and

$$U(t)x = T(t)x + \int_0^t U(s)CT(t-s)x\,ds$$

for every $x \in X$ and $t \geq 0$.

In fact, we not only have an estimate for the perturbed semigroup $(U(t))_{t\geq 0}$, but we can express it by the so-called *Dyson-Phillips Series*.

Theorem 1.36. *The strongly continuous semigroup $(U(t))_{t\geq 0}$ generated by $D := A + C$, where A is the generator of a strongly continuous semigroup $(T(t))_{t\geq 0}$ on a Banach space X and $C \in \mathcal{L}(X)$, can be obtained as*

$$U(t) = \sum_{n=0}^{\infty}(U_n)(t), \tag{1.7}$$

where $U_0(t) = T(t)$ and

$$U_{n+1}(t)x = \int_0^t T(t-s)CU_n(s)x\,ds \tag{1.8}$$

for all $x \in X$ and $t \geq 0$. Moreover, the series in Equation (1.7) converges in the operator norm uniformly for t in compact intervals of $\mathbb{R}_+$.

1.4.2   The Perturbation Theorem of Miyadera and Voigt

Now we quote the Miyadera and Voigt Perturbation Theorem. In this theorem, the perturbation C is bounded only from the domain of the generator $D(A)$ endowed with the graph norm $\|x\|_A := \|x\| + \|Ax\|$. Since a generator $(A, D(A))$ is a closed operator, the space $X_1 = ((D(A), \|\cdot\|_A)$ becomes a Banach space.

Theorem 1.37. *Let $(A, D(A))$ be the generator of a strongly continuous semigroup $(T(t))_{t\geq 0}$ on a Banach space X and let $C \in \mathcal{L}((D(A), \|\cdot\|_A), X)$. Assume that there exist constants $t_0 > 0$, $0 \leq q < 1$ such that*

$$\int_0^{t_0} \|CT(s)x\|\,ds \leq q\|x\| \qquad for\ all\ x \in D(A). \tag{1.9}$$

Then $(A + C, D(A))$ generates a strongly continuous semigroup $(U(t))_{t \geq 0}$ on X, which satisfies the Variation of Parameters Formulas

$$U(t)x = T(t)x + \int_0^t T(t-s)CU(s)x\,ds \qquad and \qquad (1.10)$$

$$U(t)x = T(t)x + \int_0^t U(t-s)CT(s)x\,ds \qquad and \qquad (1.11)$$

$$\int_0^{t_0} \|CU(s)x\|\,ds \leq \frac{q}{1-q}\|x\| \qquad for\ x \in D(A)\ and\ t \geq 0. \qquad (1.12)$$

We note that for the perturbed semigroup $(U(t))_{t \geq 0}$ we again have a Dyson-Phillips Series analogous to the one in the bounded perturbation case (see Equation (1.7)). However, due to the unboundedness of C, we need some more notation to state this series. Let $\mathcal{L}_s(X)$ be the set of all bounded linear operators on the Banach space X endowed with the strong operator topology, i.e., the local convex topology generated by the seminorms

$$p_x(T) := \|Tx\|, \quad T \in \mathcal{L}(X), x \in X$$

(see, e.g., [39, Chapter IV]). Let

$$\mathcal{X} := C(\mathbb{R}_+, \mathcal{L}_s(X))$$

be the space of all strongly continuous functions from $\mathbb{R}_+$ to $\mathcal{L}_s(X)$, and let $V : \mathcal{X} \to \mathcal{X}$ be the abstract Volterra operator defined by

$$(VF)(t)x := \int_0^t F(t-s)CT(s)x\,ds \quad for \quad x \in D(A), t \geq 0, \qquad (1.13)$$

and by

$$(VF)(t)x := \lim_{n \to \infty} (VF)(t)x_n \quad for \quad x \in X, t \geq 0, \qquad (1.14)$$

for $F \in C(\mathbb{R}_+, \mathcal{L}_s(X))$, where $(x_n) \subset D(A)$ is a sequence such that $\lim_{n \to \infty} x_n = x$. Then the semigroup $(U(t))_{t \geq 0}$ generated by $(A+C, D(A))$ in Theorem 1.37 is given by the series

$$U(t)x = \sum_0^\infty (V^n T)(t)x \quad for\ all \quad x \in X, t \geq 0, \qquad (1.15)$$

converging uniformly on compact subsets of $\mathbb{R}_+$.

1.4.3   The Perturbation Theorem of Desch and Schappacher

Finally, we quote a result on multiplicative perturbation proved by Desch and Schappacher.

Theorem 1.38.   *Let* $(A, D(A))$ *be the generator of a strongly continuous semigroup* $(T(t))_{t \geq 0}$ *on a Banach space* X *and let* $C \in \mathcal{L}(X)$. *Moreover, assume that there exist* $t_0 > 0$ *and* $1 \leq p < \infty$ *such that*

$$\int_0^{t_0} T(t_0 - r)Cf(r)\, dr \in D(A) \tag{1.16}$$

for all functions $f \in L^p([0, t_0], X)$. *Then* $(Id + C)A$ *and* $A(Id + C)$ *are generators on* X.

1.5   Regularity Properties of Perturbed Semigroups

In Section 1.3, we introduced regularity properties of strongly continuous semigroups and in Section 1.4, we showed how to obtain new semigroups from known ones. It is therefore natural to ask whether regularity properties persist under perturbations. In general this is not the case, but for some classes of semigroups and/or under additional assumptions this is true. First we consider analytic semigroups.

Definition 1.39.   Let $A : D(A) \subseteq X \to X$ be a linear operator on the Banach space X. A linear operator $C : D(C) \subseteq X \to X$ is called *A-bounded* if $D(A) \subseteq D(C)$ and if there exist constants $a, c \in \mathbb{R}_+$ such that

$$\|Cx\| \leq a\|Ax\| + c\|x\| \tag{1.17}$$

for all $x \in D(A)$. The *A-bound* of C is

$$a_0 := \inf\{a \geq 0 \; : \; \text{there exist } c \geq 0 \text{ such that (1.17) holds}\}.$$

Example 1.40.   If $C \in \mathcal{L}(X)$ is a bounded operator, then it is A-bounded with A-bound $a_0 = 0$.

Using the above definition, we can formulate the following result for analytic semigroups.

Theorem 1.41.   *Let the operator* $(A, D(A))$ *generate an analytic semigroup on a Banach space* X. *Then there exists a constant* $\alpha > 0$ *such that* $(A + C, D(A))$ *generates an analytic semigroup for every* A-*bounded operator* C *having* A-*bound* $a_0 < \alpha$.

In applications this theorem is usually applied if the operator C is relatively A-bounded with A-bound 0. There is an important criterium helping to decide this condition, which is an immediate consequence of Proposition 1.28.

Theorem 1.42. *Let $(A, D(A))$ be the generator of an analytic semigroup and let $\delta > \omega_0(A)$. If $(C, D(C))$ is a closed operator and there is a $0 \leq \vartheta < 1$ such that*

$$D\left((-A + \delta)^\vartheta\right) \subset D(C),$$

then C is A-bounded with A-bound 0.

Example 1.43. On the Hilbert space $H := L^2[0, 1]$, consider the operator

$$A := \frac{d^2}{dx^2}, \qquad D(A) := \{f \in H^2[0, 1] \ : \ f(0) = f(1) = 0\}.$$

Then $(A, D(A))$ generates an analytic semigroup on H (solving the one-dimensional heat equation on $[0, 1]$). Let $C := \frac{d}{dx}$ be the first (weak) derivative, then C is A-bounded with A-bound $a_0 = 0$, because $D((-A)^{\frac{1}{2}}) = H_0^1[0, 1]$. Therefore, $(A + C, D(A))$ generates an analytic semigroup on H.

As a special case, we obtain a bounded perturbation theorem for analytic semigroups.

Corollary 1.44. *Let the operator $(A, D(A))$ generate an analytic semigroup on a Banach space X and $C \in \mathcal{L}(X)$. Then $(A + C, D(A))$ generates an analytic semigroup on X.*

Now we turn our attention to norm continuity and compactness. We will do this for a special class of Miyadera-Voigt perturbations. From now on $(A, D(A))$ is the generator of a strongly continuous semigroup $(T(t))_{t \geq 0}$ on a Banach space X, $X_1^A := (D(A), \| \cdot \|_A)$ and $C \in \mathcal{L}(X_1^A, X)$, i.e., C is relatively bounded by A. The following hypothesis will be used, strengthening slightly the assumptions in Theorem 1.37.

Hypothesis 1.45. *Assume that there exists a function $q : \mathbb{R}_+ \to \mathbb{R}_+$ such that $\lim_{t \searrow 0} q(t) = 0$ and*

$$\int_0^t \|C\, T(s)x\| \, ds \leq q(t) \, \|x\| \tag{1.18}$$

for every $x \in D(A)$ and $t \geq 0$.

By Theorem 1.37, we know that $(A + C, D(A))$ generates a strongly continuous semigroup $(U(t))_{t \geq 0}$ on X given by the Dyson-Phillips Series (1.15), i.e.,

$$U(t)x = \sum_{0}^{\infty} (V^n T)(t)x \quad \text{for all} \quad x \in X, t \geq 0,$$

where V is the abstract Volterra operator defined in Equations (1.13) and (1.14).

Lemma 1.46. *Let $\varepsilon > 0$. If there exists a function $\tilde{q} : (0, \varepsilon) \to \mathbb{R}_+$ such that $\lim_{t \searrow 0} \tilde{q}(t) = 0$ and*

$$\int_0^t \|C\, T(s)x\|\, ds \leq \tilde{q}(t)\, \|x\|$$

for every $x \in D(A)$ and $t \in [0, \varepsilon)$, then Hypothesis 1.45 is satisfied.

Proof. Let $\nu \in (0, \varepsilon)$. For every $t \geq 0$ there exists a natural number $n \in \mathbb{N}$ and a real number $\eta \in [0, \nu)$ such that $t = n\nu + \eta$. Now we have

$$\int_0^t \|CT(s)x\|\, ds = \sum_{k=1}^{n} \int_{(k-1)\nu}^{k\nu} \|CT(s)x\|\, ds + \int_{n\nu}^{n\nu+\eta} \|CT(s)x\|\, ds$$

$$= \sum_{k=1}^{n} \int_0^{\nu} \|CT(s)T((k-1)\nu)x\|\, ds$$

$$+ \int_0^{\eta} \|CT(s)T(n\nu)x\|\, ds$$

$$\leq \sum_{k=1}^{n} \tilde{q}(\nu)\|T((k-1)\nu)\|\, \|x\| + \tilde{q}(\eta)\|T(n\nu)\|\, \|x\|$$

$$= \left(\sum_{k=1}^{n} \tilde{q}(\nu)\|T((k-1)\nu)\| + \tilde{q}(\eta)\|T(n\nu)\| \right) \|x\|$$

for all $x \in D(A)$. Hence, it is enough to choose

$$q(t) := \begin{cases} \tilde{q}(t) & \text{for } t \in (0, \varepsilon), \\ \left(\sum_{k=1}^{n} \tilde{q}(\nu)\|T((k-1)\nu)\| + \tilde{q}(\eta)\|T(n\nu)\| \right) & \text{for } t \geq \varepsilon. \end{cases}$$

$\square$

Now we can prove results on the permanence of eventual norm continuity under perturbations satisfying Condition (1.18).

Proposition 1.47. *If $(T(t))_{t\geq 0}$ is immediately norm continuous and C satisfies Hypothesis 1.45, then the perturbed semigroup $(U(t))_{t\geq 0}$ is also immediately norm continuous.*

Proof. We show by induction that $V^n T$ is immediately norm continuous for every $n \in \mathbb{N}$. For $n = 0$ this is obvious. Let us now suppose that $V^n T$ is immediately norm continuous. Let $t > 0$, $x \in D(A)$, and $h > 0$. We have

$$\|(V^{n+1}T)(t+h)x - (V^{n+1}T)(t)x\|$$

$$\leq \int_t^{t+h} \|V^n T(t+h-s)CT(s)x\|\, ds$$

$$+ \int_0^t \|((V^n T)(t+h-s) - (V^n T)(t-s))CT(s)x\|\, ds$$

$$\leq \int_0^h M\|CT(t+s)x\|\, ds$$

$$+ \int_0^t \|(V^n T)(t+h-s) - (V^n T)(t-s)\|\, \|CT(s)x\|\, ds$$

$$\leq M\, q(h)\|T(t)x\|$$

$$+ q(t) \sup_{s\in[0,t]} \|(V^n T)(t+h-s) - (V^n T)(t-s)\|\, \|x\|$$

$$= \left(M\, q(h)\|T(t)\| + q(t) \sup_{s\in[0,t]} \|(V^n T)(t+h-s) - (V^n T)(t-s)\| \right) \|x\|,$$

where $M := \sup_{s\in[0,h]} \|(V^n T)(s)\|$. Since $D(A)$ is dense in X, the estimate

$$\|(V^{n+1}T)(t+h)x - (V^{n+1}T)(t)x\|$$

$$\leq \left(M\, q(h)\|T(t)\| + q(t) \sup_{s\in[0,t]} \|(V^n T)(t+h-s) - (V^n T)(t-s)\| \right) \|x\|$$

holds for all $x \in X$. Moreover, the right-hand side of the above inequality converges to 0 as $h \searrow 0$ uniformly for $\|x\| \leq 1$ by Hypothesis 1.45 and since $V^n T$ is assumed to be norm continuous. Similarly, one can show that

$$\lim_{h\to 0^-} \|(V^{n+1}T)(t+h)x - (V^{n+1}T)(t)x\| = 0$$

uniformly for $\|x\| \leq 1$ and for all $t > 0$.

We have shown that $V^{(n+1)}T$ is immediately norm continuous. Since the Dyson-Phillips Series converges uniformly on compact intervals, we deduce that $(U(t))_{t\geq 0}$ is immediately norm continuous.   □

Therefore, the result of R. S. Phillips from [165] on the stability of immediate norm continuity is also true for certain unbounded perturbations. The stability of eventual norm continuity is more involved and we need additional assumptions as in [153, Theorem 6.1].

Lemma 1.48. *If $(T(t))_{t\geq 0}$ is norm continuous for $t > \alpha > 0$, then $V^n T$ is norm continuous for $t > (n+1)\alpha$ for every $n \in \mathbb{N}$.*

Proof. The proof is by induction on $n \in \mathbb{N}$. The assertion holds for $n = 0$. Assume now that it is true for some n, i.e., $V^n T$ is norm continuous for $t > (n+1)\alpha$. We show that it also holds for $n+1$.

Take $t > (n+2)\alpha$, $x \in D(A)$, and $1 > h > 0$. We have

$$\|(V^{n+1}T)(t+h)x - (V^{n+1}T)(t)x\|$$

$$\leq \int_t^{t+h} \|(V^n T)(t+h-s)\| \, \|CT(s)x\| \, ds$$

$$+ \left\| \int_0^t ((V^n T)(t+h-s) - (V^n T)(t-s))CT(s)x \, ds \right\|$$

$$\leq K \int_0^h \|CT(s)T(t)x\| \, ds$$

$$+ \left\| \int_0^t ((V^n T)(t+h-s) - (V^n T)(t-s))CT(s)x \, ds \right\|,$$

where $K := \sup_{s \in [0,t+1]} \|(V^n T)(s)\|$. By Condition (1.18), the first integral converges to 0 as $h \searrow 0$ uniformly for $\|x\| \leq 1$. So it remains to estimate the second integral. As in the first step, we have

$$\left\| \int_0^t ((V^n T)(t+h-s) - (V^n T)(t-s))CT(s)x \, ds \right\|$$

$$\leq \int_0^{t-(n+1)\alpha} \|(V^n T)(t+h-s) - (V^n T)(t-s)\| \, \|CT(s)x\| \, ds$$

$$+ \left\| \int_{t-(n+1)\alpha}^t ((V^n T)(t+h-s) - (V^n T)(t-s))CT(s)x \, ds \right\|$$

$$\leq q(t-(n+1)\alpha) \sup_{s \in [0,t-(n+1)\alpha]} \|(V^n T)(t+h-s) - (V^n T)(t-s)\| \, \|x\|$$

$$+ \left\| \int_{t-(n+1)\alpha}^t ((V^n T)(t+h-s) - (V^n T)(t-s))CT(s)x \, ds \right\|.$$

The first term of the above inequality converges to 0 as $h \searrow 0$ uniformly for $\|x\| \leq 1$ by the induction assumption and by Hypothesis 1.45. Now we estimate the second term as

$$\left\| \int_{t-(n+1)\alpha}^{t} ((V^n T)(t+h-s) - (V^n T)(t-s))CT(s)x \, ds \right\|$$

$$= \left\| \int_{t-(n+1)\alpha-h}^{t-h} (V^n T)(t-s)CT(s+h)x \, ds \right.$$

$$\left. - \int_{t-(n+1)\alpha}^{t} (V^n T)(t-s)CT(s)x \, ds \right\|$$

$$\leq \int_{t-(n+1)\alpha-h}^{t-(n+1)\alpha} \|(V^n T)(t-s)CT(s+h)x\| \, ds$$

$$+ \int_{t-(n+1)\alpha}^{t-h} \|(V^n T)(t-s)\, C(T(s+h) - T(s))x\| \, ds$$

$$+ \int_{t-h}^{t} \|(V^n T)(t-s)CT(s)x\| \, ds$$

$$\leq K \int_{0}^{h} \|CT(s)\, T(t-(n+1)\alpha)x\| \, ds$$

$$+ \int_{t-(n+1)\alpha}^{t-h} \|(V^n T)(t-s)\, C(T(s+h) - T(s))x\| \, ds$$

$$+ K \int_{0}^{h} \|CT(s)\, T(t-h)x\| \, ds$$

$$\leq q(h)\, K \, \|T(t-(n+1)\alpha)\| \, \|x\|$$

$$+ \int_{t-(n+1)\alpha}^{t-h} \|(V^n T)(t-s)\, C(T(s+h) - T(s))x\| \, ds$$

$$+ q(h)\, K \, \|T(t-h)\| \, \|x\|.$$

By Condition (1.18), the first and last terms converge to 0 uniformly for $\|x\| \leq 1$ as $h \searrow 0$. For the second term, we have

$$\int_{t-(n+1)\alpha}^{t-h} \|(V^n T)(t-s)\, C(T(s+h) - T(s))x\| \, ds$$

$$\leq K \int_{t-(n+1)\alpha}^{t} \|C(T(s+h) - T(s))x\| \, ds$$

$$= M \int_{0}^{(n+1)\alpha} \|CT(s)(T(t-(n+1)\,\alpha+h) - T(t-n\,\alpha))x\| \, ds$$

$$\leq M\, q((n+1)\alpha)\, \|(T(t-(n+1)\alpha+h) - T(t-(n+1)\alpha))\, x\|.$$

This converges to 0 uniformly for $\|x\| \leq 1$ as $h \searrow 0$ by the norm continuity of the semigroup $(T(t))_{t \geq 0}$ for $t > \alpha$. In an analogous way, one can show the left continuity for $t > 0$. Since $D(A)$ is dense in X, the assertion follows. $\qquad\qquad\qquad\qquad\qquad\qquad\qquad\qquad\qquad\qquad\qquad\qquad\quad$ $\square$

Corollary 1.49. *If $(T(t))_{t \geq 0}$ is norm continuous for $t > \alpha$, C satisfies Condition (1.18) and there exists $n \in \mathbb{N}$ such that $V^n T$ is norm continuous for $t > 0$, then the perturbed semigroup $(U(t))_{t \geq 0}$ is norm continuous for $t > n\alpha$.*

Proof. In the same way as in the proof of Proposition 1.47, we can prove that $V^m T$ is norm continuous for $t > 0$ for every $m \geq n$.

From Proposition 1.48, we have that $V^{(n-1)}T$ is norm continuous for $t > n\alpha$. Since the Dyson-Phillips Series in Equation (1.15) converges uniformly on compact intervals, we have that $(U(t))_{t \geq 0}$ is norm continuous for $t > n\alpha$. $\qquad\qquad\qquad\qquad\qquad\qquad\qquad\qquad\qquad\qquad\qquad\qquad\quad$ $\square$

1.6   Notes and References

In Sections 1.1, 1.2, and 1.3, we followed Engel and Nagel [72, Chapter II]. More references on the theory of strongly continuous semigroups are Belleni-Morante [24], Davies [50], Clément et al. [36], Fattorini [76], Goldstein [87], Hille and Phillips [110], Krein [127], and Pazy [164]. The results on fractional powers are taken from [164, Section 2.6] (see also [198, Section 2.3]). Theorem 1.27 is taken from [198, Lemma 2.3.5]. Proposition 1.28 is taken from [164, Corollary 2.6.11]. Theorem 1.29 can be found in Kato [120, Theorem V.3.24] or Weidmann [214, Theorem 5.39]. The last statement of the theorem is a consequence of the polar decomposition; see Kato [120, Section VI.2.7]. For applications of fractional powers to parabolic problems, see also Lunardi [135]. Maximal regularity is considered in the references Amann [2], Dore [64], Prüß [169], Weis [217], and Weis and Kunstmann [218]. The results on interpolation spaces are taken from the monographs Amann [2] and Lunardi [135].

Section 1.4 is taken from [72, Chapter III]. For the original version of the Miyadera-Voigt Perturbation Theorem, see [145] and [208]. For the original version of the Desch-Schappacher Perturbation Theorem, see [58]. Equivalent results in the context of dual semigroups are due to the Dutch school (see Clement et al. [35, 36] and van Neerven [158] for a systematic treatment of dual semigroups).

The results on perturbation of analytic semigroups of Section 1.5 are taken from [72, Section III.2], while the results on the permanence of norm continuity are taken from [166]. The case of bounded perturbation is studied in [153]. Estimates like Condition (1.18) were introduced in this abstract setting by J. Voigt [210] (see also Brendle, Nagel, and Poland [27] for applications to spectral theory).

Chapter 2

Spectral Theory and Asymptotics of Semigroups

As we have seen in Chapter 1, strongly continuous semigroups arise as solutions of abstract Cauchy problems and hence of (partial) differential equations (even with delay, as we will see in Chapter 3). Therefore, the study of the long-term behaviour of a semigroup is a most important topic. In particular, the characterization of asymptotic properties of a semigroup, such as hyperbolicity and uniform exponential stability, by means of spectral properties of its generator is a very powerful tool.

It is the aim of this chapter to recall some of these results such as the Spectral Mapping Theorem, Gearhart's stability and hyperbolicity theorem, and related extensions like the Weis-Wrobel Theorem on Banach spaces.

2.1   Spectrum, Stability, and Hyperbolicity

We saw in Chapter 1 that the generator of a strongly continuous semigroup is always a closed operator. Therefore, we start this section by recalling some spectral theory results on closed operators. We already defined spectrum, resolvent set, and resolvent of a closed operator $(A, D(A))$ on a Banach space X in Definition 1.11. Now we recall some useful properties.

Proposition 2.1. *Let $(A, D(A))$ be a linear, closed operator on a Banach space X. Then the resolvent set $\rho(A)$ is open in $\mathbb{C}$ (and therefore the spectrum $\sigma(A)$ is closed in $\mathbb{C}$) and the resolvent map $\lambda \mapsto R(\lambda, A)$ is analytic.*

Before introducing the main spectral theoretical notions we use in this book, we need the notion of a Fredholm operator.

Definition 2.2. Let $(T, D(T))$ be a closed operator in the Banach space X. We say that T is *Fredholm* if $\text{rg}\,(T)$ is closed and both $\ker T$ and $X/\text{rg}\,(T)$ are finite-dimensional.

As a next step, we look at the fine structure of the spectrum.

Definition 2.3. Let $(A, D(A))$ be a linear, closed operator on a Banach space X. We define the following sets:

(i) $P\sigma(A) := \{\lambda \in \mathbb{C} \ : \ \lambda - A \text{ is not injective}\}$ the point spectrum of (A). Moreover, each $\lambda \in P\sigma(A)$ is called eigenvalue, and each $0 \neq x \in D(A)$ satisfying $(\lambda - A)x = 0$ is called eigenvector of A (corresponding to λ).

(ii) $A\sigma(A) := \{\lambda \in \mathbb{C} \ : \ \lambda - A \text{ is not injective or } \text{rg}(\lambda - A) \text{ is not closed in } X\}$ the approximate point spectrum of A.

(iii) $R\sigma(A) := \{\lambda \in \mathbb{C} \ : \ \text{rg}(\lambda - A) \text{ is not dense in } X\}$ the residual spectrum of A.

(iv) $\sigma_{\text{ess}}(A) := \{\lambda \in \mathbb{C} \ : \ (\lambda - A) \text{ is not Fredholm}\}$ the essential spectrum[1] of A.

(v) $s(A) := \sup\{\Re\lambda \ : \ \lambda \in \sigma(A)\}$ the spectral bound of A.

The following proposition motivates the name *approximative point spectrum*.

Proposition 2.4. *Let $(A, D(A))$ be a linear, closed operator on a Banach space X. The following properties hold:*

(i) $P\sigma(A) \subset A\sigma(A)$.

(ii) $\sigma(A) = A\sigma(A) \cup R\sigma(A)$.

(iii) For a complex number $\lambda \in \mathbb{C}$, one has $\lambda \in A\sigma(A)$, i.e., λ is an approximate eigenvalue if and only if there exists a sequence $(x_n)_{n\in\mathbb{N}} \subset D(A)$, called an approximate eigenvector, such that $\|x_n\| = 1$ for all $n \in \mathbb{N}$ and $\lim_{n\to\infty} \|Ax_n - \lambda x_n\| = 0$.

The approximate point spectrum has the important property containing the boundary of the spectrum.

---

[1]In some monographs, the notion of essential spectrum may slightly differ from ours, for example, in [120]. In the presentation, we follow [85] and [72].

Theorem 2.5. *Let $(A, D(A))$ be a linear, closed operator on a Banach space X and denote with $\partial\sigma(A)$ the boundary of its spectrum in the complex plane. Then*

$$\partial\sigma(A) \subset A\sigma(A).$$

A very important result is the following.

Theorem 2.6.   (Spectral Mapping Theorem for the Resolvent.) *Let $(A, D(A))$ be a linear, closed operator on a Banach space X with nonempty resolvent set $\rho(A)$. Then the following properties hold.*

(i) $\sigma(R(\lambda_0, A)) \setminus \{0\} = (\lambda_0 - \sigma(A))^{-1} := \{\frac{1}{\lambda_0 - \mu} \ : \ \mu \in \sigma(A)\}$ *for every* $\lambda_0 \in \rho(A)$.

(ii) *Analogous statements hold for the point, approximate point, and residual spectra of A and $R(\lambda_0, A)$.*

As an important consequence of Theorem 2.6 in combination with the Riesz-Schauder Theory for compact operators (see, e.g., Yosida [226, X.5]) one obtains the following corollary.

Corollary 2.7. *If the operator A has compact resolvent, then*

$$\sigma(A) = P\sigma(A).$$

Example 2.8. (Multiplication Operators.) Let Ω be a locally compact metric space, $q : \Omega \to \mathbb{C}$ a continuous function. Define the multiplication operator

$$Af = q \cdot f$$

with domain $D(A) = \{f \in X \ : \ qf \in X\}$ as in Example 1.6. Then one easily sees that A is bounded if and only if q is a bounded function and A is invertible if and only if there exists $\varepsilon > 0$ such that $|q(s)| > \varepsilon$ for all $s \in \Omega$. Combining these, one obtains that for multiplication operators

$$\sigma(A) = \overline{q(\Omega)}.$$

It follows that for a multiplication operator

$$\sigma(A) = A\sigma(A).$$

Furthermore, for $\lambda \in \rho(A)$ the important identity

$$\|R(\lambda, A)\| = \frac{1}{d(\lambda, \sigma(A))} \tag{2.1}$$

is true.

An analogous result holds on the spaces $L^p(\Omega, \mu)$, $1 \le p < \infty$, (Ω, μ) σ-finite measure space, with $q : \Omega \to \mathbb{C}$ a measurable function.

Multiplication operators play an important role because every selfadjoint operator is unitary equivalent to a multiplication operator.

Theorem 2.9. (Spectral Theorem for Normal Operators.) *Let H be a separable Hilbert space and let $(A, D(A))$ be a normal operator, i.e., $A^*A = AA^*$. Then there exists (Ω, μ) σ-finite measure space, $q : \Omega \to \mathbb{C}$ a measurable function, and $U : H \to L^2(\Omega, \mu)$ unitary such that*

$$A = U^{-1} M_q U,$$

where M_q denotes the multiplication operator with the function q. Especially, for normal operators Identity (2.1) is valid.

Now we restrict our attentions to generators of semigroups. If $T \in \mathcal{L}(X)$ is a linear, bounded operator on a Banach space X, we denote its spectral radius by

$$r(T) := \sup\{|\lambda| \; : \; \lambda \in \sigma(T)\}.$$

Proposition 2.10. *For the spectral bound $s(A)$ of a generator A and for the growth bound $\omega_0(A)$ of the generated semigroup $(T(t))_{t \ge 0}$, one has*

$$-\infty \le s(A) \le \omega_0(A) < +\infty \tag{2.2}$$

and

$$\omega_0(A) = \inf_{t \ge 0} \frac{1}{t} \log \|T(t)\| = \lim_{t \to \infty} \frac{1}{t} \log \|T(t)\| = \frac{1}{t_0} \log r(T(t_0))$$

for each $t_0 > 0$. Thus, the spectral radius of the semigroup operator $T(t)$ is given by

$$r(T(t)) = e^{\omega_0(A)t} \qquad \text{for all } t \ge 0.$$

In the applications, one would like to determine the growth bound of a semigroup by the location of the spectrum of its generator. Unfortunately Inequality (2.2) does not prove to be too helpful. In general, the spectral bound of a generator can be strictly smaller than the growth bound of the semigroup, however, if we assume some more regularity on the semigroup we can show equality in Equation (2.2). A possible, rather general criteria is the following Spectral Mapping Theorem.

Theorem 2.11. (Spectral Mapping Theorem.) *Let $(T(t))_{t\geq 0}$ be a strongly continuous semigroup on a Banach space X with generator $(A, D(A))$. Moreover, assume that $(T(t))_{t\geq 0}$ is eventually norm continuous. Then the Spectral Mapping Theorem*

$$\sigma(T(t)) \setminus \{0\} = e^{t\sigma(A)} \qquad \text{holds for all } t \geq 0. \qquad \text{(SMT)}$$

As a corollary, we obtain the desired equality.

Corollary 2.12. *Let $(A, D(A))$ be the generator of an eventually norm continuous semigroup on a Banach space X. Then*

$$s(A) = \omega_0(A).$$

We recall two notions that are important in the analysis of the asymptotic behavior of solutions. In infinite dimensions, in contrast to the finite-dimensional case, one has to distinguish between the case where all solutions decay exponentially and the case where only the classical ones do.

Definition 2.13. We say that a semigroup $T := (T(t))_{t\geq 0}$ with generator $(A, D(A))$ is

(i) *uniformly exponentially stable* if $\omega_0(A) < 0$, and

(ii) *exponentially stable* if there exists $\varepsilon > 0$ such that

$$\lim_{t\to\infty} e^{\varepsilon t} \|T(t)x\| = 0$$

for all $x \in D(A)$.

In other words, using the previous characterization of the growth bound via the spectral radius of the semigroup operators, one obtains the following characterization of uniform exponential stability.

Proposition 2.14. *Let $(A, D(A))$ be the generator of a strongly continuous semigroup $(T(t))_{t \geq 0}$ on a Banach space X. The semigroup is uniformly exponentially stable if and only if for one (and hence all) $t > 0$, the inequality $r(T(t)) < 1$ holds, i.e., $\sigma(T(t)) \subset \mathbb{D}$.*

Here $\mathbb{D} := \{\lambda \in \mathbb{C} : |\lambda| < 1\}$ denotes the unit disk. Applying the spectral mapping theorem for eventually norm continuous semigroups, one arrives at the following significant, Ljapunov-type characterization.

Corollary 2.15. *Let $(A, D(A))$ be the generator of an eventually norm continuous semigroup $(T(t))_{t \geq 0}$ on a Banach space X. Then, the semigroup is uniformly exponentially stable if and only if $s(A) < 0$.*

In applications, especially to nonlinear equations, the so-called "splitting theorems" are also of importance.

Definition 2.16. A semigroup $(T(t))_{t \geq 0}$ in a Banach space X is called *hyperbolic* if X can be written as a direct sum $X = X_s \oplus X_u$ of $(T(t))_{t \geq 0}$-invariant, closed subspaces such that the restricted semigroups $(T_s(t))_{t \geq 0}$ on X_s and $(T_u(t))_{t \geq 0}$ on X_u satisfy the following conditions:

(i) The semigroup $(T_s(t))_{t \geq 0}$ on X_s is uniformly exponentially stable.

(ii) The operators $T_u(t)$ are invertible on X_u and the semigroup $(T_u^{-1}(t))_{t \geq 0}$ is uniformly exponentially stable.

It is easy to see that a strongly continuous semigroup $(T(t))_{t \geq 0}$ is hyperbolic if and only if there exists a projection P and constants $N, \delta > 0$ such that each $T(t)$ commutes with P, satisfies $T(t)\ker P = \ker P$, and

$$\|T(t)x\| \leq Ne^{-\delta t}\|x\| \qquad \text{for } t \geq 0 \text{ and } x \in \text{rg } P,$$

$$\|T(t)x\| \geq \frac{1}{N}e^{+\delta t}\|x\| \qquad \text{for } t \geq 0 \text{ and } x \in \ker P.$$

Therefore, we will often write that a semigroup $(T(t))_{t \geq 0}$ is hyperbolic with projection P and constants $N \geq 1, \delta > 0$. Using this notation, $X_s = \text{rg } P$, $X_u = \ker P$.

Hyperbolicity, just as uniform exponential stability, can be characterized by the location of the spectrum of the semigroup operators.

Proposition 2.17. *Let $(A, D(A))$ be the generator of a strongly continuous semigroup $(T(t))_{t \geq 0}$ on a Banach space X. The semigroup is hyperbolic if and only if for one (and hence all) $t > 0$, one has $\sigma(T(t)) \cap \Gamma = \emptyset$. A projection P yielding the desired decomposition can be obtained as*

$$P = \frac{1}{2\pi i} \oint_{\Gamma} R(\lambda, T(t)) d\lambda.$$

Here Γ denotes the unit circle, i.e., $\Gamma = \{\lambda \in \mathbb{C} : |\lambda| = 1\}$ and the integral is taken in the positive direction. A useful property of hyperbolicity is that it is stable under small perturbation.

Proposition 2.18. *Let $(T(t))_{t \geq 0}$ be a hyperbolic semigroup with projection P and constants N, δ. Assume that $(S(t))_{t \geq 0}$ is a strongly continuous semigroup and that there exists $t_0 > 0$ such that*

$$\|T(t_0) - S(t_0)\| < \frac{(1 - e^{-\delta t_0})^2}{4N^2}. \tag{2.3}$$

Then $(S(t))_{t \geq 0}$ is hyperbolic and for its projection P_S the equalities $\dim \operatorname{rg} P = \dim \operatorname{rg} P_S$, $\dim \ker P = \dim \ker P_S$ hold.

Proof. Take $\lambda \in \mathbb{C}$ with $|\lambda| = 1$. The first we have to show is that $\lambda \in \rho(S(t_0))$. We start doing this by estimating the resolvent of $T(t_0)$. Since

$$(\lambda - T(t_0))x = (\lambda - T_s(t_0))Px \oplus (\lambda - T_u(t_0))(I - P)x$$
$$= \lambda(1 - \lambda^{-1} T_s(t_0))Px \oplus (-T_u(t_0))(1 - \lambda T_u^{-1}(t_0))(I - P)x,$$

we obtain using the notation $Q = I - P$

$$R(\lambda, T(t_0)) = \sum_{n=0}^{\infty} \frac{T_s(nt_0)}{\lambda^{n-1}} \oplus - \sum_{n=1}^{\infty} \lambda^{n-1} (T_u(nt_0))^{-1}.$$

Hence,

$$\|R(\lambda, T(t_0))\| \leq \frac{2N}{1 - e^{-\delta t_0}}.$$

Using the identity

$$\begin{aligned}\lambda - S(t_0) &= \lambda - T(t_0) - (S(t_0) - T(t_0)) \\ &= (\lambda - T(t_0))[I - R(\lambda.T(t_0)(S(t_0) - T(t_0))], \end{aligned}$$

we obtain by Inequality (2.3) that it is invertible and the inverse is given by the Neumann series

$$R(\lambda, S(t_0)) = \left(\sum_{n=0}^{\infty} [R(\lambda, T(t_0)(S(t_0) - T(t_0))]^n \right) R(\lambda, T(t_0)).$$

This shows that the semigroup $(S(t))_{t \geq 0}$ is hyperbolic. It remains to show the assertion on the dimension of the projections. To achieve this goal, it is enough to show that $\|P - P_S\| < 1$ by a well-known statement in elementary functional analysis: see, for example, [85, Lemma II.4.3]. By the series representation of $R(\lambda, S(t_0))$, we obtain that

$$R(\lambda, S(t_0)) - R(\lambda, T(t_0)) = R(\lambda, S(t_0))(S(t_0) - T(t_0))R(\lambda, T(t_0)),$$

and hence

$$\|P - P_S\| \leq \frac{1}{2\pi} \oint_{\Gamma} \|R(\lambda, S(t_0)) - R(\lambda, T(t_0))\| d\lambda < 1$$

by Inequality (2.3). $\square$

Applying the Spectral Mapping Theorem for eventually norm continuous semigroups, one attains an important characterization of hyperbolicity.

Corollary 2.19. *Let $(A, D(A))$ be the generator of an eventually norm continuous semigroup $(T(t))_{t \geq 0}$ on a Banach space X. Then the semigroup is hyperbolic if and only if $\sigma(A) \cap i\mathbb{R} = \emptyset$.*

If one allows more regularity and assumes that the semigroup is compact, as a consequence of standard spectral theory, there is always a natural splitting of the space into three subspaces.

Theorem 2.20. *Assume that $(A, D(A))$ generates a compact semigroup in the Banach space X. Then there exist subspaces X_S, X_U, and X_C, which are invariant under the semigroup such that $X = X_S \oplus X_C \oplus X_U$, $\dim X_C < \infty$, $\dim X_U < \infty$, and*

(i) the semigroup $T_S(t) = T(t)|_{X_S}$ is uniformly exponentially stable,

(ii) the semigroup $T_U(t) = T(t)|_{X_U}$ is invertible and the semigroup $T_U^{-1}(t)$ is uniformly exponentially stable, and

(iii) the semigroup $T_C(t) = T(t)|_{X_C}$ is a group being polynomially bounded in both time directions and hence having growth bound 0 in both directions.

The subspaces X_S, X_U, and X_C are usually referred to as the corresponding *stable*, *unstable*, and *center* manifolds.

Another important property of eventually norm continuous semigroups is that its resolvent has to satisfy the following growth condition.

Theorem 2.21. *Let $(A, D(A))$ be the generator of an eventually norm continuous semigroup $(T(t))_{t\geq 0}$ on a Banach space X. Then, for every $a \in \mathbb{R}$ with $a > \omega_0(A)$, one has*

$$\lim_{r \to \pm\infty} \|R(a + ir, A)\| = 0.$$

As a consequence, the spectrum of the generator is bounded along all imaginary axes.

Theorem 2.22. *Let $(A, D(A))$ be the generator of an eventually norm continuous semigroup $(T(t))_{t\geq 0}$ on a Banach space X. Then, for every $b \in \mathbb{R}$ the set*

$$\{\lambda \in \sigma(A) \ : \ \Re\lambda \geq b\}$$

is bounded.

2.2   Gearhart's Theorem

In this section, we present an important stability theorem for strongly continuous semigroups on Hilbert spaces. The growth bound of a semigroup can be given with the help of the so-called abscissa of uniform boundedness of the resolvent of its generator.

Definition 2.23. Let $(A, D(A))$ be the generator of a strongly continuous semigroup $(T(t))_{t\geq 0}$ on a Banach space X. The abscissa of uniform boundedness of A is defined by

$$s_0(A) := \inf\left\{\omega \in \mathbb{R} \ : \ \{\Re\lambda > \omega\} \subset \varrho(A) \text{ and } \sup_{\Re\lambda > \omega} \|R(\lambda, A)\| < \infty\right\}.$$
$$(2.4)$$

Between the defined quantities for semigroups, the following inequality holds:

$$-\infty \leq s(A) \leq s_0(A) \leq \omega_0(A) < \infty. \qquad (2.5)$$

There are even important examples, not only trivial ones for the analysis of differential equations, where strict inequality holds.

Moreover, $s_0(A)$ never becomes a minimum as shown in the following lemma.

Lemma 2.24. *Let $(A, D(A))$ be the generator of a strongly continuous semigroup on a Banach space X. Let $\alpha \in \mathbb{R}$ such that*

$$\sup_{\Re\lambda > \alpha} \|R(\lambda, A)\| < +\infty.$$

Then there exists $\varepsilon > 0$ such that

$$\sup_{\Re\lambda > \alpha - \varepsilon} \|R(\lambda, A)\| < +\infty.$$

Proof. Let $M := \sup_{\Re\lambda > \alpha} \|R(\lambda, A)\| < +\infty$ and let $0 < \mu < \frac{1}{M}$. Then we have

$$R(\lambda - \mu, A) = R(\lambda, A)\,(1 - \mu\,R(\lambda, A))^{-1}$$
$$= R(\lambda, A) \sum_{n=0}^{\infty} (\mu\,R(\lambda, A))^n \qquad (2.6)$$

for all $\Re\lambda > \alpha$ where the Neumann series converges in $\mathcal{L}(X)$ since $0 < \mu < \frac{1}{M}$. Let $\varepsilon := \frac{1}{2M}$. Then by Equation (2.6) we obtain

$$\|R(\lambda - \varepsilon, A)\| \leq 2\,\|R(\lambda, A)\| \leq 2\,M$$

for all $\Re\lambda > \alpha$, and

$$\sup_{\Re\lambda > \alpha - \varepsilon} \|R(\lambda, A)\| \leq 2\,M < +\infty.$$

$\square$

Corollary 2.25. *For the generator $(A, D(A))$ of a strongly continuous semigroup on a Banach space X its resolvent satisfies*

$$\sup_{\Re\lambda > s_0(A)} \|R(\lambda, A)\| = \infty.$$

We recall Gearhart's Theorem.

Theorem 2.26. (Gearhart.) *Let H be a Hilbert space and let A be a generator of a strongly continuous semigroup on H. The semigroup $(T(t))_{t\geq 0}$ is hyperbolic if and only if $i\mathbb{R} \subset \rho(A)$ and $\sup\{\|R(i\omega, A)\| : \omega \in \mathbb{R}\} < \infty$. Further,*

$$s_0(A) = \omega_0(A). \qquad (2.7)$$

By the definition of uniform exponential stability, we obtain immediately the following characterization.

Corollary 2.27. *Let H be a Hilbert space and let A be a generator of a strongly continuous semigroup $(T(t))_{t\geq 0}$ on H. Then $(T(t))_{t\geq 0}$ is uniformly exponentially stable if and only if $s_0(A) < 0$.*

In order to apply Gearhart's Theorem 2.26 to our delay equations in the following chapters of this book, we will need the following lemma. It is a maximum principle result for bounded holomorphic functions on halfplanes.

Lemma 2.28. *Let X be a Banach space, $\alpha \in \mathbb{R}$ and $\varepsilon > 0$. Let f be a bounded holomorphic function on the right halfplane $\{\lambda \in \mathbb{C} : \Re\lambda > \alpha - \varepsilon\}$ with values in X. Then*

$$\sup_{\Re\lambda \geq \alpha} \|f(\lambda)\| = \sup_{s \in \mathbb{R}} \|f(\alpha + is)\|. \tag{2.8}$$

In particular, $\|f(\lambda)\| \leq \sup_{s \in \mathbb{R}} \|f(\alpha + is)\|$ for all $\Re\lambda \geq \alpha$.

Proof. Without loss of generality, we may assume $\alpha = 0$.

Let $f : \{\lambda \in \mathbb{C} : \Re\lambda > -\varepsilon\} \to X$ be bounded and holomorphic, and let $w \in \mathbb{R}$ such that $w < -\varepsilon$.

Let $g_w(\lambda) := \frac{w}{w-\lambda} f(\lambda)$ for $\Re\lambda > -\varepsilon$. Then g_w is holomorphic on the halfplane $\{\lambda \in \mathbb{C} : \Re\lambda > -\varepsilon\}$ and

$$\lim_{\Re\lambda > -\varepsilon \text{ and } |\lambda| \to \infty} \|g_w(\lambda)\| = 0.$$

It follows from the Maximum Principle that

$$\sup_{\Re\lambda \geq 0} \|g_w(\lambda)\| = \sup_{s \in \mathbb{R}} \|g_w(is)\|. \tag{2.9}$$

Therefore, since $|\frac{w}{w-\lambda}| \leq 1$ for all $\Re\lambda \geq 0$, we have

$$\|g_w(\lambda)\| \leq \|f(\lambda)\| \tag{2.10}$$

for all $\Re\lambda \geq 0$.

Hence, by Equations (2.9) and (2.10), we obtain

$$\|g_w(\lambda)\| \leq \sup_{s \in \mathbb{R}} \|g_w(is)\| \leq \sup_{s \in \mathbb{R}} \|f(is)\|$$

for all $\Re\lambda \geq 0$.

Finally, taking the limit as $w \to -\infty$, we conclude that

$$\|f(\lambda)\| = \lim_{w \to -\infty} \|g_w(\lambda)\| \leq \sup_{s \in \mathbb{R}} \|f(is)\|$$

for all $\Re\lambda \geq 0$. $\qquad\qquad\qquad\qquad\qquad\qquad\qquad\qquad\qquad\qquad\square$

Finally, we close this section with a generalization of Gerahart's Theorem to Banach spaces. Different generalizations using Fourier multipliers can be found in Section 5.2.

We define

$$\omega_\alpha(A) = \inf\{a \in \mathbb{R} : \|T(t)(\gamma - A)^{-\alpha}\| \le M_a e^{at}, \quad t \ge 0\}$$

for $\alpha \in [0,1]$ and some fixed $\gamma > \omega_0(A)$. One has

$$s(A) = \le \omega_1(A) \le \omega_\alpha(A) \le \omega_\beta(A) \le \omega_0(A)$$

for $1 \ge \alpha \ge \beta \ge 0$, and strict inequalities may occur: see, e.g., [159, Examples 1.2.4 and 4.2.9]. Recall that, using our previously introduced terminology, the semigroup generated by $(A, D(A))$ is exponentially stable if and only if $\omega_1(A) < 0$ and it is uniformly exponentially stable if and only if $\omega_0(A) < 0$.

A Banach space X has *Fourier type* $q \in [1,2]$ if the Fourier transform maps $L^q(\mathbb{R}, X)$ continuously into $L^{q'}(\mathbb{R}, X)$, where $1/q + 1/q' = 1$.[2] Clearly, any Banach space has at least Fourier type 1 and Hilbert spaces have Fourier type 2. If X has Fourier type q, then $L^s(\Omega, X)$ also has Fourier type q if $q \le s \le q'$ (Ω is a σ-finite measure space). The *Weis-Wrobel Theorem* now says that

$$\omega_{\frac{1}{q} - \frac{1}{q'}}(A) \le s_0(A) \tag{2.11}$$

if X has Fourier type q. This covers both the stability case in Gearhart's Theorem and the inequality $\omega_1(C) \le s_0(C)$ valid for every Banach space. The number $1/q - 1/q'$ in Equation (2.11) cannot be improved in general, see [159, Example 4.2.9]. However, if (SMT) holds, then all quantities $\omega_\alpha(A)$, $\alpha \ge 0$, $s(A)$, $s_0(A)$ coincide.

2.3   Notes and References

In this chapter, we mainly followed Engel and Nagel [72, Chapter IV and V]. Proposition 2.18 is a special case of Schnaubelt [186, Proposition 2.3].

Equation (2.7) was first proved by L. Gearhart [82] for contraction semigroups in Hilbert spaces. The general form appeared more or less independently in Monauni [146], Prüß [168], Greiner [92], Herbst [107], Howland [113], and F. Huang [114]. One can find generalizations in the literature as well: see, for example, S. Z. Huang [116], Kaashoek and Verduyn Lunel [117], Luo and Feng [136], Latushkin and Shvidkoy [130], and Hieber [109].

---

[2]Here we use the notation $\frac{1}{\infty} = 0$.

Lemma 2.28 is a special case of the Phragmen-Lindelöf Theorem (see Conway [40, Theorem VI.4.1]). The proof in this special case is due to M. Benes.

For general results on the asymptotic behavior of semigroups, we refer to the monograph by van Neerven [159].

Part II
Well-Posedness

Chapter 3

The Delay Semigroup

In this chapter, we present a systematic semigroup approach to linear partial differential equations with delay.

In Section 3.1, we associate an operator $(\mathcal{A}, D(\mathcal{A}))$ on the Banach space $X \times L^p([-1, 0], X)$ to the abstract delay equation on a Banach space X. There is a one-to-one correspondence between the solutions of the abstract delay equation and the solutions of the abstract Cauchy problem corresponding to the associated operator $\mathcal{A}$ (see Theorem 3.12). Thus the question of well-posedness of the delay equation reduces to the question of whether or not the operator $(\mathcal{A}, D(\mathcal{A}))$ generates a strongly continuous semigroup.

Bearing this in mind, we study in Sections 3.2, 3.3, and 3.4 the spectrum and the generator property of this operator. The main idea here is to write $\mathcal{A}$ as a perturbation of a simpler operator and then apply the Miyadera-Voigt Perturbation Theorem to obtain a sufficient condition for $\mathcal{A}$ to be a generator.

3.1   The Semigroup Approach

In this section, we will see that the appropriate setting for linear delay differential equations is that of an abstract Cauchy problem in an appropriate Banach space.

Consider first the ordinary differential equation

$$u'(t) = A\,u(t), \qquad t \geq 0, \tag{3.1}$$

for a matrix $A \in \mathcal{L}(\mathbb{C}^n)$. Everybody who has had a basic Ordinary Differential Equations course knows that the fundamental solution to Equation (3.1) is given by the exponential function $t \mapsto e^{tA}$. More precisely, for every $x \in \mathbb{C}^n$ the function $u(t) := e^{tA}x$ is the unique solution of Equation (3.1) with initial value x.

Let us now modify Equation (3.1) slightly by considering

$$u'(t) = A u(t - \tau), \qquad t \geq 0, \tag{3.2}$$

where $\tau > 0$. Do we still find an exponential function solving this equation? Of course, the function $t \mapsto e^{tA}$ does not work anymore and there is no other matrix $B \in \mathcal{L}(\mathbb{C}^m)$ such that $t \mapsto e^{tB}$ is a fundamental solution. Nevertheless, the answer is still "yes" provided we look at it in the right setting.

To do so we have to change our finite-dimensional viewpoint into an infinite-dimensional one.

Take a Banach space X and consider a function $u : [-\tau, \infty) \to X$. For each $t \geq 0$, we call the function

$$u_t : [-\tau, 0] \ni \sigma \mapsto u(t + \sigma) \in X \tag{3.3}$$

history segment with respect to $t \geq 0$.

The *history function* of u is then the function

$$h_u : t \mapsto u_t \tag{3.4}$$

on $\mathbb{R}_+$.

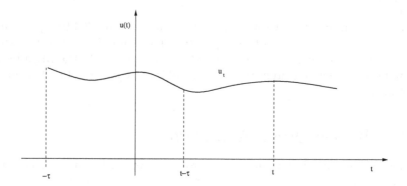

Figure 3.1. The history segment.

A delay differential equation is an equation of the form

$$u'(t) = \frac{d}{dt} u(t) = \varphi(u(t), u_t), \tag{3.5}$$

where $\varphi(\cdot, \cdot)$ is an X-valued mapping. The explanation for this terminology follows.

In many concrete situations (see the examples later), the derivative $u'(t)$ actually depends on $u(t)$ and on $u(t - \tau)$ for some fixed $\tau > 0$ (often with τ normalized to $\tau = 1$) and one has to study differential equations of the form
$$u'(t) = \Psi(u(t), u_t(-\tau)) \tag{3.6}$$
for some function Ψ from $X \times X$ into X. Thus, interpreting t as time, values of u have an effect on u' with a certain delay τ. If we now define φ as
$$\varphi(u(t), u_t) := \Psi(u(t), u_t(-\tau)),$$
we arrive at Equation (3.5).

We will show in this section that this point of view allows us to read delay equations as vector-valued ordinary differential equations.

We will discuss this, throughout this text, in the situation where the mapping φ in Equation (3.5) is the sum of a linear operator B acting on X and a linear operator Φ (the delay operator) acting on a space of functions (the history space) in which the history segments of u must lie. We are now going to make this precise by formulating our standing hypotheses, the notation, and the terminology that we will use throughout the book.

Hypothesis 3.1. (The Standing Hypotheses.) *Assume that*

(H$_1$) X *is a Banach space;*

(H$_2$) $B : D(B) \subseteq X \longrightarrow X$ *is a closed, densely defined, linear operator;*

(H$_3$) Z *is a Banach space such that $D(B) \overset{d}{\hookrightarrow} Z \overset{d}{\hookrightarrow} X$;*[1]

(H$_4$) $1 \leq p < \infty$, $f \in L^p([-1,0], Z)$ *and $x \in X$;*

(H$_5$) $\Phi : W^{1,p}([-1,0], Z) \longrightarrow X$ *is a bounded linear operator, called the delay operator; and*

(H$_6$) $\mathcal{E}_p := X \times L^p([-1,0], Z)$.

Under these hypotheses, and for given elements $x \in X$ and $f \in L^p([-1,0], Z)$, the following initial value problem will be called an (abstract) delay equation (with the history parameter $1 \leq p < \infty$)

$$(\text{DE}_p) \qquad \begin{cases} u'(t) = Bu(t) + \Phi u_t, & t \geq 0, \\ u(0) = x, \\ u_0 = f. \end{cases}$$

---

[1] By $\overset{d}{\hookrightarrow}$ we mean densely, continuously embedded.

Remark 3.2. Since we are mainly interested in the underlying semigroup structure of delay equations and for the sake of simplicity, we assume, throughout this book, for the delay $\tau = 1$. However, all the results of the book can be easily reformulated for a general delay $\tau > 0$.

Here is the natural notion of a classical solution to (DE$_p$).

Definition 3.3. We say that a function $u : [-1, \infty) \longrightarrow X$ is a *classical solution* of (DE$_p$) if

(i) $u \in C([-1, \infty), X) \cap C^1([0, \infty), X)$,

(ii) $u(t) \in D(B)$ and $u_t \in W^{1,p}([-1, 0], Z)$ for all $t \geq 0$,

(iii) u satisfies (DE$_p$) for all $t \geq 0$.

The following is a simple observation that uses a well-known fact on shift semigroups on L^p-spaces. Through its (obvious) corollary, this result will be the key to an interpretation of (DE$_p$) as an abstract Cauchy problem.

Lemma 3.4. *Let $u : [-1, \infty) \to Z$ be a function that belongs to $W^{1,p}_{loc}([-1, \infty), Z)$. Then the history function $h_u : t \mapsto u_t$ of u is continuously differentiable from $\mathbb{R}_+$ into $L^p([-1, 0], Z)$ with derivative*

$$\frac{d}{dt} h_u(t) = \frac{d}{d\sigma} u_t.$$

Proof. Let $(A, D(A))$ be the generator of the left shift group $(T(t))_{t \geq 0}$ on the space $L^p(\mathbb{R}, Z)$, i.e., $D(A) = W^{1,p}(\mathbb{R}, Z)$ and, analogously as in Example 1.7, $A = \frac{d}{d\sigma}$.

Let $t \in \mathbb{R}$ and $T > 1$. We can extend $u_{|[t-T, t+T]}$ to a function $v \in W^{1,p}(\mathbb{R}, Z) = D(A)$, so $\frac{d}{dt} T(t)v = AT(t)v$. Note that $(T(s)v)(\sigma) = u(s + \sigma) = u_s(\sigma) = h_u(s)(\sigma)$ for $\sigma \in [-1, 0]$ and $|s - t| < T - 1$. So we have

$$0 = \lim_{h \to 0} \left\| \frac{T(t+h)v - T(t)v}{h} - AT(t)v \right\|^p_{L^p(\mathbb{R}, X)}$$

$$\geq \lim_{h \to 0} \left\| \frac{h_u(t+h) - h_u(t)}{h} - \frac{d}{d\sigma} u_t \right\|^p_{L^p([-1, 0], X)},$$

which implies

$$\frac{d}{dt} h_u(t) = \frac{d}{d\sigma} u_t.$$

Moreover, the map $t \mapsto \frac{d}{d\sigma} u_t = \frac{d}{dt} h_u(t)$ is continuous from $\mathbb{R}_+$ into $L^p([-1,0], Z)$ since the map $t \mapsto AT(t)v = T(t)Av$ is continuous from $\mathbb{R}$ into $L^p(\mathbb{R}, Z)$. $\square$

By means of Lemma 3.4, we can now transform classical solutions of (DE_p) into classical solutions of an abstract Cauchy problem.

Corollary 3.5. *Let* $u : [-1, \infty) \longrightarrow X$ *be a classical solution of* (DE_p). *Then the function*

$$\mathcal{U} : t \mapsto \begin{pmatrix} u(t) \\ u_t \end{pmatrix} \in \mathcal{E}_p \tag{3.7}$$

from $\mathbb{R}_+$ *into* $\mathcal{E}_p$ *is continuously differentiable with derivative*

$$\dot{\mathcal{U}}(t) = \mathcal{A}\mathcal{U}(t),$$

where

$$\mathcal{A} := \begin{pmatrix} B & \Phi \\ 0 & \frac{d}{d\sigma} \end{pmatrix},$$

where $\frac{d}{d\sigma}$ *denotes the distributional derivative, with domain*

$$D(\mathcal{A}) := \left\{ \begin{pmatrix} x \\ f \end{pmatrix} \in D(B) \times W^{1,p}([-1,0], Z) : f(0) = x \right\}.$$

Thus every classical solution u *of* (DE_p) *yields a classical solution of the abstract Cauchy problem*

$$(ACP_p) \qquad \begin{cases} \mathcal{U}'(t) = \mathcal{A}\,\mathcal{U}(t), & t \geq 0, \\ \mathcal{U}(0) = \begin{pmatrix} x \\ f \end{pmatrix}, \end{cases}$$

on $\mathcal{E}_p$.

Therefore, every solution of (DE_p) gives us a solution of (ACP_p). Now we show that (DE_p) and (ACP_p) are actually equivalent in the sense that conversely every classical solution $t \mapsto \mathcal{U}(t)$ of (ACP_p) is of the form

$$\mathcal{U}(t) = \begin{pmatrix} u(t) \\ u_t \end{pmatrix},$$

where the function u is a classical solution of (DE_p). We fix the Banach space setting for (ACP_p) by adding the following to our standing hypotheses:

(H_7) $(\mathcal{A}, D(\mathcal{A}))$ is the operator on $\mathcal{E}_p$ defined as

$$\mathcal{A} := \begin{pmatrix} B & \Phi \\ 0 & \frac{d}{d\sigma} \end{pmatrix}, \tag{3.8}$$

with domain

$$D(\mathcal{A}) := \left\{ \begin{pmatrix} x \\ f \end{pmatrix} \in D(B) \times W^{1,p}([-1,0], Z) : f(0) = x \right\}. \tag{3.9}$$

Since we want to apply Theorem 1.22, we need to show the closedness of the operator $(\mathcal{A}, D(\mathcal{A}))$.

Lemma 3.6. *Under Hypotheses (H_1)–(H_7), the operator $(\mathcal{A}, D(\mathcal{A}))$ is closed and densely defined on $\mathcal{E}_p$.*

Proof. First we prove the closedness. Let $\begin{pmatrix} x_n \\ f_n \end{pmatrix} \subset D(\mathcal{A})$ be a sequence such that $\begin{pmatrix} x_n \\ f_n \end{pmatrix}$ converges to $\begin{pmatrix} x \\ f \end{pmatrix} \in \mathcal{E}_p$ and $\mathcal{A}\begin{pmatrix} x_n \\ f_n \end{pmatrix} = \begin{pmatrix} Bx_n + \Phi f_n \\ \frac{d}{d\sigma} f_n \end{pmatrix}$ converges to $\begin{pmatrix} y \\ g \end{pmatrix} \in \mathcal{E}_p$.

In particular, the sequence (f_n) converges to f in the norm topology of the Sobolev space $W^{1,p}([-1,0], Z)$. Hence, we have that $f \in W^{1,p}([-1,0], Z)$ and $\frac{d}{d\sigma} f = g$. Since the operator $\Phi : W^{1,p}([-1,0], Z) \to X$ is bounded, we have that $\Phi f_n \longrightarrow \Phi f$.

Moreover, by the closedness of B, we have $x \in D(B)$ and $Bx - \Phi f = y$. Finally, since the space $W^{1,p}([-1,0], Z)$ is continuously embedded into $C([-1,0], Z)$, the sequence $x_n = f_n(0)$ converges to $f(0)$. Hence, $f(0) = x$ and $\begin{pmatrix} x \\ f \end{pmatrix} \in D(\mathcal{A})$, $\mathcal{A}\begin{pmatrix} x \\ f \end{pmatrix} = \begin{pmatrix} y \\ g \end{pmatrix}$ and the operator $\mathcal{A}$ is closed.

Now we prove the density of $D(\mathcal{A})$. Let $\begin{pmatrix} y \\ g \end{pmatrix} \in \mathcal{E}_p$ and $\varepsilon > 0$. Since $D(B)$ and $W_0^{1,p}([-1,0], Z)$ are dense in X and $L^p([-1,0], Z)$, respectively, we can find $x \in D(B)$ and $\tilde{g} \in W_0^{1,p}([-1,0], Z)$ such that

$$\|x - y\| < \varepsilon \text{ and } \|\tilde{g} - g\|_p < \varepsilon.$$

Let now $h \in W^{1,p}([-1,0], Z)$ such that $h(0) = x$ and let $k \in W_0^{1,p}([-1,0], Z)$ such that $\|k - h\|_p < \varepsilon$. Finally, let $f := \tilde{g} + h - k$. We obtain $f \in W^{1,p}([-1,0], Z)$, $f(0) = x$, and

$$\left\| \begin{pmatrix} y \\ g \end{pmatrix} - \begin{pmatrix} x \\ f \end{pmatrix} \right\| \leq \left\| \begin{pmatrix} y-x \\ 0 \end{pmatrix} \right\| + \left\| \begin{pmatrix} 0 \\ g-\tilde{g} \end{pmatrix} \right\| + \left\| \begin{pmatrix} 0 \\ h-k \end{pmatrix} \right\| < 3\varepsilon.$$

Therefore, the domain $D(\mathcal{A})$ is dense.   □

In view of Lemma 3.6, we can formulate the following corollary that is a straightforward consequence of Theorem 1.22.

Corollary 3.7. *The abstract Cauchy problem (ACP_p) associated to the operator $(\mathcal{A}, D(\mathcal{A}))$ on the space $\mathcal{E}_p$ is well-posed if and only if $(\mathcal{A}, D(\mathcal{A}))$ is the generator of a strongly continuous semigroup $(\mathcal{T}(t))_{t \geq 0}$ on $\mathcal{E}_p$.*

In this case, the classical and mild solutions of (ACP_p) are given by the functions $\mathcal{U}(t) := \mathcal{T}(t)\begin{pmatrix} x \\ f \end{pmatrix}$ for $t \geq 0$.

Now we introduce the following notation.

Definition 3.8. By $\pi_1 : \mathcal{E}_p \longrightarrow X$, we denote the canonical projection from $\mathcal{E}_p$ onto X.

Similarly, by $\pi_2 : \mathcal{E}_p \longrightarrow L^p([-1,0], Z)$ we denote the canonical projection from $\mathcal{E}_p$ onto $L^p([-1,0], Z)$.

Proposition 3.9. *For every classical solution $\mathcal{U}$ of (ACP$_p$), the function*

$$u(t) := \begin{cases} (\pi_1 \circ \mathcal{U})(t) & \text{if } t \geq 0, \\ f(t) & \text{if } t \in [-1,0) \end{cases} \tag{3.10}$$

is a classical solution of (DE$_p$) and $(\pi_2 \circ \mathcal{U})(t) = u_t$ for all $t \geq 0$.

Proof. Since $\mathcal{U}$ is a classical solution of (ACP$_p$), it follows that $\pi_2 \circ \mathcal{U}$ is in $C^1(\mathbb{R}_+, L^p([-1,0], Z))$ and is a classical solution of the problem

$$\begin{cases} \frac{d}{dt}(\pi_2 \circ \mathcal{U})(t) = \frac{d}{d\sigma}(\pi_2 \circ \mathcal{U})(t), & t \geq 0, \\ (\pi_2 \circ \mathcal{U})(t)(0) = z(t), & t \geq 0, \\ (\pi_2 \circ \mathcal{U})(0) = f \end{cases} \tag{3.11}$$

in the space $L^p([-1,0], Z)$. In particular since $L^p([-1,0], Z) \overset{d}{\hookrightarrow} L^p([-1,0], X)$, the function $(\pi_2 \circ \mathcal{U})$ is in $C^1(\mathbb{R}_+, L^p([-1,0], X))$ and is a classical solution of the problem in Equation (3.11) in the space $L^p([-1,0], X)$.

Now we observe that by definition

$$u_t(\sigma) = u(t+\sigma) = \begin{cases} (\pi_1 \circ \mathcal{U})(t+\sigma) & \text{for } t+\sigma \geq 0, \\ f(t+\sigma) & \text{for } t+\sigma < 0, \end{cases}$$

where $z \in C^1([0,\infty), X)$, $f \in W^{1,p}([-1,0], Z) \overset{d}{\hookrightarrow} W^{1,p}([-1,0], X)$, and $f(0) = x = z(0)$ by assumption. Hence, $u \in W^{1,p}_{loc}([-1,\infty), X)$. We can extend u to a function in $W^{1,p}_{loc}(\mathbb{R}, X)$ and, by Lemma 3.4, we have

$$\frac{d}{dt} h_u(t) = \frac{d}{d\sigma} u_t \qquad \text{for all } t \geq 0$$

in the space $L^p([-1,0], X)$. Moreover, by definition of u_t we have

$$u_t(0) = u(t) = z(t) \qquad \text{for all } t \geq 0,$$

and

$$u_0 = f.$$

Hence, the map $t \mapsto u_t$ is also a classical solution of the problem in Equation (3.11) in the space $L^p([-1,0], X)$.

Now define $w(t) := u_t - (\pi_2 \circ \mathcal{U})(t)$ for $t \geq 0$. Then w is a classical solution of the problem

$$\begin{cases} \frac{d}{dt}w(t) = \frac{d}{d\sigma}w(t) & \text{for } t \geq 0, \\ w(t)(0) = 0 & \text{for } t \geq 0, \\ w(0) = 0 \end{cases} \tag{3.12}$$

in the space $L^p([-1,0], X)$. Since Equation (3.12) is the abstract Cauchy problem associated to the generator of the (nilpotent) left shift semigroup on $L^p([-1,0], X)$ of Example 1.34 with initial value 0, we have that $w(t) = 0$ for all $t \geq 0$. Therefore, $u_t = (\pi_2 \circ \mathcal{U})(t) \in W^{1,p}([-1,0], Z)$ and $\mathcal{U}(t) = \left(\begin{smallmatrix} u(t) \\ u_t \end{smallmatrix}\right)$ for all $t \geq 0$, and u is a classical solution of (DE$_p$). $\square$

The equivalence of (DE$_p$) and (ACP$_p$) established above enables us to use methods and results of semigroup theory in order to deal with (DE$_p$). This is our main idea.

At present, we transfer the notions of well-posedness and of mild solution, known from abstract Cauchy problems and semigroups (see Section 1.2), to (DE$_p$).

Definition 3.10.

(i) The Problem (DE$_p$) is called *well-posed* if (ACP$_p$) is well-posed (this means if $(\mathcal{A}, D(\mathcal{A}))$ generates a strongly continuous semigroup on $\mathcal{E}_p$).

(ii) Suppose (DE$_p$) is well-posed and let $(\mathcal{T}(t))_{t \geq 0}$ be the semigroup generated by $(\mathcal{A}, D(\mathcal{A}))$ on $\mathcal{E}_p$. Then for every $x \in X$ and every $f \in L^p([-1,0], Z)$ the function u defined by Equation (3.10) is called a *mild solution* of (DE$_p$).

Proposition 3.11. *Let u be a mild solution of (DE$_p$). Then u satisfies* $\int_0^t u(s)\,ds \in D(B)$, $\int_0^t u_s\,ds \in W^{1,p}([-1,0], Z)$, *and the integral equation*

$$u(t) = \begin{cases} x + B \int_0^t u(s)\,ds + \Phi \int_0^t u_s\,ds & \text{for } t \geq 0, \\ f(t) & \text{for a.e. } t \in [-1,0). \end{cases} \tag{3.13}$$

Proof. (1) First we show that

$$u_t = \pi_2\left(\mathcal{T}(t)\left(\begin{smallmatrix} x \\ f \end{smallmatrix}\right)\right) \tag{3.14}$$

for every $\left(\begin{smallmatrix} x \\ f \end{smallmatrix}\right) \in \mathcal{E}_p$ and every $t \geq 0$.

For $\left(\begin{smallmatrix}x\\f\end{smallmatrix}\right) \in D(\mathcal{A})$ the Identity (3.14) holds by Proposition 3.9. Take now $\left(\begin{smallmatrix}x\\f\end{smallmatrix}\right) \in \mathcal{E}_p$ and a sequence $\left(\begin{smallmatrix}x_n\\f_n\end{smallmatrix}\right)$ in $D(\mathcal{A})$ converging to $\left(\begin{smallmatrix}x\\f\end{smallmatrix}\right)$. Since the semigroup $(\mathcal{T}(t))_{t\geq 0}$ is strongly continuous, the sequence $\mathcal{T}(t)\left(\begin{smallmatrix}x_n\\f_n\end{smallmatrix}\right)$ converges to $\mathcal{T}(t)\left(\begin{smallmatrix}x\\f\end{smallmatrix}\right)$ in $\mathcal{E}_p$ uniformly for t in compact subsets of $[0,\infty)$.

Now, let

$$u_n(t) := \begin{cases} \pi_1\left(\mathcal{T}(t)\left(\begin{smallmatrix}x_n\\f_n\end{smallmatrix}\right)\right) & \text{if } t \geq 0, \\ f_n(t) & \text{if } t \in [-1,0). \end{cases}$$

Since $\left(\begin{smallmatrix}x_n\\f_n\end{smallmatrix}\right) \in D(\mathcal{A})$, by Proposition 3.9 we have $(u_n)_t = \pi_2\left(\mathcal{T}(t)\left(\begin{smallmatrix}x_n\\f_n\end{smallmatrix}\right)\right)$. For $t \geq 1$, we have that

$$(u_n)_t(\sigma) = u_n(t+\sigma) = \pi_1\left(\mathcal{T}(t+\sigma)\left(\begin{smallmatrix}x_n\\f_n\end{smallmatrix}\right)\right) \tag{3.15}$$

converges to $\pi_1\left(\mathcal{T}(t+\sigma)\left(\begin{smallmatrix}x\\f\end{smallmatrix}\right)\right) = u_t(\sigma)$ uniformly for $\sigma \in [-1,0]$. Hence, $(u_n)_t$ converges to u_t in $L^p([-1,0],X)$ and we have

$$u_t = \lim_{n\to\infty}(u_n)_t = \lim_{n\to\infty} \pi_2\left(\mathcal{T}(t)\left(\begin{smallmatrix}x_n\\f_n\end{smallmatrix}\right)\right) = \pi_2\left(\mathcal{T}(t)\left(\begin{smallmatrix}x\\f\end{smallmatrix}\right)\right).$$

In particular, $u_t \in L^p([-1,0],Z)$.

Let now $0 \leq t < 1$. By Corollary 1.22 and Equation (3.10), we have

$$(u_n)_t(\sigma) = \begin{cases} \pi_1(\mathcal{T}(t+\sigma)\left(\begin{smallmatrix}x_n\\f_n\end{smallmatrix}\right) & \text{for } \sigma \in [-t,0], \\ f_n(t+\sigma) & \text{for } \sigma \in [-1,-t). \end{cases}$$

This formula in conjunction with the calculation given in Equation (3.15) implies that $(u_n)_t(\sigma)$ converges to $\pi_1\left(\mathcal{T}(t+\sigma)\left(\begin{smallmatrix}x_n\\f_n\end{smallmatrix}\right)\right)$ uniformly for $\sigma \in [-t,0]$. Moreover, by assumption, $(u_n)_t$ converges to $f(t+\cdot)$ in $L^p([-1,-t),Z)$. Hence, $(u_n)_t$ converges to u_t in $L^p([-1,0],X)$ and, by the same argument as above, $u_t = \pi_2\left(\mathcal{T}(t)\left(\begin{smallmatrix}x\\f\end{smallmatrix}\right)\right)$.

(2) Take the first component of the identity

$$\mathcal{T}(t)\left(\begin{smallmatrix}x\\f\end{smallmatrix}\right) - \left(\begin{smallmatrix}x\\f\end{smallmatrix}\right) = \mathcal{A}\int_0^t \mathcal{T}(s)\left(\begin{smallmatrix}x\\f\end{smallmatrix}\right)\,ds, \qquad t \geq 0,$$

to obtain Equation (3.13). □

Theorem 3.12. *The following assertions are equivalent:*

(i) (DE_p) is well-posed.

(ii) For every $\left(\begin{smallmatrix}x\\f\end{smallmatrix}\right) \in D(\mathcal{A})$,

(a) there is a unique (classical) solution $u(x,f,\cdot)$ of (DE_p) and

(b) the solutions depend continuously on the initial values, that is, if a sequence $\left(\begin{smallmatrix} x_n \\ f_n \end{smallmatrix}\right)$ in $D(\mathcal{A})$ converges to $\left(\begin{smallmatrix} x \\ f \end{smallmatrix}\right) \in D(\mathcal{A})$ in the space $\mathcal{E}_p = X \times L^p([-1,0], Z)$, then $u(x_n, f_n, t)$ converges to $u(x, f, t)$ in X uniformly for t in compact intervals.

Proof. First we show $(b) \Rightarrow (a)$. Assume that for every $\left(\begin{smallmatrix} x \\ f \end{smallmatrix}\right) \in D(\mathcal{A})$, Equation (DE$_p$) has a unique solution u. Then Proposition 3.5 guarantees that for every $\left(\begin{smallmatrix} x \\ f \end{smallmatrix}\right) \in D(\mathcal{A})$ the abstract Cauchy problem (ACP$_p$) has a classical solution that is unique by Proposition 3.9. It is easy to see that these solutions depend continuously on the initial values. Finally, by Lemma 3.6, $(\mathcal{A}, D(\mathcal{A}))$ is closed and densely defined. Therefore, it generates a strongly continuous semigroup on $\mathcal{E}_p$ by Theorem 1.22

Conversely, if $\mathcal{A}$ is a generator, we have by Corollary 3.7 and Proposition 3.9 that for every initial value $\left(\begin{smallmatrix} x \\ f \end{smallmatrix}\right) \in D(\mathcal{A})$ there is a unique solution u of (DE$_p$) that is given by Equation (3.10). This implies that the solutions depend continuously on the initial values. $\qquad \square$

Definition 3.13. In the case that (DE$_p$) is well-posed, we call the semigroup $(\mathcal{T}(t))_{t \geq 0}$ generated by $(\mathcal{A}, D(\mathcal{A}))$ on $\mathcal{E}_p$ the *delay semigroup* corresponding to (DE$_p$).

The considerations of this section suggest that it may be worthwhile to study the delay semigroup systematically in order to obtain information on the solutions of (DE$_p$). We will do so in Part III.

Let us discuss some examples.

Example 3.14. Consider a diffusion process on an open, bounded domain $\Omega \subset \mathbb{R}^n$ with smooth boundary. Assume that there is no dispersion at the boundary, i.e., take Neumann boundary conditions. Assume moreover that there is a delayed feedback depending on the flux. Then the equation modeling this system becomes the following partial differential equation with delay.

$$\begin{cases} \partial_t u(t,s) = \Delta u(t,s) + \sum_{i=1}^n c_i \partial_i u(t - h_i, s), & t \geq 0, \ s \in \Omega, \\ \frac{\partial u}{\partial \nu}(t,s) = 0, & t \geq 0, \ s \in \partial\Omega, \\ u(t,s) = f(t,s), & t \in [-1,0], \ s \in \Omega, \end{cases} \quad (3.16)$$

for some constants $c_i \in \mathbb{R}$ and $h_i \in [0,1]$. Moreover, we assume that there exists $1 \leq p < \infty$ such that $f(t, \cdot) \in L^2(\Omega)$ for a.e. $t \in [-1,0]$ and the map $t \mapsto f(t, \cdot)$ belongs to $L^p([-1,0], L^2(\Omega))$. In order to write Equation (3.16) as an abstract delay equation, we introduce

- the Hilbert space $X := L^2(\Omega)$ with its usual scalar product,

- the operator B to be the Laplace operator with Neumann boundary conditions, i.e., $D(B) := \{g \in H^1(\Omega) \ : \ \Delta g \in L^2(\Omega) \text{ and } \frac{\partial g}{\partial \nu} = 0 \text{ on } \partial\Omega\}$ and $Bf := \Delta f$,

- the space $Z := H^1(\Omega)$, [2]

- the *delay operator* $\Phi : W^{1,p}([-1,0], Z) \to X$ defined as

$$\Phi f := \sum_{i=1}^{n} c_i \partial_i f(-h_i),$$

and

- $x := f(\cdot, 0)$.

Example 3.15. A diffusion equation with delayed reaction term (see Wu [222, Section 2.1]),

$$\begin{cases} \partial_t w(t,s) = \Delta w(t,s) + c \int_{-1}^{0} w(t+\tau, s)\, dg(\tau), \\ \qquad\qquad\qquad\qquad s \in \Omega, \ t \geq 0, \\ w(t,s) = 0, \qquad\qquad\quad s \in \partial\Omega, \ t \geq 0, \\ w(t,s) = f(t,s), \qquad\quad (s,t) \in \Omega \times [-1, 0], \end{cases} \qquad (3.17)$$

where c is a constant, $\Omega \subset \mathbb{R}^n$ an open set, and $g : [-1, 0] \to [0, 1]$, is a function of bounded variation. (In Example 5.10, this function will be the Cantor function, which is singular and has total variation 1.) The integral on the right-hand side is understood as a Riemann-Stieltjes Integral. Moreover, we assume that f belongs to $L^2([-1, 0] \times \Omega)$.

In order to write this problem as an abstract delay equation, we take

- the Hilbert space $X := L^2(\Omega)$,

- the operator $(B, D(B))$ as the variational Laplacian with Dirichlet boundary conditions,

- $Z := X = L^2(\Omega)$, and

- the operator $\Phi : W^{1,2}([-1,0], X) \to X$ defined as $\Phi f := c \int_{-1}^{0} f(\tau)\, dg(\tau)$.

All general assumptions are satisfied and so we can rewrite this reaction diffusion equation as an abstract delay equation (DE$_2$).

---

[2] In the sequel, if $p = 2$ we will usually write $H^1(\Omega)$ instead of $W^{1,2}(\Omega)$.

The previous example could be extended without modification to the case where c, the reaction coefficient, is not a constant but a bounded function.

Now we consider a population equation with age structure and delay. This equation models harvesting in a population, where the amount of harvesting done at time t is directly proportional to the population at a previous time $t - \tau$. The delay could be caused by the measuring time $\tau > 0$.

Example 3.16. A Lotka-Sharpe equation with delay,

$$\begin{cases} \partial_t u(t,a) + \partial_a u(t,a) = -\mu(a)\,u(t,a) + \nu(a)\,u(t-1,a), \\ \qquad\qquad\qquad\qquad\qquad\qquad t \geq 0,\; a \in \mathbb{R}_+, \\ u(t,0) = \int_0^{+\infty} \beta(a)\,u(t,a)\,da, \qquad\quad t \geq 0, \\ u(t,a) = f(t,a), \qquad\qquad\qquad (t,a) \in [-1,0] \times \mathbb{R}_+, \end{cases} \tag{3.18}$$

where $\mu, \nu, \beta \in L^\infty(\mathbb{R}_+)$, μ, β are positive and f is in $L^1([-1,0] \times \mathbb{R}_+)$.

To write this equation as an abstract delay equation, we choose

- the Banach space $X := L^1(\mathbb{R}_+)$,

- the operator $(B\,g)(a) := -(g')(a) - \mu(a)\,g(a)$, $a \in \mathbb{R}_+$, with domain $D(B) := \left\{ g \in W^{1,1}(\mathbb{R}_+) : g(0) = \int_0^{+\infty} \beta(a)\,g(a)\,da \right\}$, and

- the delay operator $\Phi : W^{1,p}([-1,0], X) \to X$, defined as

$$\Phi f := \nu\, f(-1).$$

With these definitions, we obtain an abstract delay equation of the form (DE_p).

Now we give an example of a second-order Cauchy problem with delay (see Part IV).

Example 3.17. A one-dimensional wave equation with delay,

$$\begin{cases} \partial_t^2 u(t,s) = \Delta u(t,s) + c_1 \partial_s u(t-h_1,s) + c_2 \partial_t u(t-h_2,s), \\ \qquad\qquad t \geq 0,\; s \in (0,1), \\ u(t,s) = f(t,s), \quad \partial_t u(t,s) = g(t,s), \quad (t,s) \in [-1,0] \times (0,1), \\ u(t,0) = u(t,1) = 0, \end{cases}$$

where the initial data verify

- $f(0,\cdot) \in H_0^1(0,1)$ and $g(0,\cdot) \in L^2(0,1)$,

- the map $t \mapsto f(t, \cdot)$ is in $L^2([-1,0], H_0^1(0,1))$, and

- the map $t \mapsto g(t, \cdot)$ is in $L^2([-1,0], L^2(0,1))$.

To write the problem in abstract form (DE$_2$), we take

- the Hilbert space $X := H_0^1(0,1) \times L^2(0,1)$,

- the operator $B := \begin{pmatrix} 0 & \mathrm{Id} \\ \Delta & 0 \end{pmatrix}$

  with domain $D(B) := \left(H_0^1(0,1) \cap H^2(0,1) \right) \times H_0^1(0,1)$,

- the space $Z := X = H_0^1(0,1) \times L^2(0,1)$,

- the function $\mathbb{R}_+ \ni t \mapsto u(t) = u(\cdot, t) \in L^2(0,1)$, and

- the operator $\Phi : W^{1,p}([-1,0], X) \to X$ defined by

$$
\Phi\left(\tfrac{g}{h}\right) := \begin{pmatrix} 0 & 0 \\ c_1 \, \partial_x \, \delta_{-h_1} & c_2 \, \delta_{-h_2} \end{pmatrix} \left(\tfrac{g}{h}\right)
$$

$$
= \begin{pmatrix} 0 \\ c_1 \, \partial_x g(-h_1) + c_2 h(-h_2) \end{pmatrix},
$$

  where δ_{-h_1} and δ_{-h_2} are the point evaluations in $-h_1$ and $-h_2$, respectively.

Remark 3.18. In Examples 3.15, 3.16, and 3.17, the space Z coincides with X, while in Example 3.14, the space $Z = H^1(\Omega)$ is strictly smaller than $X = L^2(\Omega)$.

3.2 Spectral Theory for Delay Equations

We will see in Part III that many qualitative properties of the solutions of (DE$_p$) can be characterized by the spectrum of the operator $(\mathcal{A}, D(\mathcal{A}))$. As a first step towards this goal, we calculate the resolvent $R(\lambda, \mathcal{A})$ and the resolvent set $\rho(\mathcal{A})$ of the operator $\mathcal{A}$.

Let $1 \leq p < \infty$ be fixed for the rest of this section. Before starting our investigations we introduce some notation. Here and in the following $(A_0, D(A_0))$ is the generator of the nilpotent left shift semigroup $(T_0(t))_{t \geq 0}$ on $L^p([-1,0], Z)$. For $\lambda \in \mathbb{C}$, ϵ_λ denotes the function

$$
\epsilon_\lambda(s) := e^{\lambda s} \text{ for } s \in [-1,0].
$$

Now let $\lambda \in \mathbb{C}$ and $\left(\begin{smallmatrix} y \\ g \end{smallmatrix}\right) \in \mathcal{E}_p$; we have to find $\left(\begin{smallmatrix} x \\ f \end{smallmatrix}\right) \in D(\mathcal{A})$ such that

$$(\lambda - \mathcal{A})\left(\begin{matrix} x \\ f \end{matrix}\right) = \left(\begin{matrix} (\lambda - B)x - \Phi f \\ \lambda f - f' \end{matrix}\right) = \left(\begin{matrix} y \\ g \end{matrix}\right). \tag{3.19}$$

By the Variation of Parameters Formula, we can integrate the second line of Equation (3.19) and obtain

$$f = \epsilon_\lambda\, f(0) + R(\lambda, A_0)\, g, \tag{3.20}$$

where $R(\lambda, A_0)$ is the resolvent of $(A_0, D(A_0))$, which exists because its spectrum $\sigma(A_0) = \emptyset$. Since $f(0) = x$, Equation (3.20) becomes

$$f = \epsilon_\lambda\, x + R(\lambda, A_0)\, g. \tag{3.21}$$

Hence, x has to satisfy the equation

$$(\lambda - B - \Phi_\lambda)x = \Phi R(\lambda, A_0)g + y, \tag{3.22}$$

where $\Phi_\lambda \in \mathcal{L}(Z, X)$ is defined by $\Phi_\lambda x := \Phi(e^{\lambda \cdot} x)$ for $x \in Z$.
This leads to the following proposition.

Proposition 3.19. *For $\lambda \in \mathbb{C}$ and for all $1 \le p < \infty$, we have*

$$\lambda \in \rho(\mathcal{A}) \text{ if and only if } \lambda \in \rho(B + \Phi_\lambda). \tag{3.23}$$

Moreover, for $\lambda \in \rho(\mathcal{A})$ the resolvent $R(\lambda, \mathcal{A})$ is given by

$$\left(\begin{matrix} R(\lambda, B + \Phi_\lambda) & R(\lambda, B + \Phi_\lambda)\Phi R(\lambda, A_0) \\ \epsilon_\lambda\, R(\lambda, B + \Phi_\lambda) & [\epsilon_\lambda\, R(\lambda, B + \Phi_\lambda)\Phi + Id]R(\lambda, A_0) \end{matrix}\right). \tag{3.24}$$

Proof. Let $\lambda \in \rho(B + \Phi_\lambda)$. Then the matrix defined by Equation (3.24) is a bounded operator from $\mathcal{E}_p$ to $D(\mathcal{A})$ defining the inverse of $(\lambda - \mathcal{A})$.

To see this, as a first step we show that the range of the matrix defined by Equation (3.24) is contained in $D(\mathcal{A})$. Take $\left(\begin{smallmatrix} y \\ g \end{smallmatrix}\right) \in \mathcal{E}_p$ and consider the vector

$$\left(\begin{matrix} x \\ f \end{matrix}\right) := \left(\begin{matrix} R(\lambda, B + \Phi_\lambda)y + R(\lambda, B + \Phi_\lambda)\Phi R(\lambda, A_0)g \\ \epsilon_\lambda\, R(\lambda, B + \Phi_\lambda)y + \epsilon_\lambda\, R(\lambda, B + \Phi_\lambda)\Phi R(\lambda, A_0)g + R(\lambda, A_0)g \end{matrix}\right).$$

Since $R(\lambda, B + \Phi_\lambda)$ maps into $D(B)$, we know that $x \in D(B)$. For the second component, note that the function ϵ_λ is smooth, and the range of $R(\lambda, A_0)$ is contained in $W_0^{1,p}([-1,0], Z)$. Hence, $f \in W^{1,p}([-1,0], Z)$. Since $R(\lambda, A_0)g(0) = 0$ and $(\epsilon_\lambda x)(0) = x$, it is now easy to see that $f(0) = x$. So we have $\left(\begin{smallmatrix} x \\ f \end{smallmatrix}\right) \in D(\mathcal{A})$.

Write $R_\lambda := R(\lambda, B + \Phi_\lambda)$ and compute the matrix product

$$(\lambda - \mathcal{A}) \begin{pmatrix} R(\lambda, B + \Phi_\lambda) & R(\lambda, B + \Phi_\lambda)\Phi R(\lambda, A_0) \\ \epsilon_\lambda R(\lambda, B + \Phi_\lambda) & [\epsilon_\lambda R(\lambda, B + \Phi_\lambda)\Phi + Id]R(\lambda, A_0) \end{pmatrix}$$
$$= \begin{pmatrix} (\lambda - B)R_\lambda - \Phi\epsilon_\lambda R_\lambda & (\lambda - B)R_\lambda \Phi R(\lambda, A_0) - \Phi[\epsilon_\lambda R_\lambda \Phi + Id]R(\lambda, A_0) \\ \left(\lambda - \frac{d}{d\sigma}\right)\epsilon_\lambda R_\lambda & \left(\lambda - \frac{d}{d\sigma}\right)[\epsilon_\lambda R_\lambda \Phi + Id]R(\lambda, A_0) \end{pmatrix}.$$

The identities

$$(\lambda - B)R_\lambda - \Phi(\epsilon_\lambda R_\lambda) = (\lambda - B - \Phi_\lambda)R_\lambda = Id$$

and

$$(\lambda - B)R_\lambda \Phi R(\lambda, A_0) - \Phi[\epsilon_\lambda R_\lambda \Phi + Id]R(\lambda, A_0) = \\ (\lambda - B - \Phi_\lambda)R_\lambda \Phi R(\lambda, A_0) - \Phi R(\lambda, A_0) = 0$$

hold. Moreover, we have $\left(\lambda - \frac{d}{d\sigma}\right)(\epsilon_\lambda R_\lambda) = 0$. Using this identity we also obtain

$$\left(\lambda - \frac{d}{d\sigma}\right)[\epsilon_\lambda R_\lambda \Phi + Id]R(\lambda, A_0) = \left(\lambda - \frac{d}{d\sigma}\right)R(\lambda, A_0) = Id.$$

So the operator in Equation (3.24) is a right inverse of $(\lambda - \mathcal{A})$. In an analogous way, we can prove that the operator in Equation (3.24) is also a left inverse of $(\lambda - \mathcal{A})$ and is hence an inverse of $(\lambda - \mathcal{A})$.

Conversely, if $\lambda \in \rho(\mathcal{A})$, then for every $\binom{y}{g} \in \mathcal{E}_p$ there exists a unique $\binom{x}{f} \in D(\mathcal{A})$ such that Equations (3.21) and (3.22) hold. In particular, for $g = 0$ and for every $y \in X$, there exists a unique $x \in D(B)$ such that

$$(\lambda - B - \Phi_\lambda)x = y.$$

This means that $(\lambda - B - \Phi_\lambda)$ is invertible, i.e., $\lambda \in \rho(B + \Phi_\lambda)$. $\qquad\square$

The complementary statement to Equation (3.23) in Proposition 3.19 yields

$$\lambda \in \sigma(\mathcal{A}) \text{ if and only if } \lambda \in \sigma(B + \Phi_\lambda). \tag{3.25}$$

We call Equation (3.25) the *characteristic equation* of the delay equation. An important consequence of the characteristic equation is that the spectrum of the operator $\mathcal{A}$ in $\mathcal{E}_p$ can be determined in the smaller space X. In particular, if the dimension of X is finite, we have

$$\lambda \in \sigma(\mathcal{A}) \text{ if and only if } \det(\lambda - B - \Phi_\lambda) = 0. \tag{3.26}$$

An analogous characteristic equation also holds for subsets of the spectrum such as the point spectrum $P\sigma$, approximative point spectrum $A\sigma$, the residual spectrum $R\sigma$, and the essential spectrum σ_{ess} (see Section 2.1 for the definitions).

Lemma 3.20. *For $\lambda \in \mathbb{C}$, the following hold:*

$$\lambda \in P\sigma(\mathcal{A}) \text{ if and only if } \lambda \in P\sigma(B + \Phi_\lambda),$$
$$\lambda \in A\sigma(\mathcal{A}) \text{ if and only if } \lambda \in A\sigma(B + \Phi_\lambda),$$
$$\lambda \in R\sigma(\mathcal{A}) \text{ if and only if } \lambda \in R\sigma(B + \Phi_\lambda), \text{ and }$$
$$\lambda \in \sigma_{ess}(\mathcal{A}) \text{ if and only if } \lambda \in \sigma_{ess}(B + \Phi_\lambda)$$

for all $1 \leq p < \infty$.

Proof. Let $\lambda \in \mathbb{C}$. The main idea of the proof is to show that the decomposition

$$\lambda - \mathcal{A} = \begin{pmatrix} \mathrm{Id} & -\Phi R(\lambda, A_0) \\ 0 & \mathrm{Id} \end{pmatrix} \begin{pmatrix} \lambda - B - \Phi_\lambda & 0 \\ 0 & \lambda - A_0 \end{pmatrix} \begin{pmatrix} \mathrm{Id} & 0 \\ -\epsilon_\lambda \otimes \mathrm{Id} & \mathrm{Id} \end{pmatrix}$$

holds.

Let

$$\mathcal{A}_\lambda := \begin{pmatrix} \lambda - B - \Phi_\lambda & 0 \\ 0 & \lambda - A_0 \end{pmatrix}$$

with domain $D(\mathcal{A}_\lambda) := D(B) \times D(A_0)$. Moreover, let

$$T := \begin{pmatrix} \mathrm{Id} & -\Phi R(\lambda, A_0) \\ 0 & \mathrm{Id} \end{pmatrix} \text{ and } S := \begin{pmatrix} \mathrm{Id} & 0 \\ -\epsilon_\lambda \otimes \mathrm{Id} & \mathrm{Id} \end{pmatrix},$$

where Id denotes the identity on X and $\epsilon_\lambda \otimes \mathrm{Id} : X \to L^p([-1,0], X)$ is the operator satisfying $(\epsilon_\lambda \otimes \mathrm{Id})x := \epsilon_\lambda x = e^{\lambda \cdot} x$.

Taking $\begin{pmatrix} x \\ f \end{pmatrix} \in D(\mathcal{A})$, then $S\begin{pmatrix} x \\ f \end{pmatrix} \in D(\mathcal{A}_\lambda)$, and we obtain

$$\begin{pmatrix} \mathrm{Id} & -\Phi R(\lambda, A_0) \\ 0 & \mathrm{Id} \end{pmatrix} \begin{pmatrix} \lambda - B - \Phi_\lambda & 0 \\ 0 & \lambda - A_0 \end{pmatrix} \begin{pmatrix} \mathrm{Id} & 0 \\ -\epsilon_\lambda \otimes \mathrm{Id} & \mathrm{Id} \end{pmatrix} \begin{pmatrix} x \\ f \end{pmatrix}$$

$$= \begin{pmatrix} \mathrm{Id} & -\Phi R(\lambda, A_0) \\ 0 & \mathrm{Id} \end{pmatrix} \begin{pmatrix} \lambda - B - \Phi_\lambda & 0 \\ 0 & \lambda - A_0 \end{pmatrix} \begin{pmatrix} x \\ f - \epsilon_\lambda x \end{pmatrix}$$

$$= \begin{pmatrix} \mathrm{Id} & -\Phi R(\lambda, A_0) \\ 0 & \mathrm{Id} \end{pmatrix} \begin{pmatrix} \lambda x - Bx - \Phi_\lambda x \\ (\lambda - A_0)(f - \epsilon_\lambda x) \end{pmatrix}$$

$$= \begin{pmatrix} \lambda x - Bx - \Phi_\lambda x - \Phi R(\lambda, A_0)(\lambda - A_0)(f - \epsilon_\lambda x) \\ (\lambda - A_0)(f - \epsilon_\lambda x) \end{pmatrix}$$

$$= \begin{pmatrix} \lambda x - Bx - \Phi f \\ \lambda f - f' \end{pmatrix} = (\lambda - \mathcal{A})\begin{pmatrix} x \\ f \end{pmatrix}.$$

Conversely, let $\begin{pmatrix} x \\ f \end{pmatrix} \in \mathcal{E}_p$ such that $S\begin{pmatrix} x \\ f \end{pmatrix} \in D(\mathcal{A}_\lambda)$. Then necessarily $x \in D(B)$, $f \in W^{1,p}([-1,0], X)$, and $f(0) = x$. Hence, $\begin{pmatrix} x \\ f \end{pmatrix} \in D(\mathcal{A})$.

Notice that the operators T and S are invertible on the space $\mathcal{E}_p$ with inverses given by

$$T^{-1} = \begin{pmatrix} \mathrm{Id} & \Phi R(\lambda, A_0) \\ 0 & \mathrm{Id} \end{pmatrix} \quad \text{and} \quad S^{-1} = \begin{pmatrix} \mathrm{Id} & 0 \\ \epsilon_\lambda \otimes \mathrm{Id} & \mathrm{Id} \end{pmatrix},$$

respectively. Hence the operator

$$\mathcal{A}_\lambda := \begin{pmatrix} \lambda - B - \Phi_\lambda & 0 \\ 0 & \lambda - A_0 \end{pmatrix}$$

with $D(\mathcal{A}_\lambda) := D(B) \times D(A_0)$ has the same spectral properties as $(\lambda - \mathcal{A}, D(\mathcal{A}))$. That is, the two operators $(\lambda - \mathcal{A})$ and $\mathcal{A}_\lambda$ are simultaneously injective, bounded from below, have closed range, or are Fredholm. Regarding the operator $\mathcal{A}$, these properties are equivalent to $\lambda \notin P\sigma(\mathcal{A})$, $\lambda \notin A\sigma(\mathcal{A})$, $\lambda \notin R\sigma(\mathcal{A})$, and $\lambda \notin \sigma_{ess}(\mathcal{A})$, respectively. These statements in turn are equivalent to $0 \notin P\sigma(\mathcal{A}_\lambda)$, $0 \notin A\sigma(\mathcal{A}_\lambda)$, $0 \notin R\sigma(\mathcal{A}_\lambda)$, and $0 \notin \sigma_{ess}(\mathcal{A}_\lambda)$, respectively. For example, this means for the essential spectrum that $\lambda \in \sigma_{ess}(\mathcal{A})$ if and only if $0 \in \sigma_{ess}(\mathcal{A}_\lambda)$. But using the fact that $\mathcal{A}_\lambda$ is a diagonal operator and that $\sigma(A_0) = \emptyset$, we see that $0 \in \sigma_{ess}(\mathcal{A}_\lambda)$ if and only if $\lambda \in \sigma_{ess}(B + \Phi_\lambda)$. $\qquad\square$

Example 3.21. We consider here a simple example. Let $X := \mathbb{C}$, $B = 0$, and $\Phi := c\,\delta_{-1}$, where $c \in \mathbb{C}$ is a constant and δ_{-1} is the point evaluation at $\sigma = -1$. The delay equation becomes

$$\begin{cases} u'(t) = c\,u(t-1) \text{ for } t \geq 0, \\ u(0) = x, \\ u_0 = f \end{cases}$$

and the associated operator matrix on the space $\mathbb{C} \times L^p[-1, 0]$ is

$$\mathcal{A} = \begin{pmatrix} 0 & c\,\delta_{-1} \\ 0 & \frac{d}{d\sigma} \end{pmatrix}$$

with domain

$$D(\mathcal{A}) = \left\{ \begin{pmatrix} x \\ f \end{pmatrix} \in \mathbb{C} \times W^{1,p}[-1, 0] \ : \ f(0) = x \right\}.$$

By Proposition 3.19 and Equation (3.26), the spectrum of $\mathcal{A}$ is given by

$$\sigma(\mathcal{A}) = \{\lambda \in \mathbb{C} \ : \ \lambda = c\,e^{-\lambda}\}.$$

More examples and more applications of Proposition 3.19 appear in the following chapters.

Remark 3.22. Note that all the above results hold for all $1 \leq p < \infty$; hence $\mathcal{A}$ has the same spectral properties in each $\mathcal{E}_p$.

3.3 Well-Posedness: Bounded Operators in the Delay Term

In Section 3.1, we transformed the problem of solving the partial differential equation with delay (DE_p) into the following problem:

> Determine when does $(\mathcal{A}, D(\mathcal{A}))$ generate a strongly continuous semigroup on $\mathcal{E}_p$.

We proceed to give sufficient conditions on $(B, D(B))$ and Φ such that this is true. In order to do so we start with the important special case $Z = X$, i.e., the case of bounded delay operators, and refer to Section 3.4 below for the case $Z \neq X$.

3.3.1 Ordinary Differential Equations with Delay

By this title, we mean the case when the operator B is bounded or, in particular, the space X is finite-dimensional.

We fix $1 \leq p < \infty$ for the rest of this section and recall the definition of the operator $(\mathcal{A}, D(\mathcal{A}))$ from Equations (3.8) and (3.9),

$$\mathcal{A} := \begin{pmatrix} B & \Phi \\ 0 & \frac{d}{d\sigma} \end{pmatrix}$$

and

$$D(\mathcal{A}) := \left\{ \begin{pmatrix} x \\ f \end{pmatrix} \in X \times L^p([-1,0], X) \ : \ f(0) = x \right\}.$$

Moreover, we recall that $(A_0, D(A_0))$ denotes the generator of the nilpotent left shift semigroup $(T_0(t))_{t \geq 0}$ on the space $L^p([-1,0], X)$. To be more precise,

$$D(A_0) := \{ f \in W^{1,p}([-1,0], X) \ : \ f(0) = 0 \},$$

$$A_0 f = f' := \frac{d}{d\sigma} f$$

and

$$(T_0(t)f)(\sigma) := \begin{cases} f(t + \sigma) & \text{for } t + \sigma \leq 0, \\ 0 & \text{for } t + \sigma > 0. \end{cases}$$

Moreover, let $\mathbb{I} : X \to L^p([-1,0], X)$ be the operator that maps $x \in X$ into the function identically equal to x in $L^p([-1,0], X)$, i.e., $\mathbb{I}(x)(\sigma) := x$ for all $\sigma \in [-1, 0]$.

Let us state the main result. The rest of this subsection is devoted to its proof that is due to K.-J. Engel [66] (see also [57]).

Theorem 3.23. *If $\Phi : W^{1,p}([-1,0], X) \to X$ is a bounded operator, then the operator matrix*

$$\mathcal{A} := \begin{pmatrix} B & \Phi \\ 0 & \frac{d}{d\sigma} \end{pmatrix}$$

with domain

$$D(\mathcal{A}) = \left\{ \begin{pmatrix} x \\ f \end{pmatrix} \in X \times L^p([-1,0], X) \ : \ f(0) = x \right\}$$

generates a strongly continuous semigroup on the space $\mathcal{E}_p = X \times L^p([-1,0], X)$ for all $1 \leq p < \infty$.

In particular, the corresponding delay equation is well-posed and its (mild and classical) solutions are determined by the semigroup generated by $(\mathcal{A}, D(\mathcal{A}))$ (see Section 3.1).

Proof. Let $1 \leq p < \infty$ and $\Phi \in \mathcal{L}(W^{1,p}([-1,0], X), X)$. If B is bounded, the matrix $\mathcal{A}$ can be written as

$$\mathcal{A} := \begin{pmatrix} B & \Phi \\ 0 & \frac{d}{d\sigma} \end{pmatrix} = \begin{pmatrix} 0 & \Phi \\ 0 & \frac{d}{d\sigma} \end{pmatrix} + \begin{pmatrix} B & 0 \\ 0 & 0 \end{pmatrix},$$

where the matrix $\begin{pmatrix} B & 0 \\ 0 & 0 \end{pmatrix}$ is a bounded operator on the space $\mathcal{E}_p$. Therefore, $\mathcal{A}$ is a generator if and only if the matrix $\begin{pmatrix} 0 & \Phi \\ 0 & \frac{d}{d\sigma} \end{pmatrix}$ is a generator.

Hence we may assume $B = 0$ without loss of generality.

As was seen in the proof of Lemma 3.20, the operator $\mathcal{A}$ can be written as

$$\mathcal{A} = \begin{pmatrix} \text{Id} & -\Phi A_0^{-1} \\ 0 & \text{Id} \end{pmatrix} \begin{pmatrix} \Phi_0 & 0 \\ 0 & A_0 \end{pmatrix} \begin{pmatrix} \text{Id} & 0 \\ -\mathbf{1} \otimes \text{Id} & \text{Id} \end{pmatrix}, \qquad (3.27)$$

where $\Phi_0 \in \mathcal{L}(X)$ is defined by

$$\Phi_0 x := \Phi(\mathbf{1} \otimes \text{Id}) x = \Phi(\mathbf{1} \otimes x).$$

Note that the operators

$$\begin{pmatrix} \text{Id} & -\Phi A_0^{-1} \\ 0 & \text{Id} \end{pmatrix} \quad \text{and} \quad \begin{pmatrix} \text{Id} & 0 \\ -\mathbf{1} \otimes \text{Id} & \text{Id} \end{pmatrix}$$

are bounded and invertible with inverse

$$\begin{pmatrix} \text{Id} & \Phi A_0^{-1} \\ 0 & \text{Id} \end{pmatrix} \quad \text{and} \quad \begin{pmatrix} \text{Id} & 0 \\ \mathbf{1} \otimes \text{Id} & \text{Id} \end{pmatrix},$$

respectively. Hence by similarity, $(\mathcal{A}, D(\mathcal{A}))$ is a generator on $\mathcal{E}_p$ if and only if the operator

$$
\begin{aligned}
\mathcal{B} : &= \begin{pmatrix} \mathrm{Id} & 0 \\ -\mathbb{1} \otimes \mathrm{Id} & \mathrm{Id} \end{pmatrix} \mathcal{A} \begin{pmatrix} \mathrm{Id} & 0 \\ \mathbb{1} \otimes \mathrm{Id} & \mathrm{Id} \end{pmatrix} \\
&= \begin{pmatrix} \mathrm{Id} & 0 \\ -\mathbb{1} \otimes \mathrm{Id} & \mathrm{Id} \end{pmatrix} \begin{pmatrix} \mathrm{Id} & -\Phi A_0^{-1} \\ 0 & \mathrm{Id} \end{pmatrix} \begin{pmatrix} \Phi_0 & 0 \\ 0 & A_0 \end{pmatrix}
\end{aligned}
$$

with domain $D(\mathcal{B}) := X \times D(A_0)$ is a generator.

In order to show the latter, we compute

$$
\begin{aligned}
\begin{pmatrix} \mathrm{Id} & 0 \\ -\mathbb{1} \otimes \mathrm{Id} & \mathrm{Id} \end{pmatrix} \begin{pmatrix} \mathrm{Id} & -\Phi A_0^{-1} \\ 0 & \mathrm{Id} \end{pmatrix} &= \begin{pmatrix} \mathrm{Id} & -\Phi A_0^{-1} \\ -\mathbb{1} \otimes \mathrm{Id} & \mathbb{1} \otimes \Phi A_0^{-1} + \mathrm{Id} \end{pmatrix} \\
&= : \mathrm{Id}_{\mathcal{E}_p} + \mathcal{C},
\end{aligned}
$$

where

$$
\mathcal{C} := \begin{pmatrix} 0 & -\Phi A_0^{-1} \\ -\mathbb{1} \otimes \mathrm{Id} & \mathbb{1} \otimes \Phi A_0^{-1} \end{pmatrix}
$$

is a bounded operator on $\mathcal{E}_p$. This implies that

$$
\mathcal{B} = (\mathrm{Id}_{\mathcal{E}_p} + \mathcal{C}) \begin{pmatrix} \Phi_0 & 0 \\ 0 & A_0 \end{pmatrix}.
$$

Let $S(t) := e^{t\Phi_0}$. Then the matrix

$$
\begin{pmatrix} \Phi_0 & 0 \\ 0 & A_0 \end{pmatrix}
$$

with domain $D(\mathcal{B}) = X \times D(A_0)$ generates the semigroup $(\mathcal{T}(t))_{t \geq 0}$, where

$$
\mathcal{T}(t) := \begin{pmatrix} S(t) & 0 \\ 0 & T_0(t) \end{pmatrix}.
$$

Now we use the Desch-Schappacher Perturbation Theorem 1.38: for $\binom{f_1}{f_2} \in L^p([0,1], \mathcal{E}_p)$ we have

$$
\begin{aligned}
&\int_0^1 \mathcal{T}(1-r) \, \mathcal{C} \left(\begin{smallmatrix} f_1(r) \\ f_2(r) \end{smallmatrix} \right) dr \\
&= \int_0^1 \begin{pmatrix} S(1-r) & 0 \\ 0 & T_0(1-r) \end{pmatrix} \begin{pmatrix} 0 & -\Phi A_0^{-1} \\ -\mathbb{1} \otimes \mathrm{Id} & \mathbb{1} \otimes \Phi A_0^{-1} \end{pmatrix} \left(\begin{smallmatrix} f_1(r) \\ f_2(r) \end{smallmatrix} \right) dr \\
&= \int_0^1 \begin{pmatrix} S(1-r) & 0 \\ 0 & T_0(1-r) \end{pmatrix} \begin{pmatrix} -\Phi A_0^{-1} f_2(r) \\ \mathbb{1} \otimes (\Phi A_0^{-1} f_2(r) - f_1(r)) \end{pmatrix} dr \\
&= \begin{pmatrix} -\int_0^1 S(1-r)(\Phi A_0^{-1} f_2(r)) \, dr \\ \int_0^1 T_0(1-r)(\mathbb{1} \otimes (\Phi A_0^{-1} f_2(r) - f_1(r))) \, dr \end{pmatrix}.
\end{aligned}
$$

The operator $\Phi A_0^{-1} : L^p([-1,0], X) \to X$ is bounded; hence $g := (\Phi A_0^{-1} f_2 - f_1) \in L^p([0,1], X)$ and we have that

$$\int_0^1 T_0(1-r)(\mathbb{1} \otimes g(r)) \, dr \, (\cdot)$$
$$= \int_{\cdot}^0 g(1+r) \, dr \in D(A_0).$$

Moreover, we have

$$\int_0^1 S(1-r)(\Phi A_0^{-1} f_2(r)) \, dr \in X. \tag{3.28}$$

Hence,

$$\int_0^1 \mathcal{T}(1-r) \, \mathcal{C} \left(\begin{smallmatrix} f_1(r) \\ f_2(r) \end{smallmatrix} \right) \, dr \in X \times D(A_0) = D(\mathcal{B}),$$

and the result follows by Theorem 1.38. $\square$

Example 3.24. We give here an example of a delay equation (DE_p) of *neutral type*. Let $X = \mathbb{C}$ and consider the initial value problem

$$\begin{cases} u'(t) = \int_{-1}^0 h(\sigma) u'(t+\sigma) \, d\sigma \text{ for } t \geq 0, \\ u(0) = x, \\ u_0 = f, \end{cases} \tag{3.29}$$

where

- $u : [-1, \infty) \to \mathbb{C}$ is a function,
- $1 \leq p < \infty$ and $1 < q \leq \infty$ are such that $\frac{1}{p} + \frac{1}{q} = 1$,
- $h \in L^q([-1,0])$, and
- $x \in \mathbb{C}$ and $f \in L^p([-1,0])$.

If we define $\Phi g := \int_{-1}^0 h(\sigma) \, g'(\sigma) \, d\sigma$, then the operator $\Phi : W^{1,p}([-1,0]) \to \mathbb{C}$ is bounded and we can rewrite Equation (3.29) as an abstract delay equation (with $B = 0$)

$$\begin{cases} u'(t) = \Phi u_t \text{ for } t \geq 0, \\ u(0) = x \\ u_0 = f. \end{cases} \tag{DE}$$

This equation is well-posed by Theorem 3.23. In particular, for every initial value $\left(\begin{smallmatrix} x \\ f \end{smallmatrix} \right) \in \mathbb{C} \times W^{1,p}([-1,0])$ with $f(0) = x$, there is a unique classical solution, while for each $\left(\begin{smallmatrix} x \\ f \end{smallmatrix} \right) \in \mathbb{C} \times L^p([-1,0])$, we have a unique mild solution.

3.3.2 Partial Differential Equations with Delay

Now we consider the case where the operator B is unbounded. In concrete cases, B is a differential operator and (DE_p) becomes a partial differential equation with delay (see the examples in Section 3.1).

The factoring technique used above does not work in this situation, but we can still use perturbation methods.

Let $1 \leq p < \infty$. We write

$$\mathcal{A} := \begin{pmatrix} B & \Phi \\ 0 & \frac{d}{d\sigma} \end{pmatrix}$$

with domain

$$D(\mathcal{A}) := \left\{ \begin{pmatrix} x \\ f \end{pmatrix} \in D(B) \times W^{1,p}([-1,0], X) \ : \ f(0) = x \right\}$$

as the sum $\mathcal{A}_0 + \mathcal{B}$, where

$$\mathcal{A}_0 := \begin{pmatrix} B & 0 \\ 0 & \frac{d}{d\sigma} \end{pmatrix} \tag{3.30}$$

with domain

$$D(\mathcal{A}_0) := D(\mathcal{A}) \tag{3.31}$$

and

$$\mathcal{B} := \begin{pmatrix} 0 & \Phi \\ 0 & 0 \end{pmatrix} \in \mathcal{L}(D(\mathcal{A}_0), \mathcal{E}_p). \tag{3.32}$$

The idea is to show first that $\mathcal{A}_0$ becomes a generator under appropriate assumptions, and then to apply perturbation results to show that the sum $\mathcal{A}_0 + \mathcal{B}$ is a generator as well. The first step is quite straightforward.

Theorem 3.25. *The following are equivalent:*

(i) *The operator $(B, D(B))$ generates a strongly continuous semigroup $(S(t))_{t\geq 0}$ on X.*

(ii) *The operator matrix $(\mathcal{A}_0, D(\mathcal{A}_0))$ generates a strongly continuous semigroup $(\mathcal{T}_0(t))_{t\geq 0}$ on the space $\mathcal{E}_p$ for all $1 \leq p < \infty$.*

(iii) *The operator matrix $(\mathcal{A}_0, D(\mathcal{A}_0))$ generates a strongly continuous semigroup $(\mathcal{T}_0(t))_{t\geq 0}$ on the space $\mathcal{E}_p$ for one $1 \leq p < \infty$.*

In this case, the semigroup $(\mathcal{T}_0(t))_{t\geq 0}$ is given by

$$\mathcal{T}_0(t) := \begin{pmatrix} S(t) & 0 \\ S_t & T_0(t) \end{pmatrix}, \tag{3.33}$$

where $(T_0(t))_{t\geq 0}$ is the nilpotent left shift semigroup on $L^p([-1,0],X)$ and $S_t : X \to L^p([-1,0],X)$ is defined by

$$(S_t x)(\tau) := \begin{cases} S(t+\tau)x & \text{if } -t < \tau \leq 0, \\ 0 & \text{if } -1 \leq \tau \leq -t. \end{cases}$$

Proof. We show (i)$\Rightarrow$(ii). Assume that $(B, D(B))$ generates a strongly continuous semigroup $(S(t))_{t\geq 0}$ on X. Let $1 \leq p < \infty$ and $(\mathcal{J}_0(t))_{t\geq 0}$ be the family of operators defined in Equation (3.33).

We endeavor to show that $(\mathcal{J}_0(t))_{t\geq 0}$ is a strongly continuous semigroup on $\mathcal{E}_p$.

In order to show the strong continuity of $(\mathcal{J}_0(t))_{t\geq 0}$, it suffices to show the strong continuity of the map $t \to S_t$. To this purpose, fix $x \in X$ and $0 \leq s \leq t$ and consider the limit

$$\lim_{t\to s} \|S_t x - S_s x\|_{L^p([-1,0],X)}^p = \lim_{t\to s} \int_{-1}^0 \|(S_t x)(\sigma) - (S_s x)(\sigma)\|^p \, d\sigma$$

$$= \lim_{t\to s} \int_{-s}^0 \|S(t+\sigma)\, x - S(s+\sigma)\, x\|^p \, d\sigma$$

$$+ \lim_{t\to s} \int_{-t}^{-s} \|S(t+\sigma)\, x\|^p \, d\sigma,$$

which is equal to 0 by Lebesgue's Dominated Convergence Theorem. Hence, the map $t \to S_t$ is strongly right continuous and analogously one can show that it is also strongly left continuous.

Now we show the semigroup property.[3] To do this, we compute the product

$$\begin{pmatrix} S(t) & 0 \\ S_t & T_0(t) \end{pmatrix} \begin{pmatrix} S(s) & 0 \\ S_s & T_0(s) \end{pmatrix} = \begin{pmatrix} S(t)S(s) & 0 \\ S_t S(s) + T_0(t) S_s & T_0(t) T_0(s) \end{pmatrix}$$

for $s, t \geq 0$. By definition,

$$(S_t S(s)\, x)(\tau) := \begin{cases} S(t+\tau)S(s)x & \text{if } t+\tau > 0, \\ 0 & \text{if } t+\tau \leq 0, \end{cases}$$

and

$$(T_0(t)S_s\, x)(\tau) := \begin{cases} (S_s x)(t+\tau) & \text{if } t+\tau \leq 0, \\ 0 & \text{if } t+\tau > 0. \end{cases}$$

$$= \begin{cases} S(s+t+\tau)x & \text{if } t+\tau \leq 0 \text{ and } s+t+\tau > 0 \\ 0 & \text{if } t+\tau \leq 0 \text{ and } s+t+\tau \leq 0 \\ 0 & \text{if } t+\tau > 0. \end{cases}$$

---

[3]This proof is due to Fragnelli and Nickel [81].

Combining these expressions, we obtain that

$$(S_t S(s) + T_0(t) S_s) x(\tau) = \begin{cases} S(s+t+\tau)x & \text{if } t+\tau > 0, \\ S(s+t+\tau)x & \text{if } t+\tau \leq 0 \text{ and } s+t+\tau > 0 \\ 0 & \text{if } t+\tau \leq 0 \text{ and } s+t+\tau \leq 0 \end{cases}$$

$$= (S_{s+t} x)(\tau)$$

for each $\tau \in [-1, 0]$. Hence, $(\mathcal{T}_0(t))_{t \geq 0}$ is a strongly continuous semigroup on the space $\mathcal{E}_p$.

Let $(\mathcal{C}, D(\mathcal{C}))$ be its generator and let $\Re\lambda > \omega_0(\mathcal{T}_0)$. We know that $\lambda \in \rho(\mathcal{C})$ and that the resolvent $R(\lambda, \mathcal{C})$ is given by the Laplace transform of the semigroup $(\mathcal{T}_0(t))_{t \geq 0}$, hence we may calculate

$$R(\lambda, \mathcal{C})\begin{pmatrix} x \\ f \end{pmatrix} = \int_0^\infty e^{-\lambda s} \mathcal{T}_0(s) \begin{pmatrix} x \\ f \end{pmatrix} ds$$

$$= \int_0^\infty e^{-\lambda s} \begin{pmatrix} S(s)\, x \\ S_s\, x + T_0(s)\, f \end{pmatrix} ds$$

$$= \begin{pmatrix} \int_0^\infty e^{-\lambda s} S(s)\, x\, ds \\ \int_0^\infty e^{-\lambda s} S_s\, x\, ds + \int_0^\infty e^{-\lambda s} T_0(s) f\, ds \end{pmatrix}$$

$$= \begin{pmatrix} R(\lambda, B) x \\ \int_0^\infty e^{-\lambda s} S_s\, x\, ds + R(\lambda, A_0) f \end{pmatrix}.$$

Now we compute $\int_0^\infty e^{-\lambda s} S_s\, x\, ds$ explicitly. Using Lemma A.5 from the Appendix, the equality

$$\left(\int_0^\infty e^{-\lambda s} S_s\, x\, ds \right)(\sigma) = \int_0^\infty e^{-\lambda s} (S_s\, x)(\sigma)\, ds$$

$$= \int_\sigma^\infty e^{-\lambda s} S(s+\sigma)x\, ds$$

$$= \int_0^\infty e^{-\lambda s + \lambda \sigma} S(s)x\, ds$$

$$= e^{\lambda \sigma} R(\lambda, B)\, x$$

holds, for a.e. $\sigma \in [-1, 0]$. Hence we obtain from an application of Proposition 3.19 that

$$R(\lambda, \mathcal{C}) = \begin{pmatrix} R(\lambda, B) & 0 \\ \varepsilon_\lambda \otimes R(\lambda, B) & R(\lambda, A_0) \end{pmatrix} = R(\lambda, \mathcal{A}_0)$$

and so $(\mathcal{C}, D(\mathcal{C}))$ is equal to $(\mathcal{A}_0, D(\mathcal{A}_0))$.

The implication (ii)⇒(iii) is trivial.

We show (iii)⇒(i). Assume that $(\mathcal{A}_0, D(\mathcal{A}_0))$ generates a semigroup on $\mathcal{E}_p$ for one $1 \leq p < \infty$. Then by Theorem 3.12, the delay equation with $\Phi = 0$

$$\begin{cases} u'(t) = B\,u(t) \text{ for } t \geq 0, \\ u(0) = x, \\ u_0 = f \end{cases} \tag{3.34}$$

is well-posed, that is, for all $\left(\begin{smallmatrix} x \\ f \end{smallmatrix}\right) \in D(\mathcal{A}_0) \subseteq D(B) \times W^{1,p}([-1,0], X)$ there exists a unique classical solution $u \in C([-1,\infty), X) \cap C^1([0,\infty), X)$ of Equation (3.34) and the solutions depend continuously on the initial data.

The restriction $u|_{[0,\infty)}$ is a classical solution of the abstract Cauchy problem associated to the operator $(B, D(B))$

$$\begin{cases} u'(t) = Bu(t) \text{ for } t \geq 0, \\ u(0) = x. \end{cases} \tag{3.35}$$

In particular, this abstract Cauchy problem is well-posed. Furthermore, since B is closed and densely defined by assumption, it generates a strongly continuous semigroup on X by Theorem 1.22. □

We are now ready to prove a sufficient condition for well-posedness of (DE$_p$). The following result constitutes the main theorem of this section.

Theorem 3.26. *Let $(B, D(B))$ be the generator of a strongly continuous semigroup $(S(t))_{t\geq 0}$ on X, let $1 \leq p < \infty$, and let $\Phi : W^{1,p}([-1,0], X) \to X$ be a delay operator. Assume that there exist constants $t_0 > 0$ and $0 < q < 1$ such that*

$$\int_0^{t_0} \|\Phi(S_r\,x + T_0(r)f)\| \, dr \leq q \left\|\left(\begin{smallmatrix} x \\ f \end{smallmatrix}\right)\right\| \tag{M}$$

for all $\left(\begin{smallmatrix} x \\ f \end{smallmatrix}\right) \in D(\mathcal{A})$. Then the operator $(\mathcal{A}, D(\mathcal{A}))$ is the generator of a strongly continuous semigroup on $\mathcal{E}_p$ and so (DE$_p$) is well-posed.

Proof. By Theorem 3.25, we know that $(\mathcal{A}_0, D(\mathcal{A}_0))$ generates the strongly continuous semigroup $(\mathcal{T}_0(t))_{t\geq 0}$ given in Equation (3.33). Now we apply the Miyadera-Voigt Perturbation Theorem (see Theorem 1.37) to the operator

$$\mathcal{B} := \begin{pmatrix} 0 & \Phi \\ 0 & 0 \end{pmatrix}.$$

We need to check that Condition (1.9) in Theorem 1.37 is satisfied, that is, we show that there exist $t_0 > 0$ and $0 \leq q < 1$ such that

$$\int_0^{t_0} \left\| \mathcal{B}\, \mathcal{T}_0(r)\left(\begin{smallmatrix} x \\ f \end{smallmatrix}\right) \right\| dr \leq q \left\| \left(\begin{smallmatrix} x \\ f \end{smallmatrix}\right) \right\| \tag{3.36}$$

for all $\left(\begin{smallmatrix} x \\ f \end{smallmatrix}\right) \in D(\mathcal{A}_0)$. Since

$$\int_0^{t_0} \left\| \mathcal{B}\, \mathcal{T}_0(r)\left(\begin{smallmatrix} x \\ f \end{smallmatrix}\right) \right\| dr = \int_0^{t_0} \left\| \begin{pmatrix} 0 & \Phi \\ 0 & 0 \end{pmatrix} \begin{pmatrix} S(r) & 0 \\ S_r & T_0(r) \end{pmatrix} \begin{pmatrix} x \\ f \end{pmatrix} \right\| dr$$

$$= \int_0^{t_0} \left\| \Phi(S_r x + T_0(r)f) \right\| dr, \tag{3.37}$$

we conclude that Equation (3.36) holds if and only if there exist $t_0 > 0$ and $0 \leq q < 1$ such that Condition (M) holds for all $\left(\begin{smallmatrix} x \\ f \end{smallmatrix}\right) \in D(\mathcal{A}_0)$. $\qquad\square$

Example 3.27. Let Φ be bounded from $L^p([-1,0], X)$ to X, for example,

$$\Phi f := \int_{-1}^0 h(\sigma)f(\sigma)\, d\sigma,$$

where the function h is an $\mathcal{L}(X)$-valued q-integrable function for $\frac{1}{p} + \frac{1}{q} = 1$, i.e., $h \in L^q([-1,0], \mathcal{L}(X))$. Then the perturbation $\mathcal{B}$ is bounded, and $(\mathcal{A}, D(\mathcal{A}))$ is a generator on $\mathcal{E}_p$.

3.3.3   The Main Example

We assume as usual that X is a Banach space and that $(B, D(B))$ generates a strongly continuous semigroup $(S(t))_{t \geq 0}$ on X.

This section is devoted to a large class of delay operators Φ such that the matrix $(\mathcal{A}, D(\mathcal{A}))$ becomes a generator for every semigroup generator $(B, D(B))$.

We have already seen in Section 3.3.1 that in the case that $B \in \mathcal{L}(X)$, then $(\mathcal{A}, D(\mathcal{A}))$ is a generator for all $\Phi \in \mathcal{L}(W^{1,p}([-1,0], X), X)$.

Whether the analogous statement is true in the case that B is unbounded is still an open question as of the writing of this manuscript.

However, we can presently show that $(\mathcal{A}, D(\mathcal{A}))$ is a generator for a very large class of delay operators.

Let η : $[-1,0]$ $\rightarrow$ $\mathcal{L}(X)$ be of bounded variation and let Φ : $C([-1,0],X)$ $\rightarrow$ X be the bounded linear operator given by the Riemann-Stieltjes Integral,

$$\Phi(f) := \int_{-1}^{0} d\eta\, f\,. \tag{3.38}$$

Since $W^{1,p}([-1,0],X)$ is continuously embedded in $C([-1,0],X)$, we may note that Φ defines a bounded operator from $W^{1,p}([-1,0],X)$ to X.

Example 3.28. An important special case of Equation (3.38) are the so-called "discrete" delay operators, that is, operators Φ of the form

$$\Phi(f) := \sum_{k=0}^{n} B_k f(-h_k), \quad f \in W^{1,p}([-1,0],X), \tag{3.39}$$

where $B_k \in \mathcal{L}(X)$ and $h_k \in [0,1]$ for each $k = 0,\dots,n$. In fact, we have

$$\Phi f = \int_{-1}^{0} d\eta(\sigma)\, f(\sigma)$$

for $\eta := \sum_{k=0}^{n} B_k \chi_{[-h_k,0]}$.

Now we show that all delay operators as in Equation (3.38) satisfy Condition (M) for each $1 \leq p < \infty$.

Theorem 3.29. *Let $1 \leq p < \infty$ and let Φ be as in Equation (3.38). Then the operator matrix $(\mathcal{A}, D(\mathcal{A}))$ defined by*

$$\mathcal{A} := \begin{pmatrix} B & \Phi \\ 0 & \frac{d}{d\sigma} \end{pmatrix},$$

$$D(\mathcal{A}) := \left\{ \begin{pmatrix} x \\ f \end{pmatrix} \in D(B) \times W^{1,p}([-1,0],Z) \; : \; f(0) = x \right\}$$

generates a strongly continuous semigroup, hence the corresponding delay equation (DE_p) is well-posed.

Proof. First we consider the case $1 < p < \infty$. Using Fubini's Theorem and Hölder's Inequality, we obtain for $0 < t < 1$ that

$$
\int_0^t \|\Phi(S_r\, x + T_0(r)f)\|\, dr
$$

$$
= \int_0^t \left\| \int_{-1}^{-r} d\eta(\sigma)\, f(r+\sigma) + \int_{-r}^0 d\eta(\sigma)\, S(r+\sigma)x \right\| dr
$$

$$
\leq \int_0^t \int_{-1}^{-r} \|f(r+\sigma)\|\, d|\eta|(\sigma)\, dr + \int_0^t \int_{-r}^0 \|S(r+\sigma)x\|\, d|\eta|(\sigma)\, dr
$$

$$
\leq \int_{-t}^0 \int_\sigma^0 \|f(s)\|\, ds\, d|\eta|(\sigma) + \int_{-1}^{-t} \int_\sigma^{t+\sigma} \|f(s)\|\, ds\, d|\eta|(\sigma)
$$

$$
+ \int_0^t M\|x\| |\eta|([-1,0])\, dr
$$

$$
\leq \int_{-t}^0 (-\sigma)^{1/p'} \|f\|_p\, d|\eta|(\sigma) + \int_{-1}^{-t} t^{1/p'} \|f\|_p\, d|\eta|(\sigma)
$$

$$
+ tM\|x\| |\eta|([-1,0])
$$

$$
\leq \int_{-1}^0 t^{1/p'} \|f\|_p\, d|\eta|(\sigma) + tM\|x\| |\eta|([-1,0])
$$

$$
= (t^{1/p'} \|f\|_p + tM\|x\|)\, |\eta|([-1,0]), \tag{3.40}
$$

where $\frac{1}{p} + \frac{1}{p'} = 1$,

$$
M := \sup_{r\in[0,1]} \|S(r)\|,
$$

and $|\eta|$ is the positive Borel measure on $[-1,0]$ defined by the total variation of η.

Finally, we conclude that

$$
\int_0^t \|\Phi(S_r x + T_0(r)f)\|\, dr \leq t^{1/p'} M\, |\eta|([-1,0])(\|x\| + \|f\|_p) \tag{3.41}
$$

for all $0 < t < 1$. Choose now t_0 small enough such that

$$
t_0^{1/p'} M\, |\eta|([-1,0]) < 1.
$$

Then condition (M) is satisfied with $q := t_0^{1/p'} M\, |\eta|([-1,0])$.

Now we consider the case $p = 1$.[4] On the space $\mathcal{E}_1 = X \times L^1([-1,0], X)$, we choose the equivalent norm

$$
\left\| \left(\begin{smallmatrix} x \\ f \end{smallmatrix} \right) \right\|_c := \|x\| + c\|f\|_1
$$

---

[4]This result goes back to Maniar and Voigt [139]. This proof is due to Stein, Vogt, and Voigt [197].

with $c > |\eta|([-1,0]$. Then by Equation (3.40), we obtain

$$\int_0^t \left\| \mathcal{B}\mathcal{T}_0(r)\left(\begin{smallmatrix} x \\ f \end{smallmatrix}\right) \right\|_c dr = \int_0^t \left\| \Phi(S_r x + T_0(r)f) \right\| dr$$
$$\leq tM \, |\eta|([-1,0])\|x\| + |\eta|([-1,0])\|f\|_1$$
$$\leq \max\{tM \, |\eta|([-1,0]), c^{-1}|\eta|([-1,0])\} \, \left\|\left(\begin{smallmatrix} x \\ f \end{smallmatrix}\right)\right\|_c,$$

for all $\left(\begin{smallmatrix} x \\ f \end{smallmatrix}\right) \in D(\mathcal{A}_0)$ and all $0 \leq t \leq 1$. Since

$$\max\{tM \, |\eta|([-1,0]), c^{-1}|\eta|([-1,0])\} < 1,$$

for all $0 \leq t < (M \, |\eta|([-1,0]))^{-1}$, it follows that $\mathcal{B}$ is a Miyadera-Voigt perturbation of $\mathcal{A}_0$ and, by Theorem 1.37, the operator $(\mathcal{A}, D(\mathcal{A}))$ is a generator. $\square$

Example 3.30. As a consequence, we have that

(i) the reaction-diffusion equation with delay in Example 3.15 is well-posed and

(ii) the age-structured population equation with delay in Example 3.16 is well-posed.

3.4 Well-Posedness: Unbounded Operators in the Delay Term

In this section, we give sufficient conditions for the well-posedness of the delay equation (DE_p) even in the case that $Z \subsetneq X$, i.e., in the case of unbounded delay operators. We assume that all Hypotheses (H_1)–(H_7) from Section 3.1 hold for some $1 \leq p < \infty$. Moreover, to be able to carry out the same perturbation argument as in Section 3.3.2, we must make additional assumptions on $(B, D(B))$.

To this end, we assume in this section that

$$(B, D(B)) \text{ generates an analytic semigroup } (S(t))_{t\geq0} \text{ on } X, \qquad (3.42)$$

and that for some $\delta > \omega_0(B)$, there exists $\vartheta < \frac{1}{p}$ such that

$$D((-B+\delta)^\vartheta) \overset{d}{\hookrightarrow} Z \overset{d}{\hookrightarrow} X. \qquad (3.43)$$

Remark 3.31. For $\delta, \nu > \omega_0(B)$, the equality

$$D((-B + \delta)^\theta) = D((-B + \nu)^\theta)$$

holds for every $0 < \theta < 1$ (see Theorem 1.27 (i)). Therefore, the space $D((-B + \delta)^\theta)$ is independent of the choice of $\delta > \omega_0(B)$.

In the case that $\omega_0(B) < 0$, it suffices to choose $\delta = 0$.

Recall that for an analytic semigroup $(S(t))_{t\geq 0}$ with generator $(B, D(B))$, one always has

$$S(t)X \subset D(B^n) \tag{3.44}$$

for each $t > 0$ and for each $n \in \mathbb{N}$. Furthermore, recall that the estimate

$$\|(-B + \delta)^\beta e^{-\delta t} S(t)\| \leq \frac{K_\beta}{t^\beta} \tag{3.45}$$

holds for some constants $\beta > 0$ and $K_\beta > 0$, and for all $t > 0$ (see Theorem 1.27 (iv)).

We remind that $\mathcal{A}_0$ is the operator

$$\mathcal{A}_0 := \begin{pmatrix} B & 0 \\ 0 & \frac{d}{d\sigma} \end{pmatrix} \tag{3.46}$$

with domain

$$D(\mathcal{A}_0) := \left\{ \begin{pmatrix} x \\ f \end{pmatrix} \in D(B) \times W^{1,p}([-1,0], Z) \ : \ f(0) = x \right\}, \tag{3.47}$$

and we state the first result of this section.

Proposition 3.32. *Let $(B, D(B))$ be the generator of an analytic semigroup $(S(t))_{t\geq 0}$ on X. Then $(\mathcal{A}_0, D(\mathcal{A}_0))$ generates the strongly continuous semigroup $(\mathcal{T}_0(t))_{t\geq 0}$ on $\mathcal{E}_p$ given by*

$$\mathcal{T}_0(t) := \begin{pmatrix} S(t) & 0 \\ S_t & T_0(t) \end{pmatrix}, \tag{3.48}$$

where $(T_0(t))_{t\geq 0}$ is the (nilpotent) left shift semigroup on $L^p([-1,0], Z)$ and $S_t : X \to L^p([-1,0], Z)$ is defined by

$$(S_t x)(\tau) := \begin{cases} S(t + \tau)x & \text{if } -t < \tau \leq 0, \\ 0 & \text{if } -1 \leq \tau \leq -t. \end{cases}$$

Remark 3.33. Note that Equation (3.48) is well defined due to Conditions (3.44) and (3.45) on $(B, D(B))$. It is true because the semigroup $(S(t))_{t \geq 0}$ is analytic, it maps the entire space X into the domain $D(B) \overset{d}{\hookrightarrow} Z$. Hence, $S(t + \tau)x \in Z$ for all $x \in X$, $t \geq 0$ and $-t < \tau \leq 0$. Moreover, since $D((-B + \delta)^{\vartheta}) \overset{d}{\hookrightarrow} Z$ for some $\delta > \omega_0(B)$ and $\theta < \frac{1}{p}$, we have that the function $e^{-\delta(t + \cdot)} S_t \, x$ belongs to $L^p([-1, 0], Z)$ for all $x \in X$ and all $t \geq 0$. Therefore, we also have that $S_t x \in L^p([-1, 0], Z)$ for all $x \in X$ and all $t \geq 0$.

The proof of Proposition 3.32 is analogous to that of Theorem 3.25 and is therefore omitted.

The next step is to give sufficient conditions on the operator

$$\mathcal{B} := \begin{pmatrix} 0 & \Phi \\ 0 & 0 \end{pmatrix} \in \mathcal{L}(D(\mathcal{A}_0), \mathcal{E}_p)$$

for $\mathcal{A} = \mathcal{A}_0 + \mathcal{B}$ to be the generator of a strongly continuous semigroup on $\mathcal{E}_p$. We follow again the treatment in Section 3.3.2 and apply the Miyadera-Voigt Perturbation Theorem (see Theorem 1.37).

The Miyadera-Voigt Perturbation Theorem says that $(\mathcal{A}, D(\mathcal{A}))$ generates a strongly continuous semigroup on $\mathcal{E}_p$ if there exist $t_0 > 0$ and $0 \leq q < 1$ such that

$$\int_0^{t_0} \left\| \mathcal{B} \, \mathcal{T}_0(r) \begin{pmatrix} x \\ f \end{pmatrix} \right\| dr \leq q \left\| \begin{pmatrix} x \\ f \end{pmatrix} \right\| \tag{3.49}$$

for all $\begin{pmatrix} x \\ f \end{pmatrix} \in D(\mathcal{A}_0)$. However, since

$$\int_0^{t_0} \left\| \mathcal{B} \, \mathcal{T}_0(r) \begin{pmatrix} x \\ f \end{pmatrix} \right\| dr = \int_0^{t_0} \left\| \begin{pmatrix} 0 & \Phi \\ 0 & 0 \end{pmatrix} \begin{pmatrix} S(r) & 0 \\ S_r & T_0(r) \end{pmatrix} \begin{pmatrix} x \\ f \end{pmatrix} \right\| dr$$

$$= \int_0^{t_0} \left\| \Phi(S_r x + T_0(r)f) \right\| dr \, ,$$

we conclude that Equation (3.49) holds if and only if there exist $t_0 > 0$ and $0 \leq q < 1$ such that

$$\int_0^{t_0} \left\| \Phi(S_r \, x + T_0(r)f) \right\| dr \leq q \left\| \begin{pmatrix} x \\ f \end{pmatrix} \right\| \tag{M'}$$

for all $\left(\begin{smallmatrix} x \\ f \end{smallmatrix}\right) \in D(\mathcal{A}_0)$. Let us remark that this expression is well defined because $S_r x \in W^{1,p}([-1,0], Z)$ for $x \in D(B)$ and for $r \leq 1$. We therefore obtain the following result, which is the counterpart to Theorem 3.26.

Theorem 3.34. *Suppose that Conditions (3.42), (3.43), and (M') are satisfied. Then the operator $(\mathcal{A}, D(\mathcal{A}))$ is the generator of a strongly continuous semigroup on $\mathcal{E}_p$. Hence (DE_p) is well-posed.*

We are now ready to prove that $(\mathcal{A}, D(\mathcal{A}))$ is a generator for a large class of delay operators Φ. The following result is analogous to Theorem 3.29.

Theorem 3.35. *Let $1 \leq p < \infty$ and let $\eta : [-1,0] \to \mathcal{L}(Z, X)$ be of bounded variation. Let $\Phi : C([-1,0], Z) \to X$ be the bounded linear operator given by the Riemann-Stieltjes Integral*

$$\Phi(f) := \int_{-1}^{0} d\eta \, f \, . \tag{3.50}$$

Assume also that the operator $(B, D(B))$ satisfies the Conditions (3.42) and (3.43). Then Condition (M') of Theorem 3.34 is satisfied. In particular, the operator matrix $(\mathcal{A}, D(\mathcal{A}))$ generates a strongly continuous semigroup and so (DE_p) is well-posed.

Proof. Since $W^{1,p}([-1,0], Z)$ is continuously embedded in $C([-1,0], Z)$, Φ defines a bounded operator from $W^{1,p}([-1,0], Z)$ to X.

Let $1 < p < \infty$. For each $0 < t < 1$, we obtain the estimate

$$\int_{0}^{t} \|\Phi(S_r x + T_0(r)f)\| \, dr \leq t^{1/p'} M \, |\eta|([-1,0])(\|f\|_p + \|x\|) \tag{3.51}$$

by the same calculations as in the first part of the proof of Theorem 3.29. Now choose t_0 small enough so that

$$t_0^{1/p'} M \, |\eta|([-1,0]) < 1.$$

Then Condition (M') of Theorem 3.34 is satisfied with

$$q := t_0^{1/p'} M \, |\eta|([-1,0]),$$

where $|\eta|$ denotes the positive Borel measure on $[-1,0]$ defined by the total variation of η.

The proof in the case $p = 1$ is analogous to the proof of Theorem 3.29 and is left to the reader.   □

Example 3.36. The delay equation of Example 3.14 in Section 3.1 is well-posed.

3.5   Notes and References

The study of delay equations on the state space $X \times L^p([-1,0], X)$ with $X = \mathbb{C}^n$ was started in 1966 by Coleman and Mizel [38] and then developed by many authors. All these authors first solve the delay equation by using the Variation of Parameters Formula for inhomogeneous Cauchy problems and a fixed point argument and then show that the solutions form a semigroup on the product space and its generator is the operator $(\mathcal{A}, D(\mathcal{A}))$ defined in Equations (3.8) and (3.9). The converse approach is due to Webb in 1976. In [213], he considers the operator $(\mathcal{A}, D(\mathcal{A}))$ (even for nonlinear B and Φ) on a Hilbert space X and shows that for Φ given as in Equation (3.38), this generates a (nonlinear) semigroup on the Hilbert space $X \times L^p([-1,0], \mu dx, X)$ for a suitable weight function μ. The weight μ is necessary in order to obtain a norm where the operator $(\mathcal{A}, D(\mathcal{A}))$ becomes dissipative. Then he proves that the semigroup actually gives the solutions of the delay equation.

The results of this chapter are mainly based on [18,19]. The equivalence of functional differential equations to abstract Cauchy problems on product spaces was studied by Burns, Herdman, and Stech [28] and a version of the equivalence Theorem 3.12 for neutral differential equations and $X = \mathbb{C}^n$ was proved by Kappel and Zhang [119]. See also the review article by Kappel [118]. The proof of Lemma 3.6 (density) is taken from Engel [70, Proposition 1.3], and Kramar, Mugnolo, and Nagel [123, Lemma 2.4]. Proposition 3.19 is also proved in Nakagiri [155,156]. An abstract version of Lemma 3.20 is shown in [13].

For further references on the characteristic equation (Equation (3.25)), see Diekmann et al. [62, Section I.3], Hale and Verduyn Lunel [103, Section 7.3], and Wu [222, Section 3.1] for functional differential equations, or Nagel [151,152] for abstract Cauchy problems.

The second part of the proof of Theorem 3.25 is taken from Fragnelli and Nickel [81]. Section 3.3.1 is based on [70]. The proof of the case $p = 1$ in Theorem 3.29 was first due to L. Maniar and J. Voigt [139]. We present here an easier proof due to M. Stein, H. Vogt, and J. Voigt [197].

Necessary conditions for $(\mathcal{A}, D(\mathcal{A}))$ to be a generator were proved by Kunisch and Schappacher (see [129, Proposition 4.2]).

For different approaches to the well-posedness of delay equations with bounded and unbounded delay operators, see also Di Blasio [59]; Di Blasio, Kunisch, and Sinestrari [60,61]; Diekmann et al. [62]; Favini, Pandolfi, and Tanabe [78]; Hale and Verduyn Lunel [103]; Kunisch and Mastinsek [128]; Maniar and Rhandi [137]; Mastinsek [140,141]; Prüß [169]; Rhandi [175]; Ruess [177, 178]; Nakagiri and Tanabe [157]; Tanabe [199]; Travis and Webb [201]; and Webb [212]. For a recent systematic treatment extending the perturbation argument to delays in the highest order derivatives, see [20].

For delay equations with infinite delay see Hale and Kato [102] and Hino, Murakami, and Naito [111].

For delay equations with infinite delay and nonautonomous past in L^p-spaces, see Fragnelli and Nickel [80,81].

For finite-dimensional neutral equations like in Example 3.24, see also Arino and Győri [8] and Győri and Turi [98].

For more references on the theory of operator matrices, see R. Nagel [149,150]; K.-J. Engel [68–70]; Atkinson, Langer, Mennicken, and Shkalikov [9]; and [13].

For applications of delay equations in control theory, see Delfour [52,53], Delfour and Mitter [54,55], Nakagiri [155,156], and the overview by Curtain and Zwart [43, Section 2.4].

There exist extensions of the Miyadera-Voigt Perturbation Theorem to nonautonomous equations by Räbiger, Rhandi, Schnaubelt, and Voigt [170] and to bi-continuous semigroups by Farkas [73]. These results may allow to extend the presented theory of delay equations to nonautonomous equations or to L^∞ spaces.

Part III
Asymptotic Behavior

Chapter 4

Stability via Spectral Properties

In Part II, we saw how to associate to a delay differential equation,

$$(\text{DE}_p) \qquad \begin{cases} u'(t) = Bu(t) + \Phi u_t, & t \geq 0, \\ u(0) = x, \\ u_0 = f, \end{cases}$$

an operator matrix

$$\mathcal{A} := \begin{pmatrix} B & \Phi \\ 0 & \frac{d}{d\sigma} \end{pmatrix} \quad D(\mathcal{A}) := \left\{ \begin{pmatrix} x \\ f \end{pmatrix} \in D(B) \times W^{1,p}([-1,0], Z) \; : \; f(0) = x \right\},$$

on the Banach space $\mathcal{E}_p = X \times L^p([-1,0], Z)$, $1 \leq p < \infty$, and how to discuss well-posedness of (DE_p).

Now we focus our interest on the long-term behavior of the solutions. The aim of the following chapters is to study systematically questions of this kind by semigroups methods. Due to our personal taste and interest, mainly uniform exponential stability and connected asymptotic properties are investigated.

For the key notions of semigroup theory that will be used in the following chapters, we refer the reader to Part I, where these concepts and related results have been collected (to make the book self-contained).

In this chapter, we consider the situation where the asymptotic behavior of the semigroup can be characterized by spectral properties of its generator. We consider two cases: the case where the delay semigroup is eventually norm continuous and therefore the Spectral Mapping Theorem holds, and the case where the growth bound of the semigroup only depends on the spectra of its generator and of the generator B.

In Section 4.1, we show under which assumptions the delay semigroup becomes eventually norm continuous. Then we investigate the asymptotic behavior of the delay semigroup in Section 4.2, in the case where it is eventually norm continuous, by applying the Spectral Mapping Theorem. We obtain a characteristic equation and make the investigations that you may also find in other existing monographs on delay equations. Under additional compactness assumptions, the splitting of the space $\mathcal{E}_p$ into stable, center, and unstable manifolds is shown.

In Section 4.3, we introduce the critical spectrum and state its properties. Finally in Section 4.4, we apply the properties of the critical spectrum to delay equations.

Throughout this part, we assume that our standing Hypotheses (H_1)–(H_7) of Chapter 3 are satisfied.

4.1   Regularity of the Delay Semigroup

In this section, we apply the results of Section 1.5 to give sufficient conditions such that the delay semigroup becomes eventually norm continuous (see Propositions 4.3 and 4.8) or/and the generator has compact resolvent (see Lemmas 4.5 and 4.9).

We start our investigations with the case of bounded delay operators, i.e., $Z = X$ in (H_3), and will assume for the first half of this section that the delay operator Φ satisfies the following condition being slightly stronger than Condition (M) in Section 3.3.2 on page 67.

There exists $q : \mathbb{R}_+ \longrightarrow \mathbb{R}_+$ with $\lim_{t \to 0+} q(t) = 0$ and

$$\int_0^t \left\| \Phi(S_s x + T_0(s)f) \right\| ds \leq q(t) \left\| \left(\begin{smallmatrix} x \\ f \end{smallmatrix} \right) \right\| \tag{K}$$

for all $\left(\begin{smallmatrix} x \\ f \end{smallmatrix} \right) \in D(\mathcal{A}_0)$ and $t > 0$.

As a consequence, $(\mathcal{A}, D(\mathcal{A}))$ is a generator on each $\mathcal{E}_p$, $1 \leq p < \infty$ by Theorem 3.26.

Remark 4.1. The main example of Theorem 3.29, in which the delay operator is given by

$$\Phi f := \int_{-1}^0 d\eta \, f$$

for a function $\eta : [-1,0] \to \mathcal{L}(X)$ of bounded variation, satisfies Condition (K) for all $1 < p < \infty$ (see the proof of Theorem 3.29). Unfortunately, at this moment, we do not know whether Condition (K) also holds for the case $p = 1$.

Since we will use perturbation, we recall that $(\mathcal{A}_0, D(\mathcal{A}_0))$ is the operator defined in Section 3.3.2 and Equations (3.30) and (3.31) as

$$\mathcal{A}_0 := \begin{pmatrix} B & 0 \\ 0 & \frac{d}{d\sigma} \end{pmatrix}$$

with domain

$$D(\mathcal{A}_0) := D(\mathcal{A}) = \{ \begin{pmatrix} x \\ f \end{pmatrix} \in X \times W^{1,p}([-1,0], X) \; : \; f(0) = x \}.$$

It generates the strongly continuous semigroup $(\mathcal{T}_0(t))_{t \geq 0}$ on each space $\mathcal{E}_p$, $1 \leq p < \infty$, given by

$$\mathcal{T}_0(t) := \begin{pmatrix} S(t) & 0 \\ S_t & T_0(t) \end{pmatrix}, \tag{3.33}$$

where $(S(t))_{t \geq 0}$ is the semigroup generated by the operator $(B, D(B))$, the operators S_t are defined as in Theorem 3.25, and $(T_0(t))_{t \geq 0}$ is the nilpotent left shift semigroup on $L^p([-1,0], X)$. First we show that the semigroup $(\mathcal{T}_0(t))_{t \geq 0}$ is eventually norm continuous. Due to Equation (3.33), this is relatively easy.

Proposition 4.2. *Assume $Z = X$ in* (H_3). *If $(S(t))_{t \geq 0}$ is norm continuous for $t > t_0 \geq 0$, then $(\mathcal{T}_0(t))_{t \geq 0}$ is norm continuous for $t > t_0 + 1$ for all $1 \leq p < \infty$.*

Proof. For $t \geq 1$, we have $T_0(t) = 0$ and $\mathcal{T}_0(t) = \begin{pmatrix} S(t) & 0 \\ S_t & 0 \end{pmatrix}$. So it suffices to show that the map $t \mapsto S_t$ from $[t_0 + 1, \infty)$ to $\mathcal{L}(X, L^p([-1,0], X))$ is norm continuous.

For $s, t \geq t_0 + 1$ we have

$$\lim_{s \to t} \|S_s - S_t\| = \lim_{s \to t} \sup_{\|x\| \leq 1} \left(\int_{-1}^{0} \|S(s+\sigma)x - S(t+\sigma)x\|^p \, d\sigma \right)^{\frac{1}{p}}$$

$$\leq \lim_{s \to t} \sup_{\|x\| \leq 1} \left(\int_{-1}^{0} \|S(s+\sigma) - S(t+\sigma)\|^p \, d\sigma \right)^{\frac{1}{p}} \|x\|$$

$$= \lim_{s \to t} \left(\int_{-1}^{0} \|S(s+\sigma) - S(t+\sigma)\|^p \, d\sigma \right)^{\frac{1}{p}},$$

which converges to 0 as s tends to t since $(S(t))_{t \geq 0}$ is norm continuous on (t_0, ∞) and therefore uniformly norm continuous on compact intervals. $\square$

Now we apply Corollary 1.49 to show eventual norm continuity of the delay semigroup.

Proposition 4.3. *Assume $Z = X$ in (H_3). If the semigroup $(S(t))_{t \geq 0}$ generated by $(B, D(B))$ is immediately norm continuous and Φ satisfies Condition (K), then the delay semigroup $(\mathcal{T}(t))_{t \geq 0}$ is norm continuous for $t > 1$ on each $\mathcal{E}_p$, $1 \leq p < \infty$.*

Proof. By Proposition 4.2, $(\mathcal{T}_0(t))_{t \geq 0}$ is norm continuous for $t > 1$. Now we show that $V\mathcal{T}_0$, where V is the abstract Volterra operator defined in Equations (1.13) and (1.14), is norm continuous for $t \geq 0$. In fact, for $t \geq 0$ and $\binom{x}{f} \in D(\mathcal{A}_0)$, we have

$$
V\mathcal{T}_0(t)\binom{x}{f} = \int_0^t \mathcal{T}_0(t-s)B\mathcal{T}_0(s)\binom{x}{f}\, ds
$$

$$
= \int_0^t \mathcal{T}_0(t-s)\begin{pmatrix} 0 & \Phi \\ 0 & 0 \end{pmatrix}\begin{pmatrix} S(s)x \\ S_s x + T_0(s)f \end{pmatrix} ds
$$

$$
= \int_0^t \begin{pmatrix} S(t-s) & 0 \\ S_{t-s} & T_0(t-s) \end{pmatrix}\begin{pmatrix} \Phi(S_s x + T_0(s)f) \\ 0 \end{pmatrix} ds
$$

$$
= \int_0^t \begin{pmatrix} S(t-s)\Phi(S_s x + T_0(s)f) \\ S_{t-s}\Phi(S_s x + T_0(s)f) \end{pmatrix} ds.
$$

We prove norm continuity of both components separately.

(1) Let $t \geq 0$ and $1 > h > 0$. Then we have

$$
\left\| \int_0^{t+h} S(t+h-s)\Phi(S_s x + T_0(s)f)\, ds - \int_0^t S(t-s)\Phi(S_s x + T_0(s)f)\, ds \right\|
$$

$$
\leq \left\| \int_t^{t+h} S(t+h-s)\Phi(S_s x + T_0(s)f)\, ds \right\|
$$

$$
+ \left\| \int_0^t (S(t+h-s) - S(t-s))\Phi(S_s x + T_0(s)f)\, ds \right\|
$$

$$
\leq \int_0^h \|S(h-s)\|\, \|\Phi(S_{s+t}x + T_0(s+t)f)\|\, ds
$$

$$
+ \int_0^t \|S(t+h-s) - S(t-s)\|\, \|\Phi(S_s x + T_0(s)f)\|\, ds.
$$

$$
\leq \sup_{0 \leq r \leq 1} q(h)\, \|S(r)\|\, \left\| \mathcal{T}_0(t)\binom{x}{f} \right\|
$$

$$
+ \int_0^t \|S(t+h-s) - S(t-s)\|\, \|\Phi(S_s x + T_0(s)f)\|\, ds.
$$

By Condition (K), by Lebesgue's Dominated Convergence Theorem, and by the immediate norm continuity of $(S(t))_{t\geq 0}$, we obtain that

$$\sup_{0\leq r\leq 1}\|S(r)\|q(h)\left\|\mathfrak{T}_0(t)\binom{x}{f}\right\|+\int_0^t\|S(t+h-s)-S(t-s)\|\;\|\Phi(S_s x+T_0(s)f)\|\,ds$$

tends to 0 as $h\to 0^+$ uniformly in $\binom{x}{f}\in D(\mathcal{A}_0)$, $\left\|\binom{x}{f}\right\|\leq 1$. The proof for $h\to 0^-$ is analogous.

Since $D(\mathcal{A})$ is dense in $\mathcal{E}_p$, the first component of $V\mathfrak{T}_0$ is immediately norm continuous.

(2) To prove immediate norm continuity of the second component of $V\mathfrak{T}_0$, one proceeds in a similar way. We only have to use the norm continuity of the map $t\mapsto S_t$, which was proved in Proposition 4.2.

Hence, the map $t\mapsto V\mathfrak{T}_0(t)$ is norm continuous on $\mathbb{R}_+$ and, by Corollary 1.49, we conclude that $(\mathfrak{T}(t))_{t\geq 0}$ is norm continuous for $t>1$. $\qquad\square$

We remark that the immediate norm continuity of $(B,D(B))$ is essential here and cannot be weakened to eventual norm continuity. To show this, we consider the following counterexample due to R. Shvidkoy.

Example 4.4. Consider the space $X=L^1[0,2]$, the operators

$$Bx=-x',\quad D(B):=\left\{x\in W^{1,1}[0,2]:x(0)=0\right\},$$

which is the generator of the nilpotent right-shift semigroup, and

$$Cx(s):=\begin{cases}x(s+1),&0\leq s+1\leq 1,\\ 0,&\text{otherwise.}\end{cases}$$

Notice that the functions

$$x_n(s):=\begin{cases}s\cdot e^{-2\pi ins},&0\leq s+1\leq 1,\\ e^{-2\pi ns},&\text{otherwise}\end{cases}$$

satisfy

$$(B+C)x_n=2\pi inx_n\quad\text{for each }n\in\mathbb{N}.$$

Taking $\Phi:=C\delta_{-1}$ and

$$F_n:=\binom{x_n}{e^{2\pi in\cdot}x_n},$$

we obtain that $F_n\in D(\mathcal{A})$ and

$$\mathcal{A}F_n=2\pi inF_n,\text{ i.e., }2\pi in\in P\sigma(\mathcal{A}).$$

Hence, the spectrum of $\mathcal{A}$ is not bounded along the imaginary axis. Therefore, by Theorem 2.22, the semigroup generated by $(\mathcal{A}, D(\mathcal{A}))$ is not eventually norm continuous.

Now we turn our attention to another regularity property: compactness of the resolvent. This is an important property for the study of the asymptotic behavior of the solutions (see Theorems 4.15 and 5.20).

Lemma 4.5. *Assume $Z = X$ in (H_3). If the semigroup $(S(t))_{t \geq 0}$ generated by $(B, D(B))$ is immediately compact, then $R(\lambda, B + \Phi_\lambda)$ is compact for all $\lambda \in \rho(\mathcal{A})$.*

Proof. By Theorem 1.33, we have that the resolvent $R(\mu, B)$ is compact for all $\mu \in \rho(B)$. Take $\lambda \in \rho(\mathcal{A})$. Then $R(\mu, B + \Phi_\lambda)$ exists and is compact for all $\mu \in \rho(B + \Phi_\lambda)$. This holds in particular for $\mu = \lambda$, since by Proposition 3.19 $\lambda \in \rho(B + \Phi_\lambda)$ holds.   □

Let us consider now the case of unbounded delay operators, i.e., we allow (H_3): $D(B) \overset{d}{\hookrightarrow} Z \overset{d}{\hookrightarrow} X$ in full generality. To that purpose, we need Conditions (3.42) and (3.43) from Section 3.4. In particular, the semigroup $(S(t))_{t \geq 0}$ with generator $(B, D(B))$ is analytic, hence immediately norm continuous and analogously to Proposition 4.2 we obtain the following result.

Proposition 4.6. *Under Conditions (3.42) and (3.43), the semigroup $(\mathcal{T}_0(t))_{t \geq 0}$ generated by $(\mathcal{A}_0, D(\mathcal{A}_0))$ is norm continuous for $t > 1$.*

Proof. We remark first that it follows from our Condition (3.42) on $(B, D(B))$ that the map $S : (0, +\infty) \to \mathcal{L}(X, D(B))$ is norm continuous.

The rest of the proof is analogous to the proof of Proposition 4.2 and therefore will be omitted.   □

In order to be able to carry out the same perturbation argument as for bounded delay operators, we must make an additional assumption on the delay operator Φ, which is the analogous of Condition (K) in Proposition 4.3. We will assume that Φ satisfies the following condition being slightly stronger than Condition (M') of Theorem 3.34.

There exists $q : \mathbb{R}_+ \longrightarrow \mathbb{R}_+$ with $\lim_{t \to 0+} q(t) = 0$ and

$$\int_0^t \| \Phi(S_s x + T_0(s) f) \| \, ds \leq q(t) \left\| \begin{pmatrix} x \\ f \end{pmatrix} \right\| \tag{K'}$$

for all $\begin{pmatrix} x \\ f \end{pmatrix} \in D(\mathcal{A}_0)$ and $t > 0$.

As a consequence, the operator $(\mathcal{A}, D(\mathcal{A}))$ is a generator by Theorem 3.34.

Example 4.7. Let $1 < p < \infty$ and $\Phi : W^{1,p}([-1,0], Z) \to X$ be given by the Riemann-Stieltjes Integral of a function of bounded variation $\eta : [-1,0] \to \mathcal{L}(Z, X)$, i.e.,

$$\Phi(f) = \int_{-1}^{0} d\eta \, f, \qquad f \in W^{1,p}([-1,0], Z).$$

Then Condition (K') is satisfied (see Equation (3.51) in Theorem 3.35).

Proposition 4.8. *Assume that Conditions* (3.42) *and* (3.43) *hold and that the delay operator* Φ *satisfies Condition* (K'). *Then the semigroup* $(\mathcal{T}(t))_{t\geq 0}$ *generated by* $(\mathcal{A}, D(\mathcal{A}))$ *is norm continuous for* $t > 1$.

The proof is analogous to the one of Proposition 4.3 and is omitted.

Now we prove a result analogous to Lemma 4.5. This is now more complicate because the operators Φ_λ are bounded only from Z to X.

Lemma 4.9. *Assume that Conditions* (3.42) *and* (3.43) *hold and that* B *has compact resolvent. Then* $R(\lambda, B + \Phi_\lambda)$ *is compact for all* $\lambda \in \rho(\mathcal{A})$.

Proof. We remark that $\Phi_\lambda \in \mathcal{L}(Z, X)$. By Condition (3.43), for every $\delta > \omega_0(B)$, there exists $0 < \theta < \frac{1}{p}$ such that

$$D((B+\delta)^\theta) \overset{d}{\hookrightarrow} Z \overset{d}{\hookrightarrow} X.$$

By Proposition 1.28, the operator Φ_λ is B-bounded with B-bound 0. This means that for every $a > 0$, there exists $b > 0$ such that

$$\|\Phi_\lambda x\| \leq a\|Bx\| + b\|x\| \qquad (4.1)$$

for all $x \in D(B)$. Hence, by Theorem 1.41, $B + \Phi_\lambda$ generates an analytic semigroup. In particular, $\mu \in \rho(B + \Phi_\lambda)$ for $\Re\mu$ large enough.

Moreover, $\mu \in \rho(B)$ if $\Re\mu$ is large enough, and we have

$$(\mu - B - \Phi_\lambda) = (\mathrm{Id} - \Phi_\lambda \, R(\mu, B)) \, (\mu - B).$$

If we can prove that $\|\Phi_\lambda \, R(\mu, B)\| < 1$ for $\Re\mu$ large, then

$$R(\mu, B + \Phi_\lambda) = R(\mu, B) \sum_{n=0}^{\infty} (\Phi_\lambda \, R(\mu, B))^n$$

and the assertion follows.

To this purpose, we recall that the analyticity of $(S(t))_{t\geq 0}$ implies that there exists a constant $M > 0$ such that

$$\|B\,R(\mu, B)\| \leq M \tag{4.2}$$

for all $\Re\mu$ large (see Theorem 1.24). Let now $0 < \varepsilon < 1$ and $a < \frac{\varepsilon}{2M}$. By Equation (4.1), there exist $b > 0$ such that

$$\|\Phi_\lambda\, R(\mu, B)x\| \leq a\,\|B\,R(\mu, B)x\| + b\|R(\mu, B)x\| \tag{4.3}$$

for all $x \in X$. Choose now $\Re\mu$ so large that

$$b\|R(\mu, B)\| < \frac{\varepsilon}{2}. \tag{4.4}$$

Then by Equations (4.2), (4.3), and (4.4), we obtain

$$\|\Phi_\lambda\, R(\mu, B)\| \leq \varepsilon < 1$$

for $\Re\mu$ large enough.   □

Example 4.10. We recall from Example 3.15 the following reaction-diffusion equation with delay

$$\begin{cases} \partial_t w(s,t) = \Delta w(s,t) + c\int_{-1}^0 w(s, t+\tau)\, dg(\tau), \\ \qquad\qquad\qquad\qquad\qquad s \in \Omega,\ t \geq 0, \\ w(s,t) = 0, \qquad\qquad\qquad s \in \partial\Omega,\ t \geq 0, \\ w(s,t) = f(s,t), \qquad\qquad (s,t) \in \Omega \times [-1,0], \end{cases} \tag{3.17}$$

where c is a constant, $\Omega \subset \mathbb{R}^n$ an open set with smooth boundary, and $g : [-1, 0] \to [0, 1]$ is a function of bounded variation.

In this case the generator $(B, D(B))$ is the Laplacian with Dirichlet boundary conditions; hence, it generates an analytic and in particular immediately norm continuous semigroup $(T(t))_{t\geq 0}$ on $L^2(\Omega)$. Condition (K) is satisfied by Remark 4.1.

Hence, the delay semigroup solving Equation (3.17) is norm continuous for $t > 1$. Moreover, if Ω is bounded, then the resolvent of $(B, D(B))$ is compact and by Theorem 1.33 we have that $(T(t))_{t\geq 0}$ is immediately compact. Therefore, by Lemma 4.5 the resolvent $R(\lambda, B + \Phi_\lambda)$ is compact for all $\lambda \in \rho(B + \Phi_\lambda)$.

4.2   Stability via Spectral Mapping Theorem

In this section, we apply the results of Section 4 and the Spectral Mapping Theorem to characterize uniform exponential stability of the delay semigroup. Moreover, under a compactness hypothesis, we show the existence of a stable, center, and unstable manifold for the delay semigroup.

Throughout this section, we will always assume either

$$
\begin{cases}
Z = X \text{ in } (\mathrm{H}_3), \\
(B, D(B)) \text{ generates an immediately norm continuous} \\
\quad \text{semigroup } (S(t))_{t \geq 0}, \text{ and} \\
\Phi \text{ satisfies Condition (K) of Proposition 4.3,}
\end{cases}
\tag{4.5}
$$

or that

$$
\begin{cases}
\text{Conditions (3.42) and (3.43) hold and} \\
\Phi \text{ satisfies Condition (K') of Proposition 4.8.}
\end{cases}
\tag{4.6}
$$

Theorem 4.11. *Assume either* (4.5) *or* (4.6) *holds. Then*

$$
s(\mathcal{A}) = \omega_0(\mathcal{A}).
$$

In particular, the delay semigroup is uniformly exponentially stable if and only if the implication

$$
\lambda \in \sigma(B + \Phi_\lambda) \implies \Re\lambda < 0
$$

holds.

Proof. Assume Equation (4.5) holds. It follows from Proposition 4.3 that the delay semigroup is norm continuous for $t > 1$. By Theorem 2.11, the Spectral Mapping Theorem holds and, by Corollary 2.12, the equality $s(\mathcal{A}) = \omega_0(\mathcal{A})$ holds. Using the spectral characterization of Proposition 3.19, we obtain that $\lambda \in \sigma(\mathcal{A})$ if and only if $\lambda \in \sigma(B + \Phi_\lambda)$. Using the fact that the spectrum of the generator of an eventually norm continuous semigroup is bounded along imaginary lines (see Theorem 2.22), it cannot happen that the points in the spectrum approach the imaginary axis. Hence, we obtain the desired condition for the negativity of the spectral bound.

Now assume that Equation (4.6) holds; then the proof is analogous to the previous case applying Proposition 4.8 instead of Proposition 4.3. □

Before going on with the investigation of the stability, we have to clarify a technical detail. If we are given a delay equation with an initial function $f \in L^p([-1, 0], Z)$ for all $1 \leq p < \infty$, does it really matter which $\mathcal{E}_p$ space we choose? Looking at Theorem 4.11, we immediately see that, at least concerning uniform exponential stability, it does not. Therefore, in the following, we will just speak of the delay semigroup on the space $\mathcal{E}$ if a result holds for all $\mathcal{E}_p$, $1 \leq p < \infty$.

Corollary 4.12. *Assume that either Equation (4.5) holds and the delay operator Φ is given by a function of bounded variation η as in Theorem 3.29, or Equation (4.6) holds and the delay operator Φ is given by a function of bounded variation η as in Theorem 3.35. Then the growth bound of the delay semigroup is independent of $1 < p < \infty$.*

Though we have a complete characterization of uniform exponential stability via the spectrum of the generator, it is sometimes rather complicated to use it effectively. However, the situation is much easier if we have compactness.

Corollary 4.13. *Assume either Equation (4.5) or (4.6), and $(B, D(B))$ generates an immediately compact semigroup. Then the delay semigroup is uniformly exponentially stable if and only if the implication*

$$\lambda x - Bx - \Phi_\lambda x = 0 \quad \text{for some } 0 \neq x \in D(B) \implies \Re\lambda < 0$$

holds.

Proof. By Lemma 4.5, or Lemma 4.9 respectively, we have that $\sigma(\mathcal{A}) = P\sigma(\mathcal{A})$. The result then follows immediately from Proposition 4.11. $\square$

Example 4.14. We illustrate the above corollary by the following equation modeling heat conduction in a rod.

$$\begin{cases} \partial_t u(x,t) = \partial_x^2 u(x,t) - au(x,t) - bu(x,t-r), \ x \in [0,l], \ t \geq 0, \\ u(0,t) = u(l,t) = 0, \hspace{4.7cm} t \geq 0, \\ u(x,t) = f(x,t), \ x \in [0,l], \hspace{3.3cm} t \in [-r,0], \end{cases}$$

where $l, r > 0$, $a, b \in \mathbb{R}$. Recalling the well-posedness treated in Example 3.15, we define the following spaces and operators:

- The Hilbert space $X := L^2(0,l)$.

- The operator $B := (\Delta - a)$ with Dirichlet boundary conditions, i.e., $D(B) := H_0^1(0,l) \cap H^2(0,l)$.

- The functions $\mathbb{R}_+ \ni t \mapsto u(t) = u(\cdot, t) \in L^2(0,l)$ and $u_t : [-1,0] \to L^2(0,l)$, $u_t(s) := u(t+s)$.

- $Z := X = L^2(0,l)$.

- The operator $\Phi : W^{1,p}([-1,0], X) \to X$ defined as $\Phi f := -b\delta_{-r}f = -bf(-r)$.

In order to apply Corollary 4.13, we have to find all $\lambda \in \mathbb{C}$ such that the equation

$$(\lambda + be^{-\lambda r})x - Bx = 0$$

has a solution $0 \neq x \in D(B)$. We recall that $\sigma(B) = \left\{ -\frac{n^2\pi^2}{l^2} - a : n \in \mathbb{N} \right\}$. Hence, the solution semigroup is uniformly exponentially stable if and only if all solutions $\lambda \in \mathbb{C}$ of the equation

$$\lambda + be^{-\lambda r} = -\frac{n^2\pi^2}{l^2} - a, \quad n \in \mathbb{N},$$

have negative real part.

Now we prove a decomposition result. It says that, if we assume the same compactness on the semigroup generated by $(B, D(B))$ as before, then the space $\mathcal{E}_p$ naturally decomposes into three subspaces that are invariant under the delay semigroup.

Theorem 4.15. *Assume either Equation (4.5) or (4.6), and $(B, D(B))$ generates a compact semigroup. Then there exist subspaces $\mathcal{E}_S$, $\mathcal{E}_U$, and $\mathcal{E}_C$, which are invariant under the delay semigroup such that $\mathcal{E} = \mathcal{E}_S \oplus \mathcal{E}_C \oplus \mathcal{E}_U$, $\dim \mathcal{E}_C < \infty$, $\dim \mathcal{E}_U < \infty$, and*

(i) the semigroup $T_S(t) = T(t)|_{\mathcal{E}_S}$ is uniformly exponentially stable,

(ii) the semigroup $T_U(t) = T(t)|_{\mathcal{E}_U}$ is invertible and the semigroup $T_U^{-1}(t)$ is uniformly exponentially stable, and

(iii) the semigroup $T_C(t) = T(t)|_{\mathcal{E}_C}$ is a group being polynomially bounded in both time directions and hence having growth bound 0 in both directions.

Proof. We only give the proof under Equation (4.5). The other case can be proved similarly. First, we show that under the above conditions the operator $(\mathcal{A}, D(\mathcal{A}))$ has pure point spectrum with finite-dimensional spectral subspaces. To this end, recall that $\lambda \in \sigma_{ess}(\mathcal{A})$ if and only if $\lambda \in \sigma_{ess}(B + \Phi_\lambda) = \emptyset$ since $B + \Phi_\lambda$ has compact resolvent by Lemma 4.5. Hence, by Lemma 3.20, it follows that $\sigma(\mathcal{A}) = P\sigma(\mathcal{A})$.

Further, since

$$\lambda x = Bx + \Phi_\lambda x \quad \Leftrightarrow \quad \lambda \begin{pmatrix} x \\ \epsilon_\lambda x \end{pmatrix} = \mathcal{A} \begin{pmatrix} x \\ \epsilon_\lambda x \end{pmatrix},$$

the corresponding eigenspaces have the same, finite dimension.

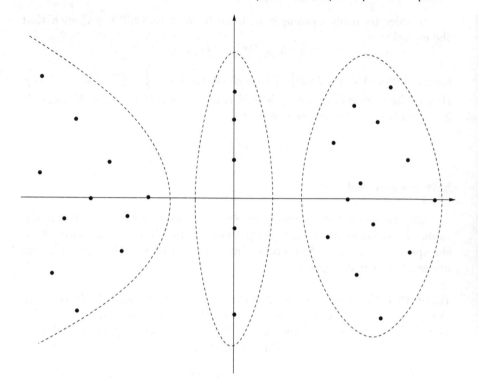

Figure 4.1. $\sigma(\mathcal{A}_S)$, $\sigma(\mathcal{A}_C)$, **and** $\sigma(\mathcal{A}_U)$.

Consider now the following decomposition of $\sigma(\mathcal{A})$:

$$\Sigma_U := \sigma(\mathcal{A}) \cap \{\lambda \in \mathbb{C} : \Re\lambda > 0\},$$
$$\Sigma_C := \sigma(\mathcal{A}) \cap i\mathbb{R},$$
$$\Sigma_S := \sigma(\mathcal{A}) \cap \{\lambda \in \mathbb{C} : \Re\lambda < 0\}.$$

By our previous considerations and Lemma 3.20, the sets Σ_U and Σ_C are finite and the corresponding spectral projections P_U and P_C are finite-dimensional by Theorem 2.20. Let us denote the corresponding spectral subspaces with $\mathcal{E}_U := P_U\mathcal{E}$, $\mathcal{E}_C := P_C\mathcal{E}$, and $\mathcal{E}_S := P_S\mathcal{E}$, where P_S is the spectral projection corresponding to Σ_S. Then, by the construction,

$$\mathcal{E} = \mathcal{E}_S \oplus \mathcal{E}_C \oplus \mathcal{E}_U.$$

All these subspaces are invariant under the delay semigroup, and the parts of the generator $\mathcal{A}$ in the corresponding subspaces have spectrum Σ_S, Σ_C, and Σ_U. Since the delay semigroup is eventually norm continuous, the spectral mapping theorem holds and the asymptotic behavior is determined by the spectrum. The assertion follows since $\mathcal{A}_C$ has purely imaginary spectrum in the finite-dimensional space $\mathcal{E}_C$, $\mathcal{A}_U$ has spectral values with strictly positive real part in the finite-dimensional space $\mathcal{E}_U$, and $\mathcal{A}_S$ has strictly negative spectrum and generates an eventually norm continuous semigroup in the space $\mathcal{E}_S$. □

The subspaces $\mathcal{E}_S$, $\mathcal{E}_U$, and $\mathcal{E}_C$ are usually referred to as the corresponding *stable*, *unstable*, and *center* manifolds.

Although for eventually norm continuous semigroups the growth bound is equal to the spectral bound, it can be very difficult to compute it. Therefore, it is sometimes useful to estimate the growth bound by resolvent estimates. We will pursue this technique systematically in Chapter 5. Here we only give an appetizer.

Theorem 4.16. *Assume either Equation (4.5) or (4.6), and $\omega_0(B) < 0$. Let $\omega_0(B) < \alpha \leq 0$. If*

$$r(\Phi_\lambda R(\lambda, B)) < 1 \qquad (4.7)$$

for $\Re\lambda \geq \alpha$, then $\omega_0(\mathcal{A}) < \alpha \leq 0$.

Proof. The condition ensures that the resolvent exists on the halfplane $\{\Re\lambda \geq \alpha\}$ and is given by the Neumann series

$$R(\lambda, B + \Phi_\lambda) = R(\lambda, B) \sum_{n=0}^{\infty} (\Phi_\lambda R(\lambda, B))^n.$$

Therefore, $s(\mathcal{A}) < \alpha$. Since the delay semigroup generated by $\mathcal{A}$ is eventually norm continuous, $\omega_0(\mathcal{A}) = s(\mathcal{A}) < \alpha$. □

Corollary 4.17. *Under the assumptions of Theorem 4.16, let $\omega_0(B) < \alpha \leq 0$. If*

$$\sup_{\omega \in \mathbb{R}} \|\Phi_{\alpha+i\omega} R(\alpha + i\omega, B)\| < 1, \qquad (4.8)$$

then $\omega_0(\mathcal{A}) < \alpha \leq 0$.

Proof. Using Lemma 2.28, it follows that $\sup_{\Re\lambda \geq \alpha} \|\Phi_\lambda R(\lambda, B)\| < 1$, hence the conditions of Theorem 4.16 are satisfied. □

Example 4.18. We consider the partial differential equation with delay

$$\begin{cases} \partial_t w(x,t) &= \Delta w(x,t) + \sum_{j=1}^n c_j \partial_j w(x,t-1), \\ & \qquad\qquad\qquad\qquad x \in \Omega, \ t \geq 0, \\ w(x,t) &= 0, \qquad\qquad\quad x \in \partial\Omega, \ t \geq 0, \\ w(x,t) &= f(x,t), \qquad\quad (x,t) \in \Omega \times [-1,0], \end{cases} \tag{P'}$$

where c is a constant and $\Omega \subset \mathbb{R}^n$ a bounded open set.

We consider $X := L^2(\Omega)$, the Dirichlet-Laplacian

$$B := \Delta_D \text{ with domain } D(B) := \left\{ u \in H_0^1(\Omega) : \Delta u \in L^2(\Omega) \right\},$$

$1 < p < 2$ arbitrary,

$$\Phi := \sum_{j=1}^n c_j \partial_j \delta_{-1},$$

$$\Phi_\lambda = e^{-\lambda} \sum_{j=1}^n c_j \partial_j ,$$

and as space

$$Z := D((-B)^{\frac{1}{2}}) = H_0^1(\Omega).$$

Here we define $\delta_{-1} : W^{1,p}([-1,0], Z) \longrightarrow Z$ by $\delta_{-1}(f) := f(-1)$. Note that $\||\nabla h\|| = \|(-B)^{\frac{1}{2}} h\|_Z$ for all $h \in Z$; see Triggiani [202, Formula (2.9)].

The well-posedness follows from the calculations in Theorem 3.35. We want to verify the stability estimate.

Using that B is a selfadjoint operator on a Hilbert space (see Kato [120, Section V.3.8]), we can compute for $\omega \in \mathbb{R}$

$$\begin{aligned} \|\Phi_{i\omega} R(i\omega, B)\| &\leq \sum_{j=1}^n |c_j| \left\|(-B)^{\frac{1}{2}} R(i\omega, B)\right\| \\ &= \sum_{j=1}^n |c_j| \left\|(-B)^{-\frac{1}{2}} B R(i\omega, B)\right\| \\ &\leq \sum_{j=1}^n |c_j| \left\|(-B)^{-\frac{1}{2}}\right\| \|B R(i\omega, B)\| \qquad\qquad (4.9) \\ &\leq \sum_{j=1}^n |c_j| \left\|(-B)^{-\frac{1}{2}}\right\| \left(|\omega| \|R(i\omega, B)\| + 1 \right) \\ &= \sum_{j=1}^n |c_j| \left\|(-B)^{-\frac{1}{2}}\right\| \left(\frac{|\omega|}{\sqrt{\omega^2 + \lambda_1^2}} + 1 \right), \end{aligned}$$

where λ_1 is the first eigenvalue of the Dirichlet-Laplacian. We obtain that

$$\sup_{\omega \in \mathbb{R}} \|\Phi_{i\omega} R(i\omega, B)\| \leq \frac{2 \sum_{j=1}^{n} |c_j|}{\sqrt{|\lambda_1|}}.$$

Thus, the solutions of (P') decay exponentially if

$$2 \sum_{j=1}^{n} |c_j| < \sqrt{|\lambda_1|}.$$

4.3   The Critical Spectrum

In this section, we give a short overview of recent results on the critical spectrum of strongly continuous semigroups due to S. Brendle, R. Nagel, and J. Poland [27, 154].

First, we introduce the critical spectrum. Let $(T(t))_{t \geq 0}$ be a strongly continuous semigroup on a Banach space X with generator $(A, D(A))$ and let $l^\infty(X)$ be the space of all bounded sequences in X endowed with the sup-norm $\| \cdot \|_\infty$. We consider the semigroup $(\tilde{T}(t))_{t \geq 0}$ given by

$$\tilde{T}(t)(x_n)_{n \in \mathbb{N}} := (T(t)x_n)_{n \in \mathbb{N}} \qquad \text{for} \quad t \geq 0$$

and the operator $(\tilde{A}, D(\tilde{A}))$ given by

$$\tilde{A}(x_n)_{n \in \mathbb{N}} := (Ax_n)_{n \in \mathbb{N}}$$

with domain

$$D(\tilde{A}) := \{(x_n)_{n \in \mathbb{N}} \in l^\infty(X) \ : \ x_n \in D(A), \ (Ax_n) \in l^\infty(X)\}.$$

Note that the semigroup $(\tilde{T}(t))_{t \geq 0}$ is strongly continuous if and only if $(T(t))_{t \geq 0}$ is uniformly continuous; that means the generator A is bounded. Moreover, for the spectra, one has $\sigma(\tilde{T}(t)) = \sigma(T(t))$ and $\sigma(\tilde{A}) = \sigma(A)$ and the resolvent is given by $R(\lambda, \tilde{A})(x_n)_{n \in \mathbb{N}} = (R(\lambda, A)x_n)_{n \in \mathbb{N}}$. Since $\|R(\lambda, A)^n\| = \|R(\lambda, \tilde{A})^n\|$ for all $n \in \mathbb{N}$, $\tilde{A}$ is a (nondensely defined) Hille-Yosida operator. Consider now the space of strong continuity

$$l_T^\infty(X) := \{(x_n)_{n \in \mathbb{N}} \in l^\infty(X) \ : \ \lim_{t \to 0} \sup_{n \in \mathbb{N}} \|T(t)x_n - x_n\| = 0\}.$$

Then we have that $l_T^\infty(X) = \overline{D(\tilde{A}_0)}$, where $\tilde{A}_0$ denotes the part of $\tilde{A}$ in $l_T^\infty(X)$. Moreover, $l_T^\infty(X)$ is a closed, $(\tilde{T}(t))_{t \geq 0}$-invariant subspace of $l^\infty(X)$. Therefore, the following definition makes sense.

On the quotient space $\hat{X} := l^\infty(X)/l_T^\infty(X)$, we define the semigroup $\hat{\mathfrak{T}} := (\hat{T}(t))_{t \geq 0}$ by

$$\hat{T}(t)\hat{x} := (T(t)x_n)_{n \in \mathbb{N}} + l_T^\infty(X)$$

for $\hat{x} := (x_n)_{n \in \mathbb{N}} + l_T^\infty(X) \in \hat{X}$.

The family $(\hat{T}(t))_{t \geq 0}$ is a semigroup of bounded operators on $\hat{X}$, however, no orbit $t \mapsto \hat{T}(t)\hat{x}$, except for $\hat{x} = 0$, is continuous.

We are now able to define the critical spectrum and the critical growth bound.

Definition 4.19. For a strongly continuous semigroup $T := (T(t))_{t \geq 0}$ on a Banach space X with generator $(A, D(A))$, we call

$$\sigma_{crit}(T(t)) := \sigma(\hat{T}(t))$$

the *critical spectrum* of $T(t)$ and denote by

$$\omega_{crit}(T) := \omega_0(\hat{T})$$

its *critical growth bound*.

Alternatively, we will also write $\omega_{crit}(A)$ for the critical growth bound.

Starting from this abstract definition, it is also possible to characterize the critical growth bound in the following way.

Proposition 4.20. *For a strongly continuous semigroup $T := (T(t))_{t \geq 0}$, the equality*

$$\omega_{crit}(T) = \inf\{v \in \mathbb{R} \ : \ \exists M > 0 \ \text{such that}$$
$$\limsup_{h \searrow 0} \|T(t + h) - T(t)\| \leq Me^{vt}\} \tag{4.10}$$

holds.

We can now formulate a version of the Spectral Mapping Theorem for arbitrary strongly continuous semigroups.

Theorem 4.21. *For a strongly continuous semigroup $(T(t))_{t \geq 0}/$, with generator $(A, D(A))$, one has*

$$\sigma(T(t)) \setminus \{0\} = e^{t\sigma(A)} \cup \sigma_{crit}(T(t)) \setminus \{0\} \qquad \text{for all } t \geq 0.$$

As a corollary of Theorem 4.21, we obtain the following characterization of the growth bound $\omega_0(T)$.

Corollary 4.22. *For a strongly continuous semigroup $(T(t))_{t\geq0}$ with generator $(A, D(A))$, one has*

$$\omega_0(T) = \max\{s(A), \omega_{crit}(T)\}.$$

Finally, we can formulate a perturbation theorem for the growth bound. We do this in the setting of Section 1.5.

Let $(A, D(A))$ be the generator of a strongly continuous semigroup $T := (T(t))_{t\geq0}$ on a Banach space X and let $C \in \mathcal{L}(X_1^A, X)$, i.e., C is relatively bounded to A. Moreover, assume that Hypothesis 1.45 is satisfied, i.e., there exist $\varepsilon > 0$ and a function $q : (0, \varepsilon) \to \mathbb{R}_+$ such that $\lim_{t\searrow0} q(t) = 0$ and

$$\int_0^t \|C\,T(s)x\|\,ds \leq q(t)\,\|x\| \tag{1.18}$$

for every $x \in D(A)$ and every $0 < t < \varepsilon$. Let $S := (S(t))_{t\geq0}$ denote the perturbed semigroup generated by the operator $(A + C, D(A))$.

Theorem 4.23. *Assume that Hypotheses 1.45 holds. Moreover, assume that $V^n\mathcal{T}$ is norm continuous for $t \geq 0$ and some $n \in \mathbb{N}$. Then the equality*

$$\omega_0(S) = \max\{s(A + C), \omega_{crit}(T)\} \tag{4.11}$$

holds.

4.4   Stability via Critical Spectrum

In this section, we are still aiming at spectral criteria for uniform exponential stability even when the semigroup is not eventually norm continuous and hence the Spectral Mapping Theorem and the identity $s(A) = \omega_0(A)$ do not hold. In this situation the *critical spectrum* and the *critical growth bound* (see Section 4.3) are a useful tool. In this section, we present a result due to S. Brendle [26].

In order to apply the critical spectrum, we only consider the special class of delay operators $\Phi : W^{1,p}([-1,0], X) \to X$, $1 \leq p < \infty$, of the form

$$\Phi f := \int_{-1}^0 \varphi(r)f(r)dr \tag{4.12}$$

for all $f \in W^{1,p}([-1,0], X)$, where $\varphi \in L^1([-1,0], \mathcal{L}(X))$ and $1 \leq p < \infty$. The well-posedness for these operators follows from a special case of Theorem 3.29 in Section 3.3.

Lemma 4.24. *There exists a function* $q : \mathbb{R}_+ \to \mathbb{R}_+$ *satisfying* $\lim_{h \searrow 0} q(h)$
$= 0$ *such that*

$$\int_0^h \|\mathcal{B}\mathcal{T}_0(r)\left(\begin{smallmatrix} x \\ f \end{smallmatrix}\right)\| \, dr \le q(h)\|\left(\begin{smallmatrix} x \\ f \end{smallmatrix}\right)\| \tag{4.13}$$

for all $\left(\begin{smallmatrix} x \\ f \end{smallmatrix}\right) \in D(\mathcal{A}_0)$ *and* $h \ge 0$.

Proof. For $1 < p < \infty$, this has been proved in Theorem 3.29, so assume
$p = 1$ and take constants $M \ge 1$ and $\omega > \omega_0(B)$ such that $\|S(t)\| \le M e^{\omega t}$
for all $t \ge 0$.

Let $\left(\begin{smallmatrix} x \\ f \end{smallmatrix}\right) \in D(\mathcal{A}_0)$ and set $S(r)x = 0$ for $r < 0$, $f(r) = 0$ and $\varphi(r) = 0$
for $r < -1$ and $r > 0$. For $h \ge 0$ and by Fubini's Theorem and a change of
variables, we have the estimate

$$\int_0^h \|\mathcal{B}\mathcal{T}_0(r)\left(\begin{smallmatrix} x \\ f \end{smallmatrix}\right)\| \, dr = \int_0^h \left\| \begin{pmatrix} 0 & \Phi \\ 0 & 0 \end{pmatrix} \begin{pmatrix} S(r) & 0 \\ S_r & T_0 \end{pmatrix} \begin{pmatrix} x \\ f \end{pmatrix} \right\| \, dr$$

$$= \int_0^h \|\Phi(S_r x + T_0(r)f\| \, dr$$

$$\le \int_0^h \int_{-1}^0 \|\varphi(\sigma)\| \, \|S(r + \sigma)x + f(r + \sigma)\| \, d\sigma \, dr$$

$$= \int_{-1}^0 \|\varphi(\sigma)\| \int_0^h \|S(r + \sigma)x + f(r + \sigma)\| \, dr \, d\sigma$$

$$= \int_{-1}^0 \|\varphi(\sigma)\| \int_\sigma^{h+\sigma} \|S(s)x + f(s)\| \, ds \, d\sigma$$

$$= \int_{-1}^h \int_{s-h}^s \|\varphi(\sigma)\| \, \|S(s)x + f(s)\|, d\sigma \, ds$$

$$\le \left(\sup_{-1 \le s \le h} \int_{s-h}^s \|\varphi(\sigma)\| \, d\sigma \right) \int_{-1}^h \|S(s)x + f(s)\| \, ds$$

$$\le \left(\sup_{-1 \le s \le h} \int_{s-h}^s \|\varphi(\sigma)\| \, d\sigma \right) \left(\int_0^h \|S(s)x\| \, ds + \|f\|_1 \right)$$

$$\le \left(\sup_{-1 \le s \le h} \int_{s-h}^s \|\varphi(\sigma)\| \, d\sigma \right) \left(M\|x\| \int_0^h e^{\omega s} \, ds + \|f\|_1 \right)$$

$$\le \left(\sup_{-1 \le s \le h} \int_{s-h}^s \|\varphi(\sigma)\| \, d\sigma \right) \left(\max\{1, \tfrac{M}{\omega}(e^{\omega h} - 1)\} \right) \|\left(\begin{smallmatrix} x \\ f \end{smallmatrix}\right)\|.$$

Hence, setting $q(h) := \sup_{-1 \le s \le h} \int_{s-h}^s \|\varphi(\sigma)\| \, d\sigma \cdot \max\{1, \tfrac{M}{\omega}(e^{\omega h} - 1)\}$ and
since $\varphi \in L^1([-1, 0], \mathcal{L}(X))$, we obtain Equation (4.13). $\square$

The next step is to determine the critical growth bound of the undelayed
semigroup $(\mathcal{T}_0(t))_{t \ge 0}$ generated by $\mathcal{A}_0$ and defined in Equation (3.33).

Theorem 4.25. *The critical growth bounds of $(S(t))_{t\geq 0}$ and $(\mathcal{T}_0(t))_{t\geq 0}$ coincide, i.e.,*

$$\omega_{\mathrm{crit}}(\mathcal{A}_0) = \omega_{\mathrm{crit}}(B). \tag{4.14}$$

Moreover, the growth bound of $(\mathcal{T}(t))_{t\geq 0}$ is given by

$$\omega_0(\mathcal{A}) = \max\{s(\mathcal{A}), \omega_{\mathrm{crit}}(B)\}. \tag{4.15}$$

Proof. We start by showing that the critical growth bounds of $(S(t))_{t\geq 0}$ and $(\mathcal{T}_0(t))_{t\geq 0}$ are equal.

For $t \geq 1$ and $h \geq 0$, we have

$$\mathcal{T}_0(t+h) - \mathcal{T}_0(t) = \begin{pmatrix} S(t+h) - S(t) & 0 \\ S_{t+h} - S_t & 0 \end{pmatrix}.$$

This obviously implies

$$\|\mathcal{T}_0(t+h) - \mathcal{T}_0(t)\| \geq \|S(t+h) - S(t)\|$$

and

$$\|\mathcal{T}_0(t+h) - \mathcal{T}_0(t)\| = \sup_{\|\binom{x}{f}\|\leq 1} \left\| \begin{pmatrix} S(t+h) - S(t) & 0 \\ S_{t+h} - S_t & 0 \end{pmatrix} \begin{pmatrix} x \\ f \end{pmatrix} \right\|$$

$$= \sup_{\|x\|\leq 1} \left\| \begin{pmatrix} (S(t+h) - S(t))x \\ (S_{t+h} - S_t)x \end{pmatrix} \right\|$$

$$\leq \sup_{\|x\|\leq 1} (\|(S(t+h) - S(t))x\| + \|(S_{t+h} - S_t)x\|_p)$$

$$\leq \sup_{\|x\|\leq 1} (\|S(t+h-1) - S(t-1)\| \, \|S(1)x\|$$

$$+ \left(\int_{-1}^{0} \|(S(t+h+\sigma) - S(t+\sigma))x\|^p \, d\sigma \right)^{\frac{1}{p}} \Bigg)$$

$$\leq \|S(t+h-1) - S(t-1)\| \, \|S(1)\|$$

$$+ \sup_{\|x\|\leq 1} \|S(t+h-1) - S(t-1)\| \left(\int_{-1}^{0} \|S(1+\sigma)x\|^p \, d\sigma \right)^{\frac{1}{p}}$$

$$\leq \|S(t+h-1) - S(t-1)\| \left(\|S(1)\| + \left(\int_{0}^{1} M e^{\omega s} \, ds \right) \right)$$

$$\leq K\|S(t+h-1) - S(t-1)\|$$

for $K = \|S(1)\| + \frac{M}{\omega}(e^\omega - 1)$. So, by Proposition 4.20, we have that $\omega_{crit}(\mathcal{A}_0) = \omega_{crit}(B)$.

Next, we show that $(\mathcal{T}(t))_{t\geq 0}$ satisfies the conditions in Theorem 4.23. The first Dyson-Phillips term for $\mathcal{T}(t)$ is, as already computed in the proof of Proposition 4.3,

$$
\begin{aligned}
V\mathcal{T}_0(t)\left(\tfrac{x}{f}\right) &= \int_0^t \mathcal{T}_0(t-s)\mathcal{B}\mathcal{T}_0(s)\left(\tfrac{x}{f}\right)ds \\
&= \int_0^t \begin{pmatrix} S(t-s)\Phi(S_s x + T_0(s)f) \\ S_{t-s}\Phi(S_s x + T_0(s)f) \end{pmatrix} ds \\
&= \int_0^t \int_{-1}^0 \begin{pmatrix} S(t-s)\varphi(r)(S(r+s)x + f(r+s)) \\ S_{t-s}\varphi(r)(S(r+s)x + f(r+s)) \end{pmatrix} dr\,ds
\end{aligned}
$$

for each $\left(\tfrac{x}{f}\right)\in D(\mathcal{A}_0)$ and $t\geq 0$. Moreover, we have

$$
\begin{aligned}
\int_h^{t+h} &\mathcal{T}_0(t+h-s)\mathcal{B}\mathcal{T}_0(s)\left(\tfrac{x}{f}\right)ds \\
&= \int_h^{t+h}\int_{-1}^0 \begin{pmatrix} S(t+h-s)\varphi(r)(S(r+s)x + f(r+s)) \\ S_{t+h-s}\varphi(r)(S(r+s)x + f(r+s)) \end{pmatrix} dr\,ds \\
&= \int_0^t\int_{-1}^0 \begin{pmatrix} S(t-s)\varphi(r)(S(r+h+s)x + f(r+h+s)) \\ S_{t-s}\varphi(r)(S(r+h+s)x + f(r+h+s)) \end{pmatrix} dr\,ds \\
&= \int_0^t\int_{-1+h}^h \begin{pmatrix} S(t-s)\varphi(r-h)(S(r+s)x + f(r+s)) \\ S_{t-s}\varphi(r-h)(S(r+s)x + f(r+s)) \end{pmatrix} dr\,ds
\end{aligned}
$$

for all $\left(\tfrac{x}{f}\right)\in D(\mathcal{A}_0)$, $h>0$, and $t\geq 0$. Subtracting these identities, we obtain

$$
\begin{aligned}
\int_h^{t+h} &\mathcal{T}_0(t+h-s)\mathcal{B}\mathcal{T}_0(s)\left(\tfrac{x}{f}\right)ds - \int_0^t \mathcal{T}_0(t-s)\mathcal{B}\mathcal{T}_0(s)\left(\tfrac{x}{f}\right)ds \\
&= \int_0^t\int_{-1}^h \begin{pmatrix} S(t-s)[\varphi(r-h)-\varphi(r)][S(r+s)x + f(r+s)] \\ S_{t-s}[\varphi(r-h)-\varphi(r)][S(r+s)x + f(r+s)] \end{pmatrix} dr\,ds
\end{aligned}
$$

for all $\left(\tfrac{x}{f}\right)\in D(\mathcal{A}_0)$, $h>0$, and $t\geq 0$. Hence, its norm can be estimated as

$$
\begin{aligned}
\left\| \int_h^{t+h} \mathcal{T}_0(t+h-s)\mathcal{B}\mathcal{T}_0(s)\left(\tfrac{x}{f}\right)ds - \int_0^t \mathcal{T}_0(t-s)\mathcal{B}\mathcal{T}_0(s)\left(\tfrac{x}{f}\right)ds \right\| \\
\leq C\left(\int_{-1}^h \|\varphi(r-h)-\varphi(r)\|\,dr \right)\left\|\left(\tfrac{x}{f}\right)\right\|
\end{aligned}
$$

for all $\left(\begin{smallmatrix} x \\ f \end{smallmatrix}\right) \in D(\mathcal{A}_0)$, $h > 0$, and $t \geq 0$. Finally, we obtain

$$\|V\mathfrak{T}_0(t+h)\left(\begin{smallmatrix} x \\ f \end{smallmatrix}\right) - V\mathfrak{T}_0(t)\left(\begin{smallmatrix} x \\ f \end{smallmatrix}\right)\|$$

$$\leq C\left(\int_{-1}^{h} \|\varphi(r-h) - \varphi(r)\|\, dr + q(h)\right)\|\left(\begin{smallmatrix} x \\ f \end{smallmatrix}\right)\|$$

for all $\left(\begin{smallmatrix} x \\ f \end{smallmatrix}\right) \in D(\mathcal{A}_0)$, $h > 0$, and $t \geq 0$. Hence, $V\mathfrak{T}_0$ is right norm continuous for $t \geq 0$ and, in an analogous way, one can prove that it is also left norm continuous for $t > 0$. This means that the hypothesis of Theorem 4.23 is fulfilled with $n = 1$, and the identity

$$\omega_0(\mathcal{A}) = \max\{s(\mathcal{A}), \omega_{crit}(B)\}$$

follows. □

An immediate consequence of this fact is the following corollary.

Corollary 4.26. *If the delay operator Φ is of the type of Equation (4.12), then the growth bound of the semigroup $(\mathfrak{T}(t))_{t \geq 0}$ is independent of $p \geq 1$.*

Finally, we state the stability result following from Equation (4.15) and not using any regularity property of the semigroup.

Corollary 4.27. *Assume that the delay operator Φ is of the type of Equation (4.12) and that $\omega_{crit}(B) < 0$. Then*

$$\omega_0(\mathcal{A}) < 0 \iff s(\mathcal{A}) < 0.$$

4.5   Notes and References

The results of Section 4.1 are essentially taken from [18, 19]. A result analogous to Proposition 4.3 was already proved by a different technique, and for the special case $\Phi := \sum_{k=0}^{n} B_k \delta_{h_k}$, by Fischer and van Neerven [79, Proposition 3.5].

Example 4.4 is a modification of Brendle, Nagel, and Poland [27, Example 5.1] and was observed by R. Shvidkoy.

Section 4.2 considers some of the consequences of the Spectral Mapping Theorem to the asymptotics of delay equations. For stability criteria based on the Spectral Mapping Theorem and the characteristic equations, there exists an enormous amount of literature. In particular, we mention the monographs by Diekmann et al. [62], Hale and Verduyn Lunel [103], and Wu [222]. For a deep spectral analysis using compactness, see Verduyn Lunel [204, 205]. For a generalization of the decomposition result to delay equations with infinite delay, see Milota [144].

It was observed by Sinestrari [193] that in the case of general unbounded delay operators, the delay semigroup fails to have regularity properties in general. Special cases of delay operators where regularity holds were investigated in Batty [21], Di Blasio [59], and Mastinšek [143].

For more unbounded delay operators, the spectral methods cannot be used in the presented way, but have been treated in, e.g., Di Blasio, Kunisch, and Sinestrari [61], Fašanga and Prüß [74]. Also see Chapter 10, where we show how these equations can be fitted in this semigroup framework.

The results of Section 4.3 are due to S. Brendle, R. Nagel, and J. Poland, and are contained in [27, 154].

In order to better understand the failure of the Spectral Mapping Theorem for semigroups that are not eventually norm continuous, the critical spectrum was introduced by R. Nagel and J. Poland in [154]. This spectrum shows many analogies to the essential spectrum studied by many different authors (see, e.g., Engel and Nagel [72, Section IV.1.20], Kato [120, Section IV.5.6], Gohberg, Goldberg, and Kaashoek [85, Chapter XVII], or Goldberg [86, Section IV.2]). For example, a *critical growth bound* can be defined and perturbation theorems hold in analogy to the ones for the essential growth bound (see, e.g., Engel and Nagel [72, Section IV.2] and J. Voigt [209]). The critical spectrum is a much smaller set than the essential spectrum, but may also be more difficult to determine.

Section 4.4 uses the method of critical spectrum to determine the stability of delay equations. The results are due to S. Brendle [26].

Chapter 5

Stability via Perturbation

In many cases, even when the Spectral Mapping Theorem holds (see e.g., Theorem 2.11), it is easier to obtain resolvent estimates than to compute the spectral bound via the characteristic equation. Further, in a large class of applications like hyperbolic equations with delay, it is not natural to make regularity assumptions on the delay semigroup and, hence, resolvent estimates are the adequate tool to analyze the asymptotic behavior of solutions. Such resolvent estimates and their consequences for the stability and hyperbolicity of the delay semigroup are the topic of this section.

In Section 5.1, we give various conditions to ensure that the resolvent of the generator $(\mathcal{A}, D(\mathcal{A}))$ is bounded on the imaginary line $i\mathbb{R}$ or on right halfplanes. These estimates result in exponential stability by the Weis-Wrobel Theorem and uniform exponential stability and hyperbolicity in Hilbert spaces by Gearhart's Theorem. They are analyzed in detail and applied to a large collection of examples. Further examples can be found in Chapter 9, where wave equations with delay are considered.

In Section 5.2, we collect some basic facts on operator-valued Fourier multipliers, which allow us in Section 5.3 to extend the robust stability and hyperbolicity criteria to general Banach spaces.

Throughout this chapter, we assume that $Z = X$ in Hypothesis (H_3).

5.1 Stability and Hyperbolicity via Gearhart's Theorem

In this section, we consider robust stability and hyperbolicity criteria for the delay equation (DE).

Our main tools are estimates for the abscissa of uniform boundedness of the resolvent

$$s_0(\mathcal{A}) := \inf \left\{ \omega \in \mathbb{R} \ : \ \{\Re\lambda > \omega\} \subset \rho(\mathcal{A}) \text{ and } \sup_{\Re\lambda > \omega} \|R(\lambda, \mathcal{A})\| < \infty \right\}$$

(see Section 1.1). In Hilbert spaces and for $p = 2$, this gives a stability result according to Gearhart's Theorem (see Theorem 2.26). At the end of this section, we extend these results to hyperbolicity.

To this purpose, we restrict our attention to a class of delay operators Φ for which (DE) is well-posed for each generator B and which have an additional property.

Definition 5.1. We call the delay operator $\Phi \in \mathcal{L}(W^{1,p}([-1,0], X), X)$ *admissible* if

(i) the operator matrix $(\mathcal{A}, D(\mathcal{A}))$ generates a strongly continuous semigroup for all generators $(B, D(B))$ and

(ii) the function $\lambda \mapsto \Phi R(\lambda, A_0)$ is an analytic function that is bounded on each halfplane $\{\lambda \in \mathbb{C} : \Re\lambda > \omega\}$ for all $\omega \in \mathbb{R}$.

We start our series of results with an almost trivial result, which will, however, form the main idea behind the more sophisticated stability results.

Theorem 5.2. *Assume that Φ is admissible and let $\alpha \in \mathbb{R}$ such that the function*

$$\lambda \mapsto R(\lambda, B + \Phi_\lambda)$$

is a bounded analytic function on the right halfplane $\{\Re\lambda > \alpha\}$. Then $s_0(\mathcal{A}) < \alpha$.

Proof. We have to show the boundedness on the halfplane $\{\Re\lambda > \alpha\}$ of the resolvent operator given in Equation (3.24).

Under our assumptions, this is equivalent to the existence and boundedness of $R(\lambda, B + \Phi_\lambda)$ on the halfplane $\{\Re\lambda > \alpha\}$. Hence, by Lemma 2.25, we have $s_0(\mathcal{A}) < \alpha$. $\qquad\square$

It seems to be an open question whether there are examples of operators Φ satisfying condition (i) but not condition (ii) in Definition 5.1. However, we can show that all the delay operators from Theorem 3.29 are admissible.

Example 5.3. Assume that Φ is of the form of Equation (3.38), i.e.,

$$\Phi f := \int_{-1}^{0} d\eta\, f, \qquad f \in W^{1,p}([-1,0], X),$$

where $\eta : [-1, 0] \to \mathcal{L}(X)$ is of bounded variation. Then Φ is admissible.

We have already shown in Theorem 3.29 that, for such delay operators Φ, $(\mathcal{A}, D(\mathcal{A}))$ generates a strongly continuous semigroup for all generators $(B, D(B))$. Hence we only have to verify the boundedness condition (ii). This follows from the estimate

$$\|\Phi R(\lambda, A_0)f\| = \left\| \int_{-1}^{0} d\eta(\sigma) \int_{\sigma}^{0} e^{\lambda(\sigma-\tau)} f(\tau)\, d\tau \right\|$$

$$\leq \int_{-1}^{0} \int_{\sigma}^{0} \|e^{\lambda(\sigma-\tau)} f(\tau)\|\, d\tau\, d|\eta|(\sigma)$$

$$= \int_{-1}^{0} \int_{\sigma}^{0} e^{\Re\lambda(\sigma-\tau)} \|f(\tau)\|\, d\tau\, d|\eta|(\sigma)$$

$$\leq \int_{-1}^{0} \int_{\sigma}^{0} (1 + e^{-\omega}) \|f(\tau)\|\, d\tau\, d|\eta|(\sigma)$$

$$\leq \int_{-1}^{0} \int_{-1}^{0} (1 + e^{-\omega}) \|f(\tau)\|\, d\tau\, d|\eta|(\sigma)$$

$$\leq e^{-\omega} |\eta|([-1,0]) \|f\|_p$$

for every λ with $\Re\lambda > \omega$ and every $f \in L^p([-1, 0], X)$.

For this type of delay operator, we can immediately formulate a robust hyperbolicity and stability result. Unfortunately, the obtained result is not useful in applications and therefore we do not formulate it in its sharpest possible form.

Theorem 5.4. *Assume that $(B, D(B))$ generates a hyperbolic semigroup $(S(t))_{t \geq 0}$ with projection P and assume that Φ is of the form of Equation (3.38), i.e.,*

$$\Phi f := \int_{-1}^{0} d\eta\, f, \qquad f \in W^{1,p}([-1,0], X),$$

where $\eta : [-1, 0] \to \mathcal{L}(X)$ is of bounded variation and assume that $p > 1$. If $|\eta|([-1, 0])$ is sufficiently small, then the delay semigroup generated by $(\mathcal{A}, D(\mathcal{A}))$ is also hyperbolic and the dimension of its unstable subspace coincides with the dimension of the unstable subspace of $(S(t))_{t \geq 0}$.

Proof. Let us first show that the semigroup $(\mathcal{T}_0(t))_{t\geq 0}$ generated by $(\mathcal{A}_0, D(\mathcal{A}_0))$ is hyperbolic and that the dimension of its unstable subspace coincides with $\dim \ker P$.

By assumption, $(S(t))_{t\geq 0}$ is hyperbolic on X with projections P and $Q = I - P$ and constants $N, \delta > 0$. Let us denote the inverse of $S(t)$ on QX by $S_u(-t)$. We define

$$\mathcal{Q} = \begin{pmatrix} Q & 0 \\ S_u(\cdot)Q & 0 \end{pmatrix}.$$

Clearly, $\mathcal{Q}$ is a bounded projection on $\mathcal{E}_p$, $\dim \mathcal{Q}\mathcal{E}_p = \dim QX$, and $\mathcal{T}_0(t)\mathcal{Q} = \mathcal{Q}\mathcal{T}_0(t)$. The inverse of $\mathcal{T}_0(t)$, $t \geq 0$, on $\mathcal{Q}\mathcal{E}_p$ is given by

$$\mathcal{T}_u(-t) = \begin{pmatrix} S_u(-t) & 0 \\ S_u(-t+\cdot)Q & 0 \end{pmatrix}.$$

This formula implies that $\|\mathcal{T}_u(-t)\mathcal{Q}\| \leq N_1\,e^{-\delta t}$ for $t \geq 0$. For $t > r$ and $\mathcal{P} = I - \mathcal{Q}$, it holds

$$\mathcal{T}_0(t)\mathcal{P} = \begin{pmatrix} S(t) & 0 \\ S_t & 0 \end{pmatrix} \begin{pmatrix} P & 0 \\ -S_u(\cdot)Q & I \end{pmatrix} = \begin{pmatrix} S(t)P & 0 \\ S_t P & 0 \end{pmatrix}$$

so that $\|\mathcal{T}_0(t)\mathcal{P}\| \leq N_2\,e^{-\delta t}$ for $t \geq 0$ and a constant N_2. Hence, the semigroup $(\mathcal{T}_0(t))_{t\geq 0}$ is hyperbolic.

Considering now estimate (3.41) in the proof of Theorem 3.29 and combining it with (3.37) in Theorem 3.26, we obtain that there exists a constant $C > 0$ such that

$$\int_0^t \left\| \mathcal{B}\,\mathcal{T}_0(r)\left(\tfrac{x}{f}\right) \right\|\, dr \leq C|\eta|([-1,0]) \left\| \left(\tfrac{x}{f}\right) \right\|.$$

Combining this estimate with the Variation of Parameters Formula from Equation (1.11) for the Miyadera-Voigt type perturbations, the estimate

$$\|\mathcal{T}(t) - \mathcal{T}_0(t)\| \leq C'|\eta|([-1,0])$$

follows. Finally, Proposition 2.18 can be applied to finish the proof.   □

Now we turn our attention to more refined methods and obtain an estimate on the abscissa of uniform boundedness of the generator $\mathcal{A}$.

Theorem 5.5. *Assume that Φ is admissible, $s_0(B) < 0$ and let $\alpha \in (s_0(B), 0]$ such that*

$$a_{\alpha,n} := \sup_{\omega \in \mathbb{R}} \|(\Phi_{\alpha+i\omega} R(\alpha + i\omega, B))^n\| < \infty. \tag{5.1}$$

If

$$a_\alpha := \sum_{n=0}^{\infty} a_{\alpha,n} < \infty, \qquad (5.2)$$

then $s_0(\mathcal{A}) < \alpha \leq 0$. If the Banach space X has Fourier type $q \in [1,2]$, then

$$\omega_{\frac{1}{q}-\frac{1}{q'}}(\mathcal{A}) \leq \alpha.$$

Proof. We have to show the boundedness of $R(\lambda, B + \Phi_\lambda)$ on the halfplane $\{\Re\lambda > \alpha\}$. Then by Theorem 5.2 the claim follows.

Since $s_0(B) < \alpha$, by Lemma 2.28, we have that $a_{\beta,n} \leq a_{\alpha,n}$ for $\beta > \alpha$.

Defining $M_\alpha := \sup_{\Re\lambda > \alpha} \|R(\lambda, B)\|$, we obtain for all $\lambda \in \mathbb{C}$ with $\Re\lambda > \alpha$ that

$$R(\lambda, B) \sum_{n=0}^{\infty} (\Phi_\lambda R(\lambda, B))^n \in \mathcal{L}(X)$$

and

$$\left\| R(\lambda, B) \sum_{n=0}^{\infty} (\Phi_\lambda R(\lambda, B))^n \right\|$$
$$\leq M_\alpha \sum_{n=0}^{\infty} \| (\Phi_\lambda R(\lambda, B))^n \| \leq M_\alpha a_{\Re\lambda} \leq M_\alpha a_\alpha.$$

This operator is the inverse of $(\lambda - B - \Phi_\lambda)$ and is bounded on the halfplane $\{\Re\lambda > \alpha\}$. Hence, by Theorem 5.2, we have $s_0(\mathcal{A}) < \alpha$.

Finally, choose $p \in (q, q')$. Applying [84, Proposition 2.3], we obtain that $\mathcal{E}_p$ also has Fourier type q. The result now follows from the theorem of Weis-Wrobel (2.11). □

Note that the previous theorem means especially that under the Conditions (5.1) and (5.2), the inequality $\omega_1(\mathcal{A}) \leq \alpha$ holds independently of the geometry of the Banach space X.

Now we consider a direct application of Theorem 5.5 and treat the special case where $\Phi := C\delta_{-1}$ for an operator $C \in \mathcal{L}(X)$ commuting with $(B, D(B))$.

Corollary 5.6. *Assume that $\Phi = C\delta_{-1}$ for some $C \in \mathcal{L}(X)$ commuting with $(B, D(B))$, $s_0(B) < 0$ and,*

$$r(C) < \frac{1}{\sup_{\omega \in \mathbb{R}} \|R(i\omega, B)\|}.$$

Then $s_0(\mathcal{A}) < 0$.

Proof. Take $M := \sup_{\omega \in \mathbb{R}} \|R(i\omega, B)\|$ and assume that

$$r(C) \cdot M < q < 1.$$

We obtain by the definition of $r(C)$ that there exists $n_0 \in \mathbb{N}$ such that

$$\|C^n\|^{\frac{1}{n}} \cdot M < q < 1 \text{ for all } n \geq n_0.$$

This implies

$$
\begin{aligned}
a_{0,n} = \sup_{\omega \in \mathbb{R}} \|(\Phi_{i\omega} R(i\omega, B))^n\| &\leq \sup_{\omega \in \mathbb{R}} \|R(i\omega, B)^n\| \sup_{\omega \in \mathbb{R}} \|\Phi_{i\omega}^n\| \\
&\leq \|C^n\| \cdot M^n < q^n
\end{aligned}
$$

for all $n \geq n_0$.

We conclude that Equation (5.2) holds for $\alpha = 0$ and the assertion follows by Theorem 5.5. $\square$

This corollary shows that the general stability estimate presented in Theorem 5.5 is sharper than the one proved by Fischer and van Neerven [79, Corollary 3.6] or by the authors in [19, Theorem 4.2], where the condition

$$\|C\| < \frac{1}{\sup_{\omega \in \mathbb{R}} \|R(i\omega, B)\|}$$

was needed.

Now we formulate an important special case of Theorem 5.5 that will be applied to estimate the growth bound of the delay semigroup if X is a Hilbert space and $p = 2$.

Corollary 5.7. *Assume that Φ is admissible, $s_0(B) < 0$, and let $\alpha \in (s_0(B), 0]$. If*

$$\sup_{\omega \in \mathbb{R}} \|\Phi_{\alpha + i\omega} R(\alpha + i\omega, B)\| < 1,$$

or in particular if

$$\sup_{\omega \in \mathbb{R}} \|\Phi_{\alpha + i\omega}\| < \frac{1}{\sup_{\omega \in \mathbb{R}} \|R(\alpha + i\omega, B)\|}, \tag{5.3}$$

then $s_0(\mathcal{A}) < \alpha \leq 0$.

Proof. Defining $q_\alpha := \sup_{\omega \in \mathbb{R}} \|\Phi_{\alpha + i\omega} R(\alpha + i\omega, B)\| < 1$, we obtain that $a_{\alpha,n} \leq q_\alpha^n$, hence the series $\sum a_{\alpha,n}$ is convergent. $\square$

For Hilbert spaces, these results imply uniform stability estimates by Theorem 2.26. We state this only in the case of Equation (5.3), the other results being analogous.

Corollary 5.8. *Assume that X is a Hilbert space and that $p = 2$. If Φ is admissible, $\alpha \in (\omega_0(B), 0]$, and*

$$\sup_{\omega \in \mathbb{R}} \|\Phi_{\alpha + i\omega}\| < \frac{1}{\sup_{\omega \in \mathbb{R}} \|R(\alpha + i\omega, B)\|}, \tag{5.4}$$

then $\omega_0(\mathcal{A}) < \alpha \leq 0$.

Moreover, if $(B, D(B))$ is a normal operator, we can give a very useful condition implying Equation (5.3).

Corollary 5.9. *Assume that X is a Hilbert space, $p = 2$, $(B, D(B))$ is a normal operator generating a strongly continuous semigroup, and Φ is of the form of Equation (3.38), i.e., there is a function $\eta \in BV([-1, 0], \mathcal{L}(X))$ such that $\Phi(f) := \int_{-1}^{0} d\eta\, f$. If $\omega_0(B) < 0$ and*

$$|\eta|([-1, 0]) < |s(B)|, \tag{5.5}$$

then $\omega_0(\mathcal{A}) < 0$.

Proof. We show that the inequalities

$$\sup_{\omega \in \mathbb{R}} \|\Phi_{i\omega}\| \leq |\eta|([-1, 0]) < |s(B)| = \frac{1}{\sup_{\omega \in \mathbb{R}} \|R(i\omega, B)\|}$$

hold. Then the assertion follows by Corollary 5.8. For $\omega \in \mathbb{R}$, we have that

$$\|\Phi_{i\omega} y\| = \left\| \int_{-1}^{0} e^{i\omega s} d\eta(s) y \right\| \leq \|y\| \cdot \left| \int_{-1}^{0} e^{i\omega s} d|\eta|(s) \right| \leq \|y\| \cdot |\eta|([-1, 0]).$$

The right-hand side, since $(B, D(B))$ is a normal operator on a Hilbert space, can be computed as

$$\sup_{\omega \in \mathbb{R}} \|R(i\omega, B)\| = \sup_{\omega \in \mathbb{R}} \frac{1}{d\,(i\omega, \sigma(B))} = \frac{1}{\inf_{\omega \in \mathbb{R}} d\,(i\omega, \sigma(B))}$$

$$= \frac{1}{d\,(\mathbb{R}, \sigma(B))} = \frac{1}{|s(B)|}$$

(see Theorem 2.9). □

In the following, we test our results in a series of examples. All the delay operators in the following examples are admissible by Example 5.3.

Example 5.10. We consider the *diffusion equation with delayed reaction term* (see Wu [222, Section 2.1]) of Example 3.15.

$$\begin{cases} \partial_t w(x,t) = \Delta w(x,t) + c \int_{-1}^{0} w(x,t+\tau) dg(\tau), \\ \qquad\qquad\qquad\qquad\qquad\qquad x \in \Omega,\ t \geq 0, \\ w(x,t) = 0, \qquad\qquad\qquad\qquad x \in \partial\Omega,\ t \geq 0, \\ w(x,t) = f(x,t), \qquad\qquad\quad (x,t) \in \Omega \times [-1,0], \end{cases} \qquad \text{(RDD)}$$

where c is a constant, $\Omega \subset \mathbb{R}^n$ a bounded open set, $f(\cdot,t) \in L^2(\Omega)$ for all $t \geq 0$, $f(\cdot,0) \in D := \{u \in H_0^1(\Omega) : \Delta u \in L^2(\Omega)\}$, and the map $[-1,0] \ni t \mapsto f(\cdot,t) \in L^2(\Omega)$ belongs to $W^{1,2}\left([-1,0], L^2(\Omega)\right)$. The function $g : [-1,0] \to [0,1]$ is supposed to be the Cantor function (see Gelbaum and Olmsted [83, Example I.8.15]), which is singular and has total variation 1.

In order to apply our results, we take

- $X := L^2(\Omega)$,

- $B := \Delta_D$ the Dirichlet-Laplacian with usual domain $D(\Delta_D) = D$, and

- $\eta := c \cdot g \cdot Id$.

The well-posedness follows from Theorem 3.29. Next, we will verify the stability estimate of Equation (5.5). First, the expression with Φ satisfies

$$\|\Phi_\lambda y\| = |c|\|y\| \left\| \int_{-1}^{0} e^{\lambda\tau} dg(\tau) \right\| \leq |c|\|y\|.$$

The other expression, using Corollary 5.9, can be computed as

$$\sup_{\omega\in\mathbb{R}} \|R(i\omega, B)\| = \sup_{\omega\in\mathbb{R}} \frac{1}{d\left(i\omega, \sigma(B)\right)} = \frac{1}{d\left(0, \sigma(B)\right)} = \frac{1}{|\lambda_1|},$$

where λ_1 is the first eigenvalue of the Laplacian. We can summarize as follows.

Corollary 5.11. *The above reaction-diffusion equation with delay* (RDD) *is well-posed. Moreover, the solutions decay exponentially if*

$$|c| < |\lambda_1|.$$

We refer, e.g., to Davies [51, Chapter 6] for estimates on λ_1 and for further references. The same result holds for general elliptic operators as considered in [51, Section 6.3].

In the next example, the domain Ω is not bounded.

Example 5.12. Consider the diffusion equation with delayed reaction term

$$
\begin{cases}
\partial_t w(x,t) = \Delta w(x,t) + c \int_{-1}^{0} w(x, t+\tau)dh(\tau), \\
\hspace{4cm} x \in \Omega,\ t \geq 0, \\
w(x,t) = 0, \hspace{2.3cm} x \in \partial\Omega,\ t \geq 0, \\
w(x,t) = f(x,t), \hspace{1.6cm} (x,t) \in \Omega \times [-1,0],
\end{cases}
\tag{RDU}
$$

where c is a constant and $\Omega \subset \mathbb{R}^n$ ($n \geq 2$) a domain such that the Laplace operator generates an exponentially stable semigroup in $L^2(\Omega)$. We refer, e.g., to Dautray-Lions [49, Volume 2, Proposition IV.7.1.1] for the following sufficient condition:

$$
\Omega \subseteq \{x \in \mathbb{R}^n : a \leq \xi \cdot x \leq b\} \quad \text{for some } \xi \in \mathbb{R}^n \text{ and } a < b. \tag{5.6}
$$

We emphasize that in this case the Dirichlet-Laplacian in $L^2(\Omega)$ is invertible, but its resolvent is not necessarily compact.

In order to obtain classical solutions, we assume that $f(\cdot, t) \in L^2(\Omega)$ for all $t \geq 0$,

$$
f(\cdot, 0) \in \{u \in H_0^1(\Omega) : \Delta u \in L^2(\Omega)\},
$$

and the map $[-1, 0] \ni t \mapsto f(\cdot, t) \in L^2(\Omega)$ belongs to $W^{1,2}([-1,0], L^2(\Omega))$. The function $h : [-1, 0] \to [0, 1]$ is supposed to be of bounded variation, i.e., $\mathrm{Var}(h)_{[-1,0]} < \infty$.

In order to apply our results, we define

- $X := L^2(\Omega)$,

- $B := \Delta_D$ the variational Dirichlet-Laplacian with usual domain, and

- $\eta := c \cdot h \cdot Id$.

The well-posedness follows again from Theorem 3.29. Next, we verify the stability estimate of Equation (5.5). First, the expression with Φ satisfies

$$
\|\Phi_\lambda y\| = |c| \|y\| \left\| \int_{-1}^{0} e^{\lambda \tau} dh(\tau) \right\| \leq |c| \mathrm{Var}(h)_{-1}^{0} \|y\|.
$$

The other expression, using Corollary 5.9, can be computed again as

$$
\sup_{\omega \in \mathbb{R}} \|R(i\omega, B)\| = \frac{1}{|s(B)|},
$$

where $s(B) < 0$ is the spectral bound of the Dirichlet-Laplacian.

We can summarize as follows.

Corollary 5.13. *The above reaction-diffusion equation with delay* (RDU) *is well-posed. Moreover, the solutions decay exponentially if*

$$|c|\mathrm{Var}(h)_{[-1,0]} < |s(B)|.$$

Example 5.14. We consider the reaction-diffusion equation from Example 5.10, but now in the state space $\mathcal{E} := L^r(\Omega) \times L^p([-1,0], L^r(\Omega))$ for $1 \leq r < \infty$, $1 < p < \infty$. The well-posedness follows again from Theorem 3.29. We show below that the solutions decay exponentially if

$$|c| < |\lambda_1|, \tag{5.7}$$

thereby extending our result from the Hilbert space case.

To obtain Equation (5.7), we extend the result obtained in the Hilbert space case in Example 5.10 by two steps. Consider first the case $r = 2$. We know from Lemma 3.19 that the spectrum $\sigma(\mathcal{A})$ of $(\mathcal{A}, D(\mathcal{A}))$ does not depend on p and that

$$s(\mathcal{A}) = \omega_0(\mathcal{A})$$

since, by Proposition 4.3, $(\mathcal{A}, D(\mathcal{A}))$ generates an eventually norm continuous semigroup on $\mathcal{E}$. For $p = 2$ we had in Corollary 5.11 a condition implying exponential stability, i.e., $\omega_0(\mathcal{A}) < 0$. Hence we obtain that the solutions decay exponentially if Equation (5.7) holds.

For $\mathcal{E} := L^r(\Omega) \times L^p([-1,0], L^r(\Omega))$, we observe first that the operator $(B, D(B))$ has compact resolvent. From Lemma 4.5, it follows that the operator $(B + \Phi_\lambda, D(B))$ has compact resolvent in $X := L^r(\Omega)$ for all $1 \leq r < \infty$, and thus its spectrum does not depend on r; see Arendt [4, Proposition 2.6] for the details. Using a spectral mapping argument as before, we obtain that Equation (5.7) implies uniform exponential stability.

Finally, we remark that we could not simply apply Corollary 5.7 to obtain Equation (5.7). If we estimate the resolvent for the stability condition in Corollary 5.7, we obtain uniform exponential stability of the semigroup if

$$|c| < c_r,$$

where $c_r \leq |\lambda_1|$ is a constant depending on r.

Example 5.15. Consider the following first-order differential equation with delay

$$\begin{cases} \partial_t u(x,t) = -\partial_x u(x,t) - \mu u(x,t) + \int_{-1}^{0} u(x,t+s)dg(s), \\ \qquad\qquad\qquad\qquad\qquad\qquad\qquad\qquad x \in \mathbb{R},\ t \geq 0, \\ u(0,t) = 0, \qquad\qquad\qquad\qquad\qquad\qquad t \geq 0, \\ u(x,s) = f(x,s), \qquad\qquad\qquad\qquad\quad x \in \mathbb{R},\ t \in [-1,0], \end{cases}$$

where $\mu > 0$ and g is a function of bounded variation. We treat this equation in Hilbert spaces by considering

- $X := L^2(\mathbb{R})$,

- $B := -\partial_x - \mu$ with domain $D(B) = H^1(\mathbb{R})$, and

- $\eta := g \cdot Id$.

The well-posedness follows again from Theorem 3.29. Next, we verify the stability estimate of Equation (5.4). First, the expression with Φ satisfies

$$\|\Phi_\lambda y\| = \|y\| \left\| \int_{-1}^{0} e^{\lambda \tau} dg(\tau) \right\| \leq \mathrm{Var}(g)_{-1}^{0} \|y\|.$$

The other expression, using that $B = \partial_x - \mu$ is a normal operator on a Hilbert space, can be computed as

$$\sup_{\omega \in \mathbb{R}} \|R(i\omega, B)\| = \frac{1}{|\mu|}.$$

We can summarize as follows.

Corollary 5.16. *The above first-order partial differential equation with delay is well-posed. Moreover, the solutions decay exponentially if*

$$\mathrm{Var}(g)_{[-1,0]} < |\mu|.$$

Other important examples of semigroups that are not eventually norm continuous will be presented in Section 9.9.3.

Now we formulate the analogous results on hyperbolicity in Hilbert spaces.

Theorem 5.17. *Let X be a Hilbert space and consider Equation* (DE). *Assume that Φ is admissible, the semigroup $(B, D(B))$ generates a hyperbolic semigroup, and consider*

$$a_n := \sup_{\omega \in \mathbb{R}} \|(\Phi_{i\omega} R(i\omega, B))^n\| < \infty, \quad n \in \mathbb{N}. \qquad (5.8)$$

If

$$a := \sum_{n=0}^{\infty} a_n < \infty, \tag{5.9}$$

then $(\mathcal{A}, D(\mathcal{A}))$ generates a hyperbolic semigroup.

Proof. As a consequence of Gearhart's Theorem (see Theorem 2.26), the numbers a_n are defined for all $n \in \mathbb{N}$, and we only have to show the boundedness of the resolvent operator given in Equation (3.24) on the line $i\mathbb{R}$. Under our assumptions, this is equivalent to the existence and boundedness of $R(\lambda, B + \Phi_\lambda)$ for $\lambda \in i\mathbb{R}$.

Let $\lambda \in i\mathbb{R}$ and $M := \sup_{\lambda \in i\mathbb{R}} \|R(\lambda, B)\|$. We obtain that

$$R(\lambda, B) \sum_{n=0}^{\infty} (\Phi_\lambda R(\lambda, B))^n \in \mathcal{L}(X)$$

and

$$\left\| R(\lambda, B) \sum_{n=0}^{\infty} (\Phi_\lambda R(\lambda, B))^n \right\| \leq M \sum_{n=0}^{\infty} \| (\Phi_\lambda R(\lambda, B))^n \|$$

$$\leq M \sum_{n=0}^{\infty} a_n = M \cdot a.$$

This operator defines the inverse of $(\lambda - B - \Phi_\lambda)$ and remains bounded on $i\mathbb{R}$. $\qquad\square$

Using the same arguments, we can formulate the analogous version of Corollary 5.7.

Corollary 5.18. *Let X be a Hilbert space and consider Equation* (DE). *Assume that Φ is admissible and that the semigroup $(B, D(B))$ generates a hyperbolic semigroup. If*

$$\sup_{\omega \in \mathbb{R}} \|\Phi_{i\omega} R(i\omega, B)\| < 1, \tag{5.10}$$

or in particular if

$$\sup_{\omega \in \mathbb{R}} \|\Phi_{i\omega}\| < \frac{1}{\sup_{\omega \in \mathbb{R}} \|R(i\omega, B)\|}, \tag{5.11}$$

then $(\mathcal{A}, D(\mathcal{A}))$ generates a hyperbolic semigroup.

We can sharpen the previous results if we consider delay operators of the form of Equation (3.38) given by the Riemann-Stieltjes Integral $\Phi f = \int_{-1}^{0} d\eta f$ (or, more generally, satisfying condition (M) of Theorem 3.26).

Theorem 5.19. *Let X be a Hilbert space and consider Equation* (DE). *Assume that Φ is of the form of Equation* (3.38), *the semigroup $(B, D(B))$ generates a hyperbolic semigroup, and consider*

$$a_n := \sup_{\omega \in \mathbb{R}} \|(\Phi_{i\omega} R(i\omega, B))^n\| < \infty, \quad n \in \mathbb{N}. \tag{5.12}$$

If

$$a := \sum_{n=0}^{\infty} a_n < \infty, \tag{5.13}$$

then $(\mathcal{A}, D(\mathcal{A}))$ generates a hyperbolic semigroup and the dimension of its unstable subspace coincides with the dimension of the unstable subspace of $(S(t))_{t \geq 0}$.

Proof. We only have to check the assertion concerning the dimensions. To this end, we introduce the perturbations $\alpha \mathcal{B}$ for $\alpha \in [0,1]$. Since the hypotheses hold for $\alpha \mathcal{B}$, we obtain the corresponding exponentially dichotomic semigroups $\mathcal{T}_{\alpha\Phi} =: \mathcal{T}_{\alpha}$, where $\mathcal{T}_1 = \mathcal{T}_\Phi = \mathcal{T}$. In view of Proposition 2.18 and Theorem 5.4, we have to show that the projections $\mathcal{Q}_\alpha$ for $\mathcal{T}_\alpha$ depend continuously on α in operator norm.

This is the case if $\alpha \mapsto \mathcal{T}_\alpha(t) \in \mathcal{L}(\mathcal{X})$ is continuous for some $t > 0$, because of the well-known formula

$$\mathcal{P}_\alpha = \frac{1}{2\pi i} \oint_\Gamma R(\lambda, \mathcal{T}_\alpha(t)) \, d\lambda. \tag{5.14}$$

But the Miyadera-Voigt Perturbation Theorem, especially Equations (1.10) and (1.9), yield that

$$\|\mathcal{T}_\alpha(t_0) - \mathcal{T}_\beta(t_0)\| \leq |\beta - \alpha| \cdot M q,$$

where $M = \sup_{t \in [0, t_0]} \|\mathcal{T}(t)\|$. □

The above presented ideas can also be used to prove the existence of stable, center, and unstable manifolds analogous to Theorem 4.15. However, to fit in this more general situation, we have to modify slightly the definitions of manifolds. Assume that X is a Hilbert space and that it is decomposed into stable, center, and unstable manifolds for the semigroup $(S(t))_{t \geq 0}$ generated by $(B, D(B))$, i.e., $X = X_S \oplus X_C \oplus X_U$, where X_S, X_C, and X_U are invariant subspaces for the semigroup such that there are constants $M \geq 1$ and $\omega_1, \omega_2 > 0$ such that the corresponding restrictions of the semigroup $(S(t))_{t \geq 0}$ satisfy the following conditions:

- The operators $S_S(t) = S(t)|_{X_S}$ form a uniformly exponentially stable semigroup, i.e., $\|S_S(t)\| \leq M e^{-\omega_1 t}$, $t \geq 0$.

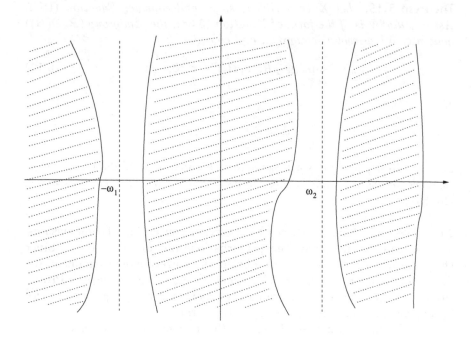

Figure 5.1. $\sigma(\mathcal{A})$

- The operators $S_U(t) = S(t)|_{X_U}$ are invertible and the semigroup $\left(S_U^{-1}(t)\right)_{t \geq 0}$ is uniformly exponentially stable, i.e., $\|S_U^{-1}(t)\| \leq M e^{-\omega_2 t}$, $t \geq 0$.

- The operators $S_C(t) = S(t)|_{S_C}$ form a group satisfying the estimate $\|S_C(t)\| \leq M e^{\omega_2 t}$, $t \geq 0$, and $\|S_C^{-1}(t)\| \leq M e^{\omega_1 t}$, $t \geq 0$.

Theorem 5.20. *Assume that the Hilbert space X can be decomposed into stable, center, and unstable manifolds for the semigroup $(S(t))_{t \geq 0}$ with the constants ω_1, ω_2 and that Φ is admissible. Define*

$$a_n := \sup_{\omega \in \mathbb{R}} \|(\Phi_{-\omega_1 + i\omega} R(i\omega, B))^n\| < \infty \qquad (5.15)$$

and

$$b_n := \sup_{\omega \in \mathbb{R}} \|(\Phi_{\omega_2 + i\omega} R(i\omega, B))^n\| < \infty. \qquad (5.16)$$

If

$$a := \sum_{n=0}^{\infty} a_n < \infty, \quad b := \sum_{n=0}^{\infty} b_n < \infty, \qquad (5.17)$$

then $\mathcal{E} = \mathcal{E}_S \oplus \mathcal{E}_C \oplus \mathcal{E}_U$, *where these subspaces are invariant under the delay semigroup and there is a constant $M_1 \geq 1$ such that the following conditions are met:*

(i) *The operators $\mathcal{T}_S(t) = \mathcal{T}(t)|_{\mathcal{E}_S}$ form a uniformly exponentially stable semigroup satisfying $\|\mathcal{T}_S(t)\| \leq M_1 e^{-\omega_1 t}$, $t \geq 0$.*

(ii) *The operators $\mathcal{T}_U(t) = \mathcal{T}(t)|_{\mathcal{E}_U}$ are invertible and the semigroup $\left(\mathcal{T}_U^{-1}(t)\right)_{t \geq 0}$ is uniformly exponentially stable such that $\|\mathcal{T}_U^{-1}(t)\| \leq M_1 e^{-\omega_2 t}$, $t \geq 0$.*

(iii) *The operators $\mathcal{T}_C(t) = T(t)|_{\mathcal{E}_C}$ form a group satisfying the estimates $\|\mathcal{T}_C(t)\| \leq M_1 e^{\omega_2 t}$, $t \geq 0$, and $\|\mathcal{T}_C^{-1}(t)\| \leq M_1 e^{\omega_1 t}$, $t \geq 0$.*

Proof. It follows from our assumptions and Theorem 5.17 that the operator $(\mathcal{A} + \omega_1, D(\mathcal{A}))$ generates a hyperbolic semigroup. Hence, a decomposition $\mathcal{E} = \mathcal{E}_S \oplus \mathcal{E}_{CU}$ holds, where these subspaces are invariant under the delay semigroup, and the restrictions for the estimates $\|\mathcal{T}_S(t)\| \leq M_1 e^{-\omega_1 t}$, $t \geq 0$, and $\|\mathcal{T}_{CU}^{-1}(t)\| \leq M_1 e^{\omega_1 t}$, $t \geq 0$, hold.

Consider now the rescaled semigroup $e^{-\omega_2 t}\mathcal{T}_{CU}(t)$. It has generator $(\mathcal{A} - \omega_2)|_{\mathcal{E}_{CU}}$ and is also hyperbolic by our assumptions. Hence, $\mathcal{E}_{CU} = \mathcal{E}_C \oplus \mathcal{E}_U$ where these subspaces are invariant under $\mathcal{T}_{CU}(t)$ and hence also under the delay semigroup. For the restrictions, the estimates $\|\mathcal{T}_C(t)\| \leq M_1 e^{\omega_2 t}$, $t \geq 0$, and $\|\mathcal{T}_U^{-1}(t)\| \leq M_1 e^{-\omega_2 t}$, $t \geq 0$, hold. $\square$

Analogous results can be formulated to the hyperbolicity concerning the dimension of the invariant subspaces (manifolds). The details are left to the reader.

5.2   Fourier Multipliers

Now we turn our attention to Fourier multipliers that are needed in Section 5.3. The main point in this section is to present the proper generalizations of Gearhart's Theorem in Banach spaces.

Definition 5.21. Let $1 \leq p < \infty$. We say that the function $m \in L^\infty(\mathbb{R}, \mathcal{L}(X))$ is an L^p-*Fourier multiplier* if the map

$$f \mapsto \mathcal{F}^{-1}(m\mathcal{F}f) = \mathcal{F}^{-1}m * f,$$

$f \in \mathcal{S}(\mathbb{R}, X)$, extends to a bounded linear operator on $L^p([-1,0], X)$. The set of all L^p-Fourier multipliers will be denoted by $\mathcal{M}_p^{\mathcal{L}(X)}$ and normed with

$$\|m\|_{\mathcal{M}_p^{\mathcal{L}(X)}} := \|\mathcal{F}^{-1}m\mathcal{F}\|_{\mathcal{L}(L^p([-1,0], X))}.$$

The following results play an important role in our investigations.
Theorem 5.22 has been proved recently by Latushkin and Shvidkoy in [130] and characterizes hyperbolicity via Fourier multipliers.

Theorem 5.22. *Let $(G, D(G))$ be the generator of a strongly continuous semigroup $(T(t))_{t\geq 0}$ on the Banach space X. Then the semigroup is hyperbolic if and only if*

$$R(i\cdot, G) \in \mathcal{M}_p^{\mathcal{L}(X)} \qquad \text{for some/all } 1 \leq p < \infty.$$

The next theorem has been proved by Clark, Latushkin, Montgomery-Smith, and Randolph [34] and Hieber [109] and characterizes the growth bound of a semigroup via Fourier multipliers.

Theorem 5.23. *Assume that $(G, D(G))$ is the generator of a strongly continuous semigroup $(T(t))_{t\geq 0}$. Then*

$$\omega_0(G) = \inf\left\{\mu > s(G) : \sup_{\alpha \geq \mu} \|R(\alpha + i\cdot, G)\|_{\mathcal{M}_p^{\mathcal{L}(X)}} < \infty\right\}$$

for some/all $1 \leq p < \infty$.

Finally, we mention a technical result that plays an important role in some proofs in this book and is usually referred to as the Hausdorff-Young Inequality (see Amann [3, Theorem 3.5]).

Theorem 5.24. (Hausdorff-Young Inequality) *Let $K \in L^1(\mathbb{R}, \mathcal{B}(X))$ and $f \in L^p(\mathbb{R}, X)$. Then $K * f \in L^p(\mathbb{R}, X)$ and*

$$\|K * f\|_{L^p} \leq \|K\|_{L^1} \cdot \|f\|_{L^p}. \tag{5.18}$$

For further references to the theory of operator-valued Fourier multipliers and their applications in semigroup theory, see Amann [3], Hieber [108, 109], and Latushkin and Shvidkoy [130].

5.3   Hyperbolicity via Fourier Multipliers

In this section, we extend the results of the previous section to Banach spaces using the powerful techniques of Fourier multipliers. The results were obtained in collaboration with E. Fašanga and R. Shvidkoy and are taken from [16].

Let us recall that in the Hilbert space case, the key assumption on the delay operator was its admissibility. This allowed us to prove the boundedness of the resolvent along imaginary lines.

Definition 5.25. We call the delay operator $\Phi \in \mathcal{L}(W^{1,p}([-1,0],X),X)$ *Fourier-admissible* if

(i) the operator $(\mathcal{A}, D(\mathcal{A}))$ is a generator for each generator $(B, D(B))$ and

(ii) the function $\Phi R(\alpha + i\cdot, A_0)$, where A_0 is the generator of the nilpotent left shift semigroup on $L^p([-1,0],X)$ (see Proposition 3.25), is an L^p-Fourier multiplier for all $\alpha \in \mathbb{R}$.

The following proposition gives a class of examples of such delay operators.

Proposition 5.26. *Assume that Φ is given by Equation (3.38) as in Theorem 3.29. Then Φ is Fourier-admissible.*

Proof. Condition (i) of Definition 5.1 is satisfied by Theorem 3.29. So we only have to prove that condition (ii) is satisfied. Let us fix an arbitrary $\alpha \in \mathbb{R}$. Denote by M the operator acting from $E := L^p(\mathbb{R}, L^p([-1,0],X))$ into $L^p(\mathbb{R},X)$ by the rule

$$(Mf)(t) := \int_{\mathbb{R}} \Phi R(\alpha + is, A_0)\hat{f}(s)e^{its}ds, \quad t \in [-1,0], \ f \in E.$$

We prove that for all f from a dense subspace $D \subset E$ and for some constant $c_{\alpha,\eta} > 0$, the inequality

$$\int_{\mathbb{R}} \|(Mf)(t)\|_X^p \, dt \le c_{\alpha,\eta}\|f\|_E^p \tag{5.19}$$

holds. We take D to be the space of all Schwartz functions from $\mathbb{R}$ to $L^p([-1,0],X)$ whose Fourier transform have compact support. This will be enough to justify the use of Fubini's Theorem in our further computations.

Before proving Equation (5.19), let us simplify the expression for $(Mf)(t)$ using Equation (3.38) and the definition of A_0. In this way, we obtain

$$(Mf)(t) = \int_{\mathbb{R}} \int_{-1}^{0} [R(\alpha + is, A_0)\hat{f}(s)](\tau)\, d\eta(\tau)e^{ist}\, ds$$

$$= \int_{\mathbb{R}} \int_{-1}^{0} \int_{\tau}^{0} e^{(\alpha+is)(\tau-r)}\hat{f}(s)(r)\, dr\, d\eta(\tau)e^{ist}\, ds$$

$$= \int_{-1}^{0} \int_{\tau}^{0} \int_{\mathbb{R}} e^{(\alpha+is)(\tau-r)+ist}\hat{f}(s)(r)\, ds\, dr\, d\eta(\tau)$$

$$= \int_{-1}^{0} \int_{\tau}^{0} e^{\alpha(\tau-r)} \int_{\mathbb{R}} e^{is(\tau-r+t)}\hat{f}(s)(r)\, ds\, dr\, d\eta(\tau)$$

$$= \int_{-1}^{0} \int_{\tau}^{0} e^{\alpha(\tau-r)} f(\tau - r + t)(r)\, dr\, d\eta(\tau).$$

Observe now that $e^{\alpha(\tau-r)}$ is bounded by some constant C for all τ and r from the domain of integration. Therefore, using this and the Hölder inequality, we obtain the estimates

$$\int_{\mathbb{R}} \|(Mf)(t)\|_X^p\, dt \leq \int_{\mathbb{R}} \int_{-1}^{0} \int_{\tau}^{0} \|f(\tau - r + t)(r)\|_X^p\, dr\, d|\eta|(\tau)\, dt$$

$$= C^p \int_{-1}^{0} \int_{\tau}^{0} \int_{\mathbb{R}} \|f(\tau - r + t)(r)\|_X^p\, dt\, dr\, d|\eta|(\tau)$$

$$= C^p \int_{-1}^{0} \int_{\tau}^{0} \int_{\mathbb{R}} \|f(t)(r)\|_X^p\, dt\, dr\, d|\eta|(\tau)$$

$$\leq C^p \int_{-1}^{0} \int_{-1}^{0} \int_{\mathbb{R}} \|f(t)(r)\|_X^p\, dt\, dr\, d|\eta|(\tau)$$

$$= C^p |\eta|([-1,0]) \int_{\mathbb{R}} \int_{-1}^{0} \|f(t)(r)\|_X^p\, dr\, dt = C_{\alpha,\eta}\|f\|_E^p,$$

which is the desired result. Here $|\eta|$ is the positive-valued Borel measure defined from η by its total variation. $\qquad\square$

Theorem 5.27. *Assume that Φ is Fourier-admissible, $(B, D(B))$ generates a hyperbolic semigroup, and let*

$$a_n := \|(\Phi_i.R(i\cdot, B))^n\|_{\mathcal{M}_p^{\mathcal{L}(X)}} < \infty. \tag{5.20}$$

If

$$a := \sum_{n=0}^{\infty} a_n < \infty, \tag{5.21}$$

then $(\mathcal{A}, D(\mathcal{A}))$ generates a hyperbolic semigroup.

Proof. By Theorem 5.22, we have to show that the resolvent operator given by Equation (3.24) is an L^p-Fourier multiplier on the line $i\mathbb{R}$.

Since Φ is Fourier-admissible, this follows if the function $R(i\cdot, B + \Phi_i.)$ is bounded as an L^p-Fourier multiplier on $i\mathbb{R}$. We remark that if m is an L^p-Fourier multiplier, then $\epsilon_{\alpha+i.} \otimes m(\cdot)$ is an L^p-Fourier multiplier between the Banach spaces $L^p(\mathbb{R}, L^p([-1,0]), X))$ and $L^p(\mathbb{R}, X)$. By Theorem 5.23, $R(i., A_0)$ is an L^p-Fourier multiplier from $L^p(\mathbb{R}, L^p([-1,0]), X))$ into itself since $(A_0, D(A_0))$ generates a uniformly exponentially stable semigroup. Define

$$M := \|R(i\cdot, B)\|_{\mathcal{M}_p^{\mathcal{L}(X)}},$$

which is finite by Theorem 5.22. Hence,

$$R(i\omega, B) \sum_{n=0}^{\infty} (\Phi_{i\omega} R(i\omega, B))^n \in \mathcal{L}(X)$$

for all $\omega \in \mathbb{R}$ and the inequality

$$\left\| R(i\cdot, B) \sum_{n=0}^{\infty} (\Phi_i.R(i\cdot, B))^n \right\|_{\mathcal{M}_p^{\mathcal{L}(X)}} \leq M \sum_{n=0}^{\infty} \| (\Phi_i.R(i\cdot, B))^n \|_{\mathcal{M}_p^{\mathcal{L}(X)}}$$

$$\leq M \sum_{n=0}^{\infty} a_n = Ma$$

holds. This operator defines the inverse of $(i\omega - B - \Phi_{i\omega})$ and remains bounded on $i\mathbb{R}$, which concludes the proof. $\qquad\square$

We can sharpen the previous result if we consider delay operators of the form of Equation (3.38) given by the Riemann-Stieltjes Integral $\Phi f = \int_{-1}^{0} d\eta f$ (or, more generally, satisfying condition (M) of Theorem 3.26). The proof is the same as in Theorem 5.19 and therefore will not be repeated here.

Theorem 5.28. *Let X be a Hilbert space and consider Equation* (DE). *Assume that Φ is of the form of Equation* (3.38), *the semigroup $(B, D(B))$ generates a hyperbolic semigroup, and consider*

$$a_n := \|(\Phi_i. R(i\cdot, B))^n\|_{\mathcal{M}_p^{\mathcal{L}(X)}} < \infty. \qquad (5.22)$$

If

$$a := \sum_{n=0}^{\infty} a_n < \infty, \qquad (5.23)$$

then $(\mathcal{A}, D(\mathcal{A}))$ generates a hyperbolic semigroup and the dimension of its unstable subspace coincides with the dimension of the unstable subspace of $(S(t))_{t\geq 0}$.

Now we turn our attention to stability questions.

Theorem 5.29. *Assume that Φ is Fourier-admissible, $\omega_0(B) < 0$, and let $\alpha \in (\omega_0(B), 0]$ such that*

$$a_{\alpha,n} := \sup_{\beta \geq \alpha} \|(\Phi_{\beta+i}. R(\beta + i\cdot, B))^n\|_{\mathcal{M}_p^{\mathcal{L}(X)}} < \infty. \qquad (5.24)$$

If

$$a_\alpha := \sum_{n=0}^{\infty} a_{\alpha,n} < \infty, \qquad (5.25)$$

then $\omega_0(\mathcal{A}) < \alpha \leq 0$.

Proof. By our assumptions and using Theorem 5.27, we have that $(\mathcal{A} - \beta)$ generates a hyperbolic semigroup for all $\beta \geq \alpha$. Thus, $\omega_0(\mathcal{A}) < \alpha \leq 0$. □

Now we formulate a useful consequence of this theorem, which will play a key role in our investigation of the "small delay" problem; see Section 7.1.

Corollary 5.30. *Assume that Φ is Fourier-admissible, $\omega_0(B) < 0$, and let $\alpha \in (\omega_0(B), 0]$. If*

$$\sup_{\beta \geq \alpha} \|\Phi_{\beta+i}. R(\beta + i\cdot, B)\|_{\mathcal{M}_p^{\mathcal{L}(X)}} < 1, \qquad (5.26)$$

then $\omega_0(\mathcal{A}) < \alpha \leq 0$.

In the following, we formulate consequences of this result. First we consider the single delay case.

Corollary 5.31. *Consider $\Phi := C\delta_{-\tau}$, where $C \in \mathcal{L}(X)$ and $\tau \in [0,1]$. Assume that $(B, D(B))$ generates a uniformly exponentially stable semigroup $(S(t))_{t \geq 0}$, let $\alpha \in (\omega_0(B), 0]$, and take $\omega \in [\omega_0(B), \alpha)$ and $M \geq 1$ such that $\|S(t)\| \leq Me^{\omega t}$. If*

$$\frac{Me^{-\alpha\tau}}{\alpha - \omega}\|C\| < 1,$$

then $\omega_0(\mathcal{A}) < \alpha \leq 0$.

Proof. It follows from the elementary properties of the Fourier transform and from the connection between the semigroup and its resolvent that for $t \in \mathbb{R}$, $\beta \geq \alpha$, and $x \in X$,

$$\mathcal{F}^{-1}(\Phi_{\beta+i\cdot}.R(\beta + i\cdot, B)x)(t) = Ce^{-\beta\tau}\mathcal{F}^{-1}(e^{-i\tau\cdot}R(i\cdot, B - \beta)x)(t) \quad (5.27)$$

$$= Ce^{-\beta\tau}\tilde{S}_\beta(t + \tau)x$$

$$= \begin{cases} Ce^{-\beta\tau}e^{-\beta(\tau+t)}S(\tau + t)x & \text{if } t + \tau \geq 0, \\ 0 & \text{if } t + \tau < 0. \end{cases}$$

Here we have used a special notation for the rescaled and extended semigroup and wrote

$$S_\beta(t) := e^{-\beta t}S(t)$$

and

$$\tilde{S}_\beta(t) := \begin{cases} S_\beta(t) & \text{if } t \geq 0, \\ 0 & \text{if } t < 0. \end{cases}$$

It follows that $\tilde{S}_\beta \in L^1(\mathbb{R}, \mathcal{L}_s(X))$. Taking $\omega \in [\omega_0(B), \alpha)$ and $M := M_\omega \geq 1$ such that $\|S(t)\| \leq Me^{\omega t}$, we obtain from Equation (5.27) using the Hausdorff-Young Inequality (see Theorem 5.24)

$$\|\Phi_{\beta+i\cdot}.R(\beta + i\cdot, B)\|_{\mathfrak{M}_p^{\mathcal{L}(X)}} \leq \|\mathcal{F}^{-1}(\Phi_{\beta+i\cdot}.R(\beta + i\cdot, B))\|_{L^1}$$

$$= \int_0^\infty \|Ce^{-\beta(s+\tau)}S(s)\| \, ds$$

$$\leq \|C\| \int_0^\infty e^{-\beta(s+\tau)}e^{\omega s} \, ds$$

$$= M\|C\|e^{-\beta\tau} \int_0^\infty e^{(\omega-\beta)s} \, ds$$

$$= \frac{Me^{-\beta\tau}}{\beta - \omega}\|C\| \leq \frac{Me^{-\alpha\tau}}{\alpha - \omega}\|C\|.$$

To finish the proof, we use Corollary 5.30.   □

The previous corollary can be considerably sharpened if we have additional commutation property.

Corollary 5.32. *Consider* $\Phi := C\delta_{-\tau}$, *where* $C \in \mathcal{L}(X)$ *commutes with* $(B, D(B))$ *and where* $\tau \in [0,1]$. *Assume that* $(B, D(B))$ *generates a uniformly exponentially stable semigroup, let* $\alpha \in (\omega_0(B), 0]$, *and take* $\omega \in [\omega_0(B), \alpha)$ *and* $M \geq 1$ *such that* $\|S(t)\| \leq Me^{\omega t}$. *If*

$$\frac{Me^{-\alpha\tau}}{\alpha - \omega}r(C) < 1,$$

then $\omega_0(\mathcal{A}) < \alpha \leq 0$.

Proof. Choosing $\frac{Me^{-\alpha\tau}}{\alpha-\omega}r(C) < q < 1$, there exists $N \in \mathbb{N}$ such that

$$\left(\frac{Me^{-\alpha\tau}}{\alpha - \omega}\right)^n \|C^n\| < q^n \quad \text{for all } n \geq N.$$

Then it follows that

$$
\begin{aligned}
a_{\alpha,n} &= \sup_{\beta \geq \alpha} \|(\Phi_{\beta+i\cdot}R(\beta + i\cdot, B))^n\|_{\mathcal{M}_p^{\mathcal{L}(X)}} \\
&= \sup_{\beta \geq \alpha} \|\Phi_{\beta+i\cdot}^n R(\beta + i\cdot, B))^n\|_{\mathcal{M}_p^{\mathcal{L}(X)}} \\
&\leq \|(e^{-\alpha\tau}C)^n\| \cdot \sup_{\beta \geq \alpha} \|e^{-i\cdot\tau}R(\beta + i\cdot, B))^n\|_{\mathcal{M}_p^{\mathcal{L}(X)}} \\
&\leq \|C^n\| e^{-\alpha\tau n} \cdot \sup_{\beta \geq \alpha} \|e^{-i\cdot\tau}R(\beta + i\cdot, B))\|_{\mathcal{M}_p^{\mathcal{L}(X)}}^n \\
&\leq \|C^n\| \left(\frac{Me^{-\alpha\tau}}{\alpha - \omega}\right)^n < q^n
\end{aligned}
$$

for $n \geq N$. The statement follows from Theorem 5.29. $\qquad\square$

Let us consider now the case of general delay operators.

Corollary 5.33. *Assume that* Φ *is of the form of Equation (3.38), assume that* $(B, D(B))$ *generates a uniformly exponentially stable semigroup* $(S(t))_{t\geq 0}$, *let* $\alpha \in (\omega_0(B), 0]$, *and take* $\omega \in [\omega_0(B), \alpha)$ *and* $M \geq 1$ *such that* $\|S(t)\| \leq Me^{\omega t}$. *If*

$$|\eta|([-1,0])\frac{M}{\alpha - \omega} < 1,$$

then $\omega_0(\mathcal{A}) < \alpha \leq 0$.

Proof. Analogously to the proof of Corollary 5.31, we have for $t \in \mathbb{R}$ and $\beta \geq \alpha$

$$\mathcal{F}^{-1}(\Phi_{\beta+i}.R(\beta+i\cdot,B)x)(t) = \mathcal{F}^{-1}\left(\int_{-1}^{0} e^{(\beta+i\cdot)s}d\eta(s)R(\beta+i\cdot,B)x\right)(t) \tag{5.28}$$

$$= \int_{\mathbb{R}}\int_{-1}^{0} e^{(\beta+iw)s}e^{iwt}\,d\eta(s)R(\beta+iw,B)x\,dw$$

$$= \int_{-1}^{0} e^{\beta s}\,d\eta(s)\int_{\mathbb{R}} e^{iwt}e^{iws}R(\beta+iw,B)x\,dw$$

$$= \int_{-1}^{0} e^{\beta s}\,d\eta(s)\mathcal{F}^{-1}\left(e^{is\cdot}R(\beta+i\cdot,B)x\right)(t)$$

$$= \int_{-1}^{0} e^{\beta s}\,d\eta(s)\tilde{S}_{\beta}(t+s)x = \Phi\left(S_{\beta,t}x\right).$$

Here $S_{\beta,t}$ are the operators defined by

$$(S_{\beta,t}\,x)(\tau) := \begin{cases} S_{\beta}(t+\tau)x, & -t < \tau \leq 0, \\ 0, & -1 \leq \tau \leq -t. \end{cases}$$

Finally, we obtain

$$\|\Phi_{\beta+i}.R(\beta+i\cdot,B)x\|_{\mathcal{M}_p^{\mathcal{L}(X)}} \leq \|\mathcal{F}^{-1}(\Phi_{\beta+i}.R(\beta+i\cdot,B)x)\|_{L^1}$$

$$= \int_0^{\infty} \|\Phi\left(S_{\beta,t}x\right)\|\,dt$$

$$= \int_0^{\infty} \left\|\int_{-1}^{0} d\eta(s)S_{\beta,t}(s)x\right\|\,dt$$

$$\leq \int_0^{\infty}\int_{-1}^{0} \|S_{\beta,t}(s)x\|\,d|\eta|(s)\,dt$$

$$= \int_{-1}^{0}\int_{-s}^{\infty} \|e^{-\beta(t+s)}S(t+s)x\|\,dt\,d|\eta|(s)$$

$$\leq |\eta|([-1,0])\frac{M}{\beta-\omega}\|x\|$$

$$\leq |\eta|([-1,0])\frac{M}{\alpha-\omega}\|x\|.$$

The assertion follows from Corollary 5.30. ∎

5.4   Notes and References

Section 5 consists of two main parts. In Section 5.1, the results are mainly taken from [19] and [12]. Stability criteria without using characteristic equations can be found, e.g., in Fischer and van Neerven [79, Corollary 3.6] and Curtain and Zwart [43, Theorem 5.1.7]. To our knowledge, the hyperbolicity and the existence of stable, center, and unstable manifolds in infinite dimensions has not been investigated until now in this context.

See also the review article Schnaubelt [188] for related well-posedness and hyperbolicity results in the nonautonomous case.

Analogous methods were extended to the stability investigation of general unbounded delay operators in Mastinsek [142]; see also Chapter 10.

The results of Section 5.3 were obtained in collaboration with E. Fašanga and R. Shvidkoy and are taken from [16].

Chapter 6

Stability via Positivity

As we have seen in Chapter 4, in order to locate the growth bound of a strongly continuous semigroup via spectral properties, additional regularity such as norm continuity or compactness is needed. In many applications, like population equations, heat equations, transport processes, or some reaction-diffusion equations, solutions with positive initial value should remain positive. This property is called positivity. In this chapter, we show that positivity opens a most elegant way to get access to spectral and asymptotic information.

In Section 6.1, we collect some basic results on positive semigroups and their asymptotic behavior. These results are applied in Section 6.2 to the delay equation to obtain stability and uniform exponential stability. Examples motivated by population equations are considered. Finally, in Section 6.3, we show how to construct the so-called modulus semigroup for a delay equation in the finite-dimensional case.

6.1 Positive Semigroups

We collect here the basic definitions and some results about positive semigroups on Banach lattices. An extensive treatment of this topic and more references can be found in Nagel et al. [148]. For our purpose, the survey in [72, Section VI.1.b] will be sufficient. Basic notions on Banach lattices will be used freely in the text; refer to the monographs Nagel et al. [148] and Schaefer [184]. Throughout this chapter, we assume that the Banach space X is even a Banach lattice (e.g., $X = L^q(\Omega)$ or $X = C_0(\Omega)$).

Definition 6.1. A strongly continuous semigroup $(T(t))_{t\geq 0}$ on a Banach lattice X is called *positive* if each operator $T(t)$ is positive, i.c., if

$$0 \leq f \in X \quad \text{implies} \quad 0 \leq T(t)f \quad \text{for all } t \geq 0.$$

Positive semigroups can be characterized through the resolvent of its generator as follows.

Theorem 6.2. *A strongly continuous semigroup* $(T(t))_{t\geq 0}$ *on a Banach lattice* X *is positive if and only if the resolvent* $R(\lambda, A)$ *of its generator* A *is positive for all sufficiently large* λ.

For positive semigroups, there is a very nice spectral and asymptotic theory. The key tool in proving these properties is the following lemma.

Lemma 6.3. *For a positive strongly continuous semigroup* $(T(t))_{t\geq 0}$ *with generator* $(A, D(A))$ *on a Banach lattice* X, *one has*

$$R(\lambda, A)f = \int_0^\infty e^{-\lambda s}T(s)f\, ds, \quad f \in X$$

for all $\Re\lambda > s(A)$. *Moreover, the following are equivalent for* $\lambda \in \rho(A)$:

(i) $0 \leq R(\lambda, A)$.

(ii) $s(A) < \lambda$.

One of the most important results in view of applications is the following.

Proposition 6.4. *Let* $(T(t))_{t\geq 0}$ *be a positive strongly continuous semigroup with generator* A *on a Banach lattice* X *with* $s(A) > -\infty$. *Then* $s(A) \in \sigma(A)$.

In particular, this means that there is always a real number in the spectrum dominating the whole spectrum.

This characterization and the representation of perturbed semigroups using the Dyson-Phillips Series leads us to the following result.

Proposition 6.5. *Let* $(A, D(A))$ *be the generator of the positive semigroup* $(T(t))_{t\geq 0}$ *in the Banach lattice* X *and let* $B \in \mathcal{L}(X)$ *be positive. Then* $C := A + B$ *generates a positive semigroup and* $s(A) \leq s(A + B)$.

It is also possible to characterize exponential stability of the semigroup via spectral properties without using the Spectral Mapping Theorem (see [72, Proposition VI.1.14]).

Proposition 6.6. *Let* $(T(t))_{t\geq 0}$ *be a positive strongly continuous semigroup with generator* G *on a Banach lattice* X. *Then the spectral bound* $s(A)$ *satisfies* $s(A) < 0$ *if and only if* $(T(t))_{t\geq 0}$ *is exponentially stable.*

In view of the above results, it is possible to investigate the stability of a semigroup by considering a positive semigroup that "dominates" it in the following sense (see Nagel et al. [148, Section C-II.4]).

Definition 6.7. Let $(T(t))_{t\geq 0}$ be a positive semigroup on a Banach lattice X and $(S(t))_{t\geq 0}$ be a semigroup. We say that

(i) $(T(t))_{t\geq 0}$ *dominates* $(S(t))_{t\geq 0}$ if

$$|S(t)f| \leq T(t)|f| \quad \text{for all} \quad f \in X, \, t > 0,$$

(ii) $(T(t))_{t\geq 0}$ is the *modulus semigroup* of $(S(t))_{t\geq 0}$ if $(T(t))_{t\geq 0}$ is dominated by any other positive semigroup dominating $(S(t))_{t\geq 0}$.

We recall a result due to Becker and Greiner [23] on the existence of the modulus semigroup.

Theorem 6.8. *Let $(S(t))_{t\geq 0}$ be a semigroup on a Banach lattice X with order continuous norm. Assume that there is a positive semigroup dominating $(S(t))_{t\geq 0}$. Then there exists the modulus semigroup of $(S(t))_{t\geq 0}$.*

The following theorem is the generalization of Theorem 2.26 for positive semigroups on L^p-spaces. Derndinger [56] proved it for $p = 1$, Greiner and Nagel [93] for $p = 2$, and Weis [215, 216] for $1 \leq p < \infty$.

Theorem 6.9. (Weis's Theorem.) *Let $(T(t))_{t\geq 0}$ be a positive semigroup with generator A on $L^p(\Omega, \mu)$, $1 \leq p < \infty$. Then*

$$s(A) = \omega_0(A).$$

Especially, the spectral bound $s(A)$ satisfies $s(A) < 0$ if and only if $(T(t))_{t\geq 0}$ is uniformly exponentially stable.

It is important to mention, however, that for positive semigroups on L^p-spaces, the Spectral Mapping Theorem does not hold and there is no analogous characterization of hyperbolicity. In fact, Montgomery-Smith [147] gave an example of a positive semigroup $(T(t))_{t\geq 0}$ with generator A on an L^2-space such that $\sigma(A) \cap i\mathbb{R} = \emptyset$, but $(T(t))_{t\geq 0}$ is not hyperbolic.

6.2 Stability via Positivity

Delay equations occur quite often in models from biology (see, e.g., Hadeler [99]), where positivity is a very natural assumption. So the corresponding semigroups should be positive.

In this section, we present the theory of positive delay semigroups on the history space $\mathcal{E}_p = X \times L^p([-1,0], X)$. This is particularly useful since in the case that X is a Lebesgue space $L^p(\Omega)$, the history space $\mathcal{E}_p$ becomes a Bochner-Lebesgue space and hence, by Weis's Theorem 6.9, the equality

$$s(\mathcal{A}) = \omega_0(\mathcal{A})$$

holds.

Assume that the Banach space X is even a Banach lattice (e.g., $X = L^q(\Omega)$ or $X = C_0(\Omega)$) and consider the usual delay equation

$$\begin{cases} u'(t) = Bu(t) + \Phi u_t, & t \geq 0, \\ u(0) = x, & \\ u_0 = f \in L^p([-1,0], X) \end{cases} \qquad (DE_p)$$

for $1 \leq p < \infty$.

It is a natural and important question to decide whether the solution $u(\cdot)$ of (DE_p) remains positive if the initial values $x \in X$ and $f \in L^p([-1,0], X)$ are positive. If (DE_p) is well-posed, this is equivalent to the fact that the solution semigroup (see Definition 3.13) is a positive semigroup on $\mathcal{E}_p$. Therefore, first we give a sufficient condition such that this delay semigroup generated by the operator $(\mathcal{A}, D(\mathcal{A}))$ is positive. It turns out that the positivity of the semigroup $(S(t))_{t \geq 0}$ generated by B and the positivity of the delay operator Φ are sufficient.

Theorem 6.10. *Assume that X is a Banach lattice, $(B, D(B))$ generates a semigroup $(S(t))_{t \geq 0}$, and $\Phi : W^{1,p}([-1,0], X) \to X$ is such that the operator $(\mathcal{A}, D(\mathcal{A}))$, defined in Equation (3.8), generates a strongly continuous semigroup $(\mathcal{T}(t))_{t \geq 0}$ for all $1 \leq p < \infty$. If $(S(t))_{t \geq 0}$ and Φ are positive, then $(\mathcal{T}(t))_{t \geq 0}$ is positive.*

Proof. We use the characterization in Theorem 6.2, i.e., we have to verify that $R(\lambda, \mathcal{A}))$ is positive if λ is sufficiently large. Using the representation in Lemma 3.19, we only have to verify that its entries are positive operators. Since $(A_0, D(A_0))$ generates the nilpotent shift, which is certainly positive, it follows that $R(\lambda, A_0)$ and $\Phi R(\lambda, A_0)$ are positive for $\lambda > 0$. Moreover, Φ_λ is a positive operator for $\lambda > 0$, and hence $R(\lambda, B + \Phi_\lambda)$ is also positive for sufficiently large λ because it is given by the Neumann series

$$R(\lambda, B + \Phi_\lambda) = R(\lambda, B) \sum_{n=0}^{\infty} (\Phi_\lambda R(\lambda, B))^n.$$

$\square$

Corollary 6.11. *Under the above assumptions, (DE_p) has positive solutions for all positive initial values for all $1 \leq p < \infty$.*

Remark 6.12. If we make the assumption that the conditions of Theorem 3.26 in Chapter 3 are satisfied, then the proof of Theorem 6.10 is a direct consequence of Rhandi [174, Theorem 1.1] (see also Arendt and Rhandi [7]). This will be our main assumption later in this section. This argument uses the Miyadera-Voigt Perturbation Theorem and the Dyson-Phillips Series.

Based on Theorem 6.10 we are now able to investigate stability by means of positivity. To do so, we have to carry out a series of technical calculations. First we analyze the operator-valued map $R : \rho \subset \mathbb{C}^2 \to \mathcal{L}(X)$ defined by

$$(\lambda, \mu) \mapsto R(\lambda, \mu) := R(\lambda, B + \Phi_\mu) \quad \text{for}$$
$$(\lambda, \mu) \in \rho := \left\{ (r, s) \in \mathbb{C}^2 : r \in \rho(B + \Phi_s) \right\}.$$

Lemma 6.13. *The set $\rho \subset \mathbb{C}^2$ is open and the mapping $R(\cdot, \cdot)$ is analytic.*

Proof. Let $(\lambda_0, \mu_0) \in \rho$ and $(\lambda, \mu) \in \mathbb{C}^2$. Then

$$(\lambda - B - \Phi_\mu) - (\lambda_0 - B - \Phi_{\mu_0}) = (\lambda - \lambda_0) + \Phi((\varepsilon_{\mu_0} - \varepsilon_\mu) \otimes I) =: \Delta_{\lambda, \mu}.$$

Since $\lim_{(\lambda, \mu) \to (\lambda_0, \mu_0)} \|\Delta_{\lambda, \mu}\| = 0$ and

$$(\lambda - B - \Phi_\mu) = (I + \Delta_{\lambda, \mu} R(\lambda_0, \mu_0))(\lambda_0 - B - \Phi_{\mu_0}), \qquad (6.1)$$

it follows that $(\lambda - B - \Phi_\mu)$ is invertible for $\|(\lambda, \mu) - (\lambda_0, \mu_0)\|$ sufficiently small meaning that the set ρ is open. It is also a consequence of Equation (6.1) that for $\|(\lambda, \mu) - (\lambda_0, \mu_0)\|$ sufficiently small, one has

$$R(\lambda, \mu) = R(\lambda_0, \mu_0) \sum_{n=0}^{\infty} [\Delta_{\lambda, \mu} R(\lambda_0, \mu_0)]^n .$$

Since this series converges uniformly on small balls and the map $\mathbb{C}^2 \ni (\lambda, \mu) \mapsto \Delta_{\lambda, \mu} \in \mathcal{L}(X)$ is analytic, one obtains the analyticity of $R(\cdot, \cdot)$ as well. $\qquad \square$

With this lemma we will now describe the behavior of the spectral bound function

$$s : \mathbb{R} \to \mathbb{R} \cup \{-\infty\} \text{ defined by } s(\lambda) := s(B + \Phi_\lambda).$$

Proposition 6.14. *Let $(B, D(B))$ generate a positive semigroup in the Banach lattice X and assume that Φ is positive. Then the spectral bound function $s(\cdot)$ is decreasing and continuous from the left on $\mathbb{R}$. If, in addition, $s(\mu_0)$ is isolated in $\sigma(B + \Phi_{\mu_0}) \cap \mathbb{R}$, then $s(\cdot)$ is even continuous in $\mu_0 \in \mathbb{R}$.*

Proof. For $\mu_0 \leq \mu_1$, we have $\Phi_{\mu_1} \leq \Phi_{\mu_0}$ and therefore $s(B + \Phi_{\mu_1}) \leq s(B + \Phi_{\mu_0})$ by Proposition 6.5. This shows that the function $s(\cdot)$ is decreasing. We also have $s(\mu) \in \sigma(B + \Phi_\mu)$ by Proposition 6.4.

To show that the function $s(\cdot)$ is left-continuous, let us assume by contradiction that

$$s(\mu_0) < s^- := \lim_{\varepsilon \searrow 0} s(\mu_0 - \varepsilon)$$

for some $\mu_0 \in \mathbb{R}$. Then $s^- \in \rho(B + \Phi_{\mu_0})$, and therefore $(s^-, \mu_0) \in \rho$. This contradicts the fact that $\rho \in \mathbb{C}^2$ is open since by Proposition 6.4 we have $s(\mu_0 - \varepsilon) \in \sigma(B + \Phi_{\mu_0 - \varepsilon})$, meaning that $(s(\mu_0 - \varepsilon), \mu_0 - \varepsilon) \notin \rho$ for all $\varepsilon > 0$ while $(s(\mu_0 - \varepsilon), \mu_0 - \varepsilon) \searrow (s^-, \mu_0)$ while $\varepsilon \to 0$.

Now let us assume that $s(\mu_0)$ is isolated in $\sigma(B + \Phi_{\mu_0}) \cap \mathbb{R}$. In order to show that $s(\cdot)$ is right continuous, we use again a contradiction argument and assume that

$$s^+ := \lim_{\varepsilon \searrow 0} s(\mu_0 + \varepsilon) < s(\mu_0).$$

Then by assumption there exists $\lambda \in \rho(B + \Phi_{\mu_0}) \cap \mathbb{R}$ satisfying

$$s^+ < \lambda < s(\mu_0).$$

In particular, $(\lambda, \mu_0) \in \rho$ and we conclude from Lemma 6.13 that

$$R(\lambda, \mu_0) = \lim_{\varepsilon \searrow 0} R(\lambda, \mu_0 + \varepsilon) \geq 0.$$

This contradicts Lemma 6.3, meaning that $s(\cdot)$ is continuous. $\qquad\square$

It is worth mentioning that the spectral bound function is always continuous at the point $\lambda_0 \in \mathbb{R}$ if B has compact resolvent or Φ_{λ_0} is compact.

Theorem 6.15. *Let B generate a positive semigroup on X and assume that Φ satisfies Equation (3.38) and is positive. Then the following assertions hold.*

(i) *If $s(B + \Phi_\lambda) < \lambda$, then $s(\mathcal{A}) < \lambda$.*

(ii) *If $s(B + \Phi_\lambda) = \lambda$, then $s(\mathcal{A}) = \lambda$.*

(iii) In addition, assume that $\sigma(B) \neq \emptyset$. If B has compact resolvent or if Φ_λ is compact for all $\lambda \in \mathbb{R}$, then the spectral bound $s(\mathcal{A})$ is the unique solution of the characteristic equation

$$\lambda = s(B + \Phi_\lambda), \qquad \lambda \in \mathbb{R}. \tag{6.2}$$

Moreover, in this case

$$s(B + \Phi_\lambda) \underset{>}{\overset{<}{=}} \lambda \quad \Longleftrightarrow \quad s(\mathcal{A}) \underset{>}{\overset{<}{=}} \lambda.$$

Proof. (i) Let $\lambda > s(\lambda)$. Then one obtains from the monotonicity of $s(\cdot)$ that

$$\mu \geq \lambda > s(\lambda) \geq s(\mu)$$

for all $\mu \geq \lambda$. This implies $\mu \in \rho(B + \Phi_\mu)$ and therefore $\mu \in \rho(\mathcal{A})$ for all $\mu \geq \lambda$. On the other hand, by Proposition 6.4 one has $s(\mathcal{A}) \in \sigma(\mathcal{A})$, hence $\lambda > s(\mathcal{A})$ as claimed.

(ii) If $\lambda = s(\lambda)$, then again by Proposition 6.4 one has $\lambda \in \sigma(B + \Phi_\lambda) = \sigma(\mathcal{A})$. On the other hand, one can show as in (i) that $\mu \in \rho(\mathcal{A})$ for all $\mu > \lambda$, which implies $\lambda = s(\mathcal{A})$.

(iii) If $\sigma(B) \neq \emptyset$, then by Proposition 6.5 the inequality $-\infty < s(B) \leq s(\lambda)$ holds for all $\lambda \in \mathbb{R}$. Moreover, by Proposition 6.14 the function $s(\cdot)$ is continuous and decreasing. Therefore, Equation (6.2) has a unique solution λ_0, which by (ii) coincides with $s(\mathcal{A})$. The last estimates are then immediate. $\qquad\square$

The following result characterizes the spectral bound.

Lemma 6.16. *Let B generate a positive semigroup on X and assume that Φ is positive. If $\sigma(B + \Phi_\lambda) \neq \emptyset$ for some $\lambda \in \mathbb{R}$, then*

$$s(\mathcal{A}) = \sup\{\lambda \in \mathbb{R} \ : \ s(B + \Phi_\lambda) \geq \lambda\}. \tag{6.3}$$

In the other case, one has $s(\mathcal{A}) = -\infty$.

Proof. If $\sigma(B + \Phi_\lambda) = \emptyset$ for all $\lambda \in \mathbb{R}$, then $s(\mathcal{A}) = -\infty$ since $\sigma(\mathcal{A}) = \emptyset$ and the second statement follows.

Assuming $\sigma(B + \Phi_\lambda) \neq \emptyset$, let us use the notation $\mu := \sup\{\lambda \in \mathbb{R} \ : \ s(B + \Phi_\lambda) \geq \lambda\}$. Then it follows from the left-continuity of $s(\cdot)$ that $s(\mu) \geq \mu$.

If $s(\mu) = \mu$, then $s(\mathcal{A}) = \mu$ by Theorem 6.15 (ii) and the statement follows.

If $s(\mu) > \mu$, then we make the proof in two steps. First, we show that in this case the inclusion

$$(\mu, s(B + \Phi_\mu)] \subset \sigma(B + \Phi_\mu) \tag{6.4}$$

holds. To do so, assume by contradiction that there exists $r \in (\mu, s(B + \Phi_\mu)] \cap \rho(B + \Phi_\mu)$. Then $(r, \mu) \in \rho$, and by the definition of μ, one obtains $r + \varepsilon > \mu + \varepsilon > s(B + \Phi_{\mu+\varepsilon})$ for all $\varepsilon > 0$. Hence,

$$R(r, \mu) = R(r, B + \Phi_\mu) = \lim_{\varepsilon \searrow 0} R(r + \varepsilon, B + \Phi_{\mu+\varepsilon}) \geq 0,$$

which by Lemma 6.3 contradicts the fact that $r \leq s(B + \Phi_\mu)$. Hence Equation (6.4) is proved and from the closedness of the spectrum we deduce $\mu \in \sigma(B + \Phi_\mu)$. Consequently, by Proposition 6.4 $\mu \in \sigma(\mathcal{A})$ and therefore $s(\mathcal{A}) \geq \mu$.

As a second step, we show that $s(\mathcal{A}) > \mu$ is not possible. To this end, we assume by contradiction that $s(\mathcal{A}) > \mu$. Then from the definition of μ it immediately follows that

$$s(B + \Phi_{s(\mathcal{A})}) < s(\mathcal{A}),$$

and hence by Theorem 6.15 (i) $s(\mathcal{A}) \in \rho(B + \Phi_{s(\mathcal{A})})$. By the spectral characterization this implies $s(\mathcal{A}) \in \rho(\mathcal{A})$, which contradicts Proposition 6.4. □

Though the preceding results are really interesting, it is in applications in general quite difficult to solve the characteristic equation. However, it can be decided much more easily whether the spectral bound $s(\mathcal{A})$ of $\mathcal{A}$ is negative.

Proposition 6.17. *Under the assumptions of Theorem 6.10, the following equivalence holds:*

$$s(\mathcal{A}) < 0 \quad \Longleftrightarrow \quad s(B + \Phi_0) < 0.$$

Proof. We can assume that there exists $\lambda \in \mathbb{R}$ such that $\sigma(B + \Phi_\lambda) \neq \emptyset$ since otherwise $s(\mathcal{A}) = s(B + \Phi_\lambda) = -\infty$ by Lemma 6.16.

Suppose first that $s(\mathcal{A}) < 0$. Then $s(B + \Phi_0) < 0$ since otherwise $s(\mathcal{A}) \geq 0$ by Lemma 6.16. This proves the implication "$\Longrightarrow$." The implication "$\Longleftarrow$" follows immediately from Theorem 6.15 (i), hence the proof is complete. □

Theorem 6.18. *Under the assumptions of Theorem 6.10, the following assertions hold.*

(i) *The delay semigroup* $(\mathfrak{T}(t))_{t\geq 0}$ *is exponentially stable if and only if the spectral bound* $s(B + \Phi_0)$ *is less than 0.*

(ii) *If the semigroup* $(S(t))_{t\geq 0}$ *generated by* $(B, D(B))$ *is immediately norm continuous and Condition (K) of Section 4.1 holds, then the semigroup* $(\mathfrak{T}(t))_{t\geq 0}$ *is uniformly exponentially stable if and only if the spectral bound* $s(B + \Phi_0)$ *is less than 0.*

Proof. (i) The assertion follows from Proposition 6.6 and Proposition 6.17.

(ii) Since $(S(t))_{t\geq 0}$ is immediately norm continuous, then, by Proposition 4.3, the delay semigroup $(\mathfrak{T}(t))_{t\geq 0}$ generated by $(\mathcal{A}, D(\mathcal{A}))$ is eventually norm continuous. Hence, the growth bound $\omega_0(\mathcal{A})$ is equal to the spectral bound $s(\mathcal{A})$ and the assertion follows by Proposition 6.17. □

Finally, we formulate a stability result for positive semigroups on Lebesgue spaces.

Theorem 6.19. *Let* $X := L^p(\Omega)$, $1 \leq p < \infty$, *and let* B *generate a positive semigroup on* X. *Let* $\Phi : W^{1,p}([-1, 0], X) \to X$ *be positive and such that* $(\mathcal{A}, D(\mathcal{A}))$ *generates a semigroup on* $\mathcal{E}_p$. *Then*

$$\omega_0(\mathcal{A}) = s(\mathcal{A}).$$

Proof. By Theorem 6.10 the semigroup generated by $(\mathcal{A}, D(\mathcal{A}))$ is positive. Moreover, by Theorem A.6, the space $\mathcal{E}_p$ is a Lebesgue space of type $L^p(\Omega)$ for an appropriate measure space Ω. Therefore, by Theorem 6.9,

$$\omega_0(\mathcal{A}) = s(\mathcal{A}).$$

□

Example 6.20. We recall the population equation from Example 3.16:

$$\begin{cases} \partial_t u(t, a) + \partial_a u(t, a) = -\mu(a)\, u(t, a) + \nu(a)\, u(t - 1, a), \\ \hspace{4cm} t \geq 0,\ a \in \mathbb{R}_+, \\ u(t, 0) = \int_0^{+\infty} \beta(a)\, u(t, a)\, da, \hspace{1.3cm} t \geq 0, \\ u(s, a) = f(s, a), \hspace{2.7cm} (s, a) \in [-1, 0) \times \mathbb{R}_+, \end{cases} \quad (6.5)$$

where $\mu, \nu, \beta \in L^\infty(\mathbb{R})$ are positive, $\mu > \nu$, and

$$\mu_\infty := \lim_{a \to \infty} \mu(a) \quad (6.6)$$

exists (see Greiner [91, Section 2]). To write this equation in the form $(DE)_1$, we choose the following operators on the space $X := L^1(\mathbb{R}_+)$:

- The operator $(Bx)(a) := -(x')(a) - \mu(a)\, x(a)$, $a \in \mathbb{R}_+$, with domain $D(B) := \left\{ x \in W^{1,1}(\mathbb{R}_+) : x(0) = \int_0^{+\infty} \beta(a)\, x(a)\, da \right\}$ generates a positive semigroup on X.

- The delay operator $\Phi : W^{1,1}([-1,0], X) \to X$ defined as

$$\Phi f := \nu\, f(-1)$$

  is a positive operator.

In Example 3.30, we have shown that Equation (6.5) is well-posed, i.e., the operator matrix $(\mathcal{A}, D(\mathcal{A}))$ associated to Equation (6.5) generates a semigroup on the Banach lattice $L^1(\mathbb{R}_+) \times L^1([-1,0], L^1(\mathbb{R}_+))$ by Theorem 3.29.

Moreover, by Theorem 6.10, $(\mathcal{A}, D(\mathcal{A}))$ generates a positive semigroup $(\mathcal{T}(t))_{t \geq 0}$. If we look closer at the state space, we see that

- the space $L^1([-1,0], L^1(\mathbb{R}_+))$ is canonically isomorphic to $L^1([-1,0] \times \mathbb{R}_+)$ by Theorem A.6 and

- the space $\mathcal{E}_1 := L^1(\mathbb{R}_+) \times L^1([-1,0] \times \mathbb{R}_+)$ with norm $\left\| \binom{x}{f} \right\|_1 := \|x\|_{L^1(\mathbb{R}_+)} + \|f\|_{L^1([-1,0] \times \mathbb{R}_+)}$ is again an L^1-space.

So we have a positive semigroup on a Banach lattice of type L^1 and by Theorem 6.9 it follows that

$$\omega_0(\mathcal{A}) = s(\mathcal{A}).$$

Moreover, by Theorem 6.18 we have that

$$\omega_0(\mathcal{A}) < 0 \quad \text{if and only if} \quad s(B + \Phi_0) < 0.$$

In our example, the operator Φ_0 is the multiplication operator by the function ν, hence

$$(B + \Phi_0)x = -x' - (\mu - \nu)\, x \quad \text{for all } x \in W^{1,1}(\mathbb{R}_+).$$

Assume now that

$$\int_0^{+\infty} \beta(a) \left(e^{\int_0^a (\mu_\infty - \mu(s) + \nu(s))\, ds} \right) da > 1. \tag{6.7}$$

Then one can show that the operator $B + \Phi_0$ satisfies the assumptions of Greiner [91, Corollary 2.4] and we have

$$s(B + \Phi_0) \gtreqless 0 \quad \text{if and only if} \quad \int_0^{+\infty} \beta(a) \left(e^{-\int_0^a (\mu(s) - \nu(s))\, ds} \right) da \gtreqless 1.$$

Hence, we obtain the following stability theorem.

Theorem 6.21. *Assume that Equations (6.6) and (6.7) hold. Then the solutions of the age-structured population equation with delay (Equation (6.5)) are uniformly exponentially stable if and only if*

$$\int_0^{+\infty} \beta(a) \left(e^{-\int_0^a (\mu(s) - \nu(s))\, ds} \right) da < 1.$$

The stability criteria valid for positive semigroups can also be used for nonpositive semigroups. In order to show stability for a delay semigroup $(\mathcal{T}(t))_{t\geq 0}$, it is actually sufficient to show it for a dominating semigroup (see also Kerscher and Nagel [121], Nagel et al. [148, C-II.4], and Definition 6.7). A sufficient condition for domination is the following.

Lemma 6.22. *Assume that X is a Banach lattice, $(B, D(B))$ generates a strongly continuous semigroup, and Φ is bounded from $W^{1,p}([-1,0], X)$ to X, satisfying the conditions of Theorem 3.26. Assume further that there exists an operator $(\hat{B}, D(\hat{B}))$ generating a positive semigroup $(\hat{S}(t))_{t\geq 0}$ dominating $(S(t))_{t\geq 0}$ and that there exists $\hat{\Phi}$ dominating Φ and also satisfying the conditions of Theorem 3.26. Then the semigroup $(\hat{\mathcal{T}}(t))_{t\geq 0}$ generated by $\hat{A}$ associated to the operators $\hat{B}$ and $\hat{\Phi}$ dominates the semigroup $(\mathcal{T}(t))_{t\geq 0}$ generated by A.*

Proof. From the Miyadera-Voigt perturbation argument we have that the delay semigroup can be written in the form

$$\mathcal{T}(t) = \sum_{k=0}^{\infty} (V^k \mathcal{T})(t), \tag{6.8}$$

where $(V^0 \mathcal{T}) := \mathcal{T}_0(t)$ was defined in Equation (3.33), and

$$(V^{k+1}\mathcal{T})(t)\left(\begin{smallmatrix} x \\ f \end{smallmatrix}\right) := \int_0^t (V^k \mathcal{T})(s)\mathcal{B}\mathcal{T}_0(t-s)\left(\begin{smallmatrix} x \\ f \end{smallmatrix}\right) ds \tag{6.9}$$

for all $\left(\begin{smallmatrix} x \\ f \end{smallmatrix}\right) \in D(A)$. The operator $\hat{A}$ defined by

$$\hat{A} := \begin{pmatrix} \hat{B} & \hat{\Phi} \\ 0 & \frac{d}{d\sigma} \end{pmatrix}, \quad D(\hat{A}) := D(A) \tag{6.10}$$

generates a positive C_0-semigroup $\hat{\mathcal{T}}(t)$ on the Banach lattice $\mathcal{E}_p$. Similarly, the semigroup generated by $\hat{A}$ can also be written in the form

$$\hat{\mathcal{T}}(t) = \sum_{k=0}^{\infty} (V^k \hat{\mathcal{T}})(t), \qquad (6.11)$$

where $\mathcal{A}_0$ is replaced by $\hat{\mathcal{A}}_0 := \begin{pmatrix} \hat{B} & 0 \\ 0 & \frac{d}{d\sigma} \end{pmatrix}$ and $\mathcal{B}$ by $\hat{\mathcal{B}} := \begin{pmatrix} 0 & \hat{\Phi} \\ 0 & 0 \end{pmatrix}$.

It follows from Theorem 6.10 that $\hat{\mathcal{T}}(t)$ is positive. We also have by induction that all the terms $(V^k \hat{\mathcal{T}})(t)$ are positive. To finish the proof, we have to show that all the terms in the Dyson-Phillips Series satisfy

$$\left| (V^k \mathcal{T})(t) \left(\tfrac{x}{f} \right) \right| \leq (V^k \hat{\mathcal{T}})(t) \left| \left(\tfrac{x}{f} \right) \right| \qquad (6.12)$$

for all $\left(\tfrac{x}{f} \right) \in D(\mathcal{A})$ and for every $k \in \mathbb{N}$.

Let us verify Equation (6.12) for $k = 0$. It follows from Proposition 3.25 and our assumptions that

$$|\mathcal{T}_0(t)\left(\tfrac{x}{f} \right)| = \begin{pmatrix} |S(t)x| \\ |S_t x + T_0(t)f| \end{pmatrix} \leq \begin{pmatrix} |S(t)x| \\ |S_t x| + |T_0(t)f| \end{pmatrix}$$

$$\leq \begin{pmatrix} \hat{S}(t)|x| \\ \hat{S}_t |x| + T_0(t)|f| \end{pmatrix} = \hat{\mathcal{T}}_0(t) \left| \left(\tfrac{x}{f} \right) \right|,$$

where we used the positivity of the shift semigroup $(T_0(t))_{t \geq 0}$. The general case can be proved by induction using that $|\Phi f| \leq \hat{\Phi}|f|$. $\qquad \square$

Example 6.23. We consider again a population equation but with a different delay operator than in Example 6.20:

$$\begin{cases} \partial_t u(t,a) + \partial_a u(t,a) = -\mu(a)\, u(t,a) + \Phi u_t, \\ \qquad\qquad\qquad\qquad\quad t \geq 0,\ a \in \mathbb{R}_+, \\ u(t,0) = \int_0^{+\infty} \beta(a)\, u(t,a)\, da, \qquad t \geq 0, \\ u(s,a) = f(s,a), \qquad\qquad\qquad (s,a) \in [-1,0) \times \mathbb{R}_+, \end{cases} \qquad (6.13)$$

where $\mu, \beta \in L^\infty(\mathbb{R}_+)$ are positive functions and

$$\Phi : W^{1,1}([-1,0], L^1(\mathbb{R}_+)) \to L^1(\mathbb{R}_+))$$

is given by

$$\Phi f := \int_{-1}^0 g(\sigma)\, f(\sigma)\, d\sigma,$$

where $g \in L^1([-1,0], L^1(\mathbb{R}_+))$.

Let $\hat{\Phi} : W^{1,1}([-1,0], L^1(\mathbb{R}_+)) \to L^1(\mathbb{R}_+))$ be given by

$$\Phi f := \int_{-1}^0 |g|(\sigma)\, f(\sigma)\, d\sigma.$$

Then we have

$$
\begin{aligned}
|\Phi f| &= \left| \int_{-1}^{0} g(\sigma) \, f(\sigma) \, d\sigma \right| \\
&\leq \int_{-1}^{0} |g|(\sigma) \, |f|(\sigma) \, d\sigma \\
&= \hat{\Phi} \, |f| \, .
\end{aligned}
$$

Hence, $\hat{\Phi}$ dominates Φ. Moreover, the semigroup generated by $(B, D(B))$ is positive and by Lemma 6.22 the solutions of the delay equation

$$
\begin{cases}
\partial_t u(t, a) + \partial_a u(t, a) = -\mu(a) \, u(t, a) + \hat{\Phi} u_t, & \\
\qquad\qquad\qquad\qquad\qquad\qquad t \geq 0, \ a \in \mathbb{R}_+, & \\
u(t, 0) = \int_0^{+\infty} \beta(a) \, u(t, a) \, da, & t \geq 0, \\
u(s, a) = f(s, a), & (s, a) \in [-1, 0) \times \mathbb{R}_+,
\end{cases} \tag{6.14}
$$

dominate those of Equation (6.13). Therefore, analogously to Example 6.20, the solutions of Equation (6.13) are uniformly exponentially stable if

$$
s(B + \hat{\Phi}_0) < 0.
$$

6.3 The Modulus Semigroup

Motivated by the stability criteria using dominating semigroups as shown in the previous example, now we characterize the modulus semigroup (see Definition 6.7) for the delay semigroup, i.e., the smallest semigroup dominating the delay semigroup. This will be done for the case when $n := \dim X < \infty$ and therefore $B = (b_{ij})$ is a real $n \times n$-matrix. The presented result is due to S. Boulite, L. Maniar, A. Rhandi, and J. Voigt [25].

We restrict our considerations to the case where $p > 1$ and Φ satisfies the conditions in Theorem 3.29, i.e., is of the form of Equation (3.38). In fact, Φ can be considered as defined by a matrix of scalar measures on $[-1, 0]$, and the measures have no mass at zero.

To obtain the existence of the modulus semigroup of $\mathcal{T}$, it suffices to show that the semigroup $\mathcal{T}$ is dominated. The existence of a minimal dominating semigroup then follows since $\mathcal{E}_p$ has order continuous norm (see Theorem 6.8).

For this purpose, let us denote by $B_\#$ the matrix in $\mathbb{R}^{n \times n}$ defined by

$$
(B_\#)_{ij} := \begin{cases} b_{ii} & \text{for } i = j, \\ |b_{ij}| & \text{for } i \neq j. \end{cases}
$$

It is known that $S_\#(t) := e^{tB_\#}$ gives the modulus semigroup of $(S(t))_{t \geq 0}$ (see Nagel et al. [148, Example C.II.4.19.]).

The operator Φ is bounded from $C([-1,0], \mathbb{R}^n)$ to $\mathbb{R}^n$, therefore Φ has a modulus defined by

$$|\Phi|f := \sup \{|\Phi g| : g \in C([-1,0], \mathbb{R}^n), |g| \leq f\}$$

for $0 \leq f \in C([-1,0], \mathbb{R}^n)$ which is a positive bounded linear operator (see Schaefer [184, Theorem IV.1.5]). This means that $|\Phi|$ is of the same form as Φ, and the operator $\tilde{\mathcal{A}}$ defined by

$$\tilde{\mathcal{A}} := \begin{pmatrix} B_\# & |\Phi| \\ 0 & \frac{d}{d\sigma} \end{pmatrix}, \quad D(\tilde{\mathcal{A}}) := D(\mathcal{A}), \tag{6.15}$$

generates a positive C_0-semigroup $(\tilde{\mathcal{T}}(t))_{t \geq 0}$ on the Banach lattice $\mathcal{E}_p$. Again, this semigroup can also be written in the form

$$\tilde{\mathcal{T}}(t) = \sum_{k=0}^{\infty} (V^k \tilde{\mathcal{T}})(t), \tag{6.16}$$

where the terms $(V^k \tilde{\mathcal{T}})(t)$ are determined by replacing $\mathcal{A}_0$ by $\tilde{\mathcal{A}}_0 := \begin{pmatrix} B_\# & 0 \\ 0 & \frac{d}{d\sigma} \end{pmatrix}$ and $\mathcal{B}$ by $\tilde{\mathcal{B}} := \begin{pmatrix} 0 & |\Phi| \\ 0 & 0 \end{pmatrix}$.

We obtain an important consequence by Lemma 6.22.

Corollary 6.24. *The semigroup $(\tilde{\mathcal{T}}(t))_{t \geq 0}$ is a positive semigroup dominating $(\mathcal{T}(t))_{t \geq 0}$.*

As noted before, this result guarantees the existence of the modulus semigroup $(\mathcal{T}_\#(t))_{t \geq 0}$ of $(\mathcal{T}(t))_{t \geq 0}$. Its generator will be denoted by $(\mathcal{A}_\#, D(\mathcal{A}_\#))$. The following lemma will be used to obtain the generator $\mathcal{A}_\#$ explicitly.

Lemma 6.25. *We have the following properties.*

(i) There exists $C > 0$ such that

$$\|\mathcal{T}(t) \left(\begin{smallmatrix} x \\ 0 \end{smallmatrix}\right) - \mathcal{T}_0(t)) \left(\begin{smallmatrix} x \\ 0 \end{smallmatrix}\right)\| \leq Ct|\eta|([-t,0])\|x\| \tag{6.17}$$

for all $x \in \mathbb{R}^n$ and $0 < t < 1$. Here $|\eta|$ is the positive Borel measure on $[-1,0]$ defined by the total variation of η as defined in Section 3.3.3.

(ii) $\frac{1}{t}\left(\pi_1 \mathcal{T}(t) \left(\begin{smallmatrix} x \\ 0 \end{smallmatrix}\right) - x\right) \to Bx$, $\frac{1}{t}\left(\pi_1 \tilde{\mathcal{T}}(t) \left(\begin{smallmatrix} x \\ 0 \end{smallmatrix}\right) - x\right) \to B_\# x$ as $t \searrow 0$ for all $x \in \mathbb{R}^n$. Here π_1 is the canonical projection from $\mathcal{E}_p = \mathbb{R}^n \times L^p([-1,0], \mathbb{R}^n)$ onto $\mathbb{R}^n$ as defined in Definition 3.8.

(iii) $\frac{1}{t}\pi_1\mathcal{T}(t)\left(\begin{smallmatrix}0\\f\end{smallmatrix}\right) \to \Phi f,\ \frac{1}{t}\pi_1\tilde{\mathcal{T}}(t)\left(\begin{smallmatrix}0\\f\end{smallmatrix}\right) \to |\Phi|f$ *as* $t \searrow 0$ *for all* $f \in W^{1,p}([-1,0],\mathbb{R}^n)$.

Proof. (i) In order to show Equation (6.17) we take $x \in \mathbb{R}^n$ and $t > 0$. For $0 < \alpha < 1 - t$, we define the function $f_\alpha \in W^{1,p}([-1,0],\mathbb{R}^n)$ by

$$f_\alpha(\theta) := \begin{cases} \frac{\theta+\alpha}{\alpha}x & \text{for } -\alpha \leq \theta < 0, \\ 0 & \text{otherwise.} \end{cases}$$

Then $\left(\begin{smallmatrix}x\\f_\alpha\end{smallmatrix}\right) \in D(\mathcal{A})$, and from the Variation of Parameters Formula, we obtain

$$\mathcal{T}(t)\left(\begin{smallmatrix}x\\f_\alpha\end{smallmatrix}\right) - \mathcal{T}_0(t))\left(\begin{smallmatrix}x\\f_\alpha\end{smallmatrix}\right) = \int_0^t \mathcal{T}(t-s)\mathcal{B}\mathcal{T}_0(t)\left(\begin{smallmatrix}x\\f_\alpha\end{smallmatrix}\right)ds.$$

Taking the constants $C_1 := \sup_{t\in[0,1]}\|\mathcal{T}(t)\|$ and $M := \sup_{t\in[0,1]}\|S(t)\|$, we estimate

$$\left\|\mathcal{T}(t)\left(\begin{smallmatrix}x\\f_\alpha\end{smallmatrix}\right) - \mathcal{T}_0(t))\left(\begin{smallmatrix}x\\f_\alpha\end{smallmatrix}\right)\right\| \leq \int_0^t \left\|\mathcal{T}(t-s)\left(\begin{smallmatrix}\Phi(S_s x + T_0(s)f_\alpha)\\0\end{smallmatrix}\right)\right\|ds$$

$$\leq C_1 \int_0^t \left\|\int_{-1}^0 d\eta(\theta)(S_s x(\theta) + T_0(s)f_\alpha(\theta))\right\|ds$$

$$\leq C_1 M \int_0^t |\eta|([-s-\alpha,0])\|x\|$$

$$\leq C_1 Mt|\eta|([-t-\alpha,0])\|x\|.$$

By letting $\alpha \searrow 0$, we obtain the desired statement.

(ii) First, it is easy to verify directly that

$$\frac{1}{t}\left(\pi_1\mathcal{T}_0(t)\left(\begin{smallmatrix}x\\0\end{smallmatrix}\right) - x\right) \to Bx \qquad \text{as } t \searrow 0 \tag{6.18}$$

for all $x \in \mathbb{R}^n$. For the original expression, we can write

$$\frac{1}{t}\left(\pi_1\mathcal{T}(t)\left(\begin{smallmatrix}x\\0\end{smallmatrix}\right) - x\right) = \frac{1}{t}\left(\pi_1\mathcal{T}(t)\left(\begin{smallmatrix}x\\0\end{smallmatrix}\right) - \pi_1\mathcal{T}_0(t)\left(\begin{smallmatrix}x\\0\end{smallmatrix}\right)\right) + \frac{1}{t}\left(\pi_1\mathcal{T}_0(t)\left(\begin{smallmatrix}x\\0\end{smallmatrix}\right) - x\right).$$

Then it follows from Equations (6.17) and (6.18) that

$$\frac{1}{t}\left(\pi_1 \mathcal{T}(t)\left(\begin{smallmatrix} x \\ 0 \end{smallmatrix}\right) - x\right) \to Bx$$

as $t \searrow 0$ for all $x \in \mathbb{R}^n$. By the same argument we obtain that

$$\frac{1}{t}\left(\pi_1 \tilde{\mathcal{T}}(t)\left(\begin{smallmatrix} x \\ 0 \end{smallmatrix}\right) - x\right) \to B_{\#}x$$

as $t \searrow 0$ for all $x \in \mathbb{R}^n$.

(iii) The statement follows directly by using (ii) and

$$\pi_1 \tilde{\mathcal{T}}(t)\left(\begin{smallmatrix} 0 \\ f \end{smallmatrix}\right) = \pi_1 \tilde{\mathcal{T}}(t)\left(\begin{smallmatrix} f(0) \\ f \end{smallmatrix}\right) - \pi_1 \tilde{\mathcal{T}}(t)\left(\begin{smallmatrix} f(0) \\ 0 \end{smallmatrix}\right).$$

Since $\left(\begin{smallmatrix} f(0) \\ f \end{smallmatrix}\right) \in D(\mathcal{A})$ for $f \in W^{1,p}([-1,0], \mathbb{R}^n)$, the assertion follows.

The other statement follows by the same argument. $\qquad\qquad\square$

The previous result allows us to obtain some first information on the generator $\mathcal{A}_{\#}$ of the modulus semigroup.

Lemma 6.26. *We have* $D(\mathcal{A}) \subseteq D(\mathcal{A}_{\#})$. *Further, there is a sequence* $(t_k) \subset (0, +\infty)$, $t_k \to 0$ *such that*

$$\lim_{k \to \infty} \frac{1}{t_k}\left(\pi_1 \mathcal{T}_{\#}(t_k)\left(\begin{smallmatrix} x \\ 0 \end{smallmatrix}\right) - x\right) \to B_{\#}x \qquad (6.19)$$

and

$$\lim_{k \to \infty} \frac{1}{t_k}\left(\pi_1 \mathcal{T}_{\#}(t_k)\left(\begin{smallmatrix} 0 \\ f \end{smallmatrix}\right) - x\right) \to |\Phi| f. \qquad (6.20)$$

Proof. By the definition of the modulus semigroup, we have the inequalities

$$\mathcal{T}(t)\left(\begin{smallmatrix} x \\ f \end{smallmatrix}\right) \leq \mathcal{T}_{\#}(t)\left(\begin{smallmatrix} x \\ f \end{smallmatrix}\right) \leq \tilde{\mathcal{T}}(t)\left(\begin{smallmatrix} x \\ f \end{smallmatrix}\right) \qquad (6.21)$$

for all $\left(\begin{smallmatrix} x \\ f \end{smallmatrix}\right) \in \mathcal{E}_{p+}$, $t \geq 0$.

Take $0 \leq \left(\begin{smallmatrix} x \\ f \end{smallmatrix}\right) \in D(\tilde{\mathcal{A}}) = D(\mathcal{A})$. Then, in view of Equation (6.21), we have

$$\frac{1}{t}\left(\mathcal{T}(t) - I\right)\left(\begin{smallmatrix} x \\ f \end{smallmatrix}\right) \leq \frac{1}{t}\left(\mathcal{T}_{\#}(t) - I\right)\left(\begin{smallmatrix} x \\ f \end{smallmatrix}\right) \leq \frac{1}{t}\left(\tilde{\mathcal{T}}(t) - I\right)\left(\begin{smallmatrix} x \\ f \end{smallmatrix}\right).$$

As $t \searrow 0$, the first and the last term of the inequality converge to $\mathcal{A}\left(\begin{smallmatrix} x \\ f \end{smallmatrix}\right)$ and $\tilde{\mathcal{A}}\left(\begin{smallmatrix} x \\ f \end{smallmatrix}\right)$, respectively.

Now we show that there exists a null sequence $(t_k) \subset (0,1]$ such that the sequence $\left(\frac{1}{t_k} (\mathcal{T}_\#(t_k) - I) \left(\begin{smallmatrix} x \\ f \end{smallmatrix} \right) \right)_k$ converges weakly. Defining

$$u(t) := \frac{1}{t} \left(\mathcal{T}(t) - I \right) \left(\begin{smallmatrix} x \\ f \end{smallmatrix} \right),$$

$$u_\#(t) := \frac{1}{t} \left(\mathcal{T}_\#(t) - I \right) \left(\begin{smallmatrix} x \\ f \end{smallmatrix} \right),$$

$$\tilde{u}(t) := \frac{1}{t} \left(\tilde{\mathcal{T}}(t) - I \right) \left(\begin{smallmatrix} x \\ f \end{smallmatrix} \right),$$

we obtain $0 \le u_\#(t) - u(t) \le \tilde{u}(t) - u(t)$ for $0 < t \le 1$, and $\lim_{t \searrow 0}(\tilde{u}(t) - u(t)) = \mathcal{A} \left(\begin{smallmatrix} x \\ f \end{smallmatrix} \right) - \tilde{\mathcal{A}} \left(\begin{smallmatrix} x \\ f \end{smallmatrix} \right)$. The set $S := \{\tilde{u}(t) - u(t) \ : \ 0 < t \le 1\} \subset \mathbb{R}^n$ is bounded, and hence relatively compact. Since $\{u_\#(t) - u(t) \ : \ 0 < t \le 1\}$ is a subset of S, there exists a null sequence (t_k) such that $(u_\#(t_k) - u(t_k))_k$ converges, and therefore $(u_\#(t_k))_k$ converges.

Observe that, for $k \in \mathbb{N}$, the pair $\left(\frac{1}{t_k} \int_0^{t_k} \mathcal{T}_\#(s) \left(\begin{smallmatrix} x \\ f \end{smallmatrix} \right) ds, u_\#(t_k) \right)$ belongs to the graph of $\mathcal{A}_\#$. Since $\mathcal{A}_\#$ is a closed operator, its graph is weakly closed. Hence, $\left(\begin{smallmatrix} x \\ f \end{smallmatrix} \right) \in D(\mathcal{A}_\#)$. Finally, $D(\tilde{\mathcal{A}}) = D(\tilde{\mathcal{A}})_+ - D(\tilde{\mathcal{A}})_+ \subset D(\mathcal{A}_\#)$ shows the first assertion.

To show Equation (6.19), let $x \in \mathbb{R}^n$. From Equation (6.21) we obtain

$$\frac{1}{t} \left(\pi_1 \mathcal{T}(t) \left(\begin{smallmatrix} x \\ 0 \end{smallmatrix} \right) - x \right) \le \frac{1}{t} \left(\pi_1 \mathcal{T}_\#(t) \left(\begin{smallmatrix} x \\ 0 \end{smallmatrix} \right) - x \right)$$
$$\le \frac{1}{t} \left(\pi_1 \tilde{\mathcal{T}}(t) \left(\begin{smallmatrix} x \\ 0 \end{smallmatrix} \right) - x \right). \tag{6.22}$$

It follows from Lemma 6.25 (ii) that the first and the third term of Equation (6.22) converge to Bx and $B_\# x$ as $t \searrow 0$, respectively.

Let us denote by $(e_1, e_2, \ldots, e_n)$ the canonical basis of $\mathbb{R}^n$. The previous statements show that the sets $\{ \frac{1}{t} \left(\pi_1 \mathcal{T}_\#(t) \left(\begin{smallmatrix} e_i \\ 0 \end{smallmatrix} \right) - e_i \right) : 0 < t \le 1 \}$ are bounded subsets of $\mathbb{R}^n$ for $i = 1, 2, \ldots, n$. Hence, there exists a null sequence $(t_k) \subset (0,1]$ such that the sequence $\left(\frac{1}{t_k} \left(\pi_1 \mathcal{T}_\#(t_k) \left(\begin{smallmatrix} e_i \\ 0 \end{smallmatrix} \right) - e_i \right) \right)_k$ is convergent for all $i = 1, 2, \ldots, n$. Consequently,

$$\hat{B}x := \lim_{k \to \infty} \frac{1}{t_k} \left(\pi_1 \mathcal{T}_\#(t_k) \left(\begin{smallmatrix} x \\ 0 \end{smallmatrix} \right) - x \right)$$

exists for all $x \in \mathbb{R}^n$. For $0 \le x \in \mathbb{R}^n$, we obviously have

$$Bx \le \hat{B}x \le B_\# x.$$

By the definition of $B_\#$ we conclude for the diagonal entries of the matrix $\hat{B}$ that

$$\hat{b}_{ii} = b_{ii} \quad i = 1, 2, \ldots, n.$$

Let p_i be the canonical projection in $\mathbb{R}^n$ onto the ith coordinate. Then, for $i \neq j$, we have

$$\left| p_i \pi_1 \mathcal{T}(t_k) \left(\begin{smallmatrix} e_j \\ 0 \end{smallmatrix} \right) \right| \leq p_i \pi_1 \mathcal{T}_\#(t_k) \left(\begin{smallmatrix} e_j \\ 0 \end{smallmatrix} \right),$$

$$\frac{1}{t_k} \left| p_i \pi_1 \mathcal{T}(t_k) \left(\begin{smallmatrix} e_j \\ 0 \end{smallmatrix} \right) \right| \leq \frac{1}{t_k} p_i \pi_1 \mathcal{T}_\#(t_k) \left(\begin{smallmatrix} e_j \\ 0 \end{smallmatrix} \right).$$

By Lemma 6.25 (ii) the terms of the last inequality converge to $|b_{ij}|$ and $\hat{b}_{ij}$, respectively. We can deduce that

$$|b_{i,j}| \leq \hat{b}_{i,j} \ \text{ for } i,j \in \{1,2,\dots,n\}, \ i \neq j.$$

Combining all this information on the entries of the matrix $\hat{B}$, we obtain that $\hat{B} = B_\#$.

Finally, we are going to show Equation (6.20). Let $f \in W^{1,p}([-1,0],\mathbb{R}^n)$. Since $\left(\begin{smallmatrix} f(0) \\ f \end{smallmatrix} \right) \in D(\mathcal{A}) \subset D(\mathcal{A}_\#)$, we obtain the existence of the limit

$$\hat{\Phi} f := \lim_{k \to 0} \frac{1}{t_k} \pi_1 \mathcal{T}_\#(t_k) \left(\begin{smallmatrix} 0 \\ f \end{smallmatrix} \right)$$

$$= \lim_{k \to 0} \frac{1}{t_k} \pi_1 \left(\mathcal{T}_\#(t_k) \left(\begin{smallmatrix} f(0) \\ f \end{smallmatrix} \right) - \left(\begin{smallmatrix} f(0) \\ f \end{smallmatrix} \right) \right) - \lim_{k \to 0} \frac{1}{t_k} \pi_1 \left(\mathcal{T}_\#(t_k) \left(\begin{smallmatrix} 0 \\ f \end{smallmatrix} \right) - f(0) \right)$$

$$= \pi_1 \mathcal{A}_\# \left(\begin{smallmatrix} f(0) \\ f \end{smallmatrix} \right) - B_\# f(0).$$

By the definition of $\hat{\Phi}$, we have that $\hat{\Phi} : W^{1,p}([-1,0],\mathbb{R}^n) \to \mathbb{R}^n$ is a positive linear operator, and by Equation (6.21), it satisfies

$$|\Phi f| \leq \hat{\Phi} f \leq |\Phi| f$$

for all $0 \leq f \in W^{1,p}([-1,0],\mathbb{R}^n)$. In order to conclude that $\hat{\Phi} = |\Phi|$, it remains to show that $0 \leq f \in W^{1,p}([-1,0],\mathbb{R}^n)$, $g \in C([-1,0],\mathbb{R}^n)$, $|g| \leq f$ implies $|\Phi g| \leq \hat{\Phi} f$.

To show this, choose a sequence $(g_k) \subset W^{1,p}([-1,0],\mathbb{R}^n)$ satisfying $\lim_{k \to \infty} \|g_k - g\|_\infty = 0$. Defining $\tilde{g}_k := (g_k \wedge f) \vee (-f)$, we obtain that $\tilde{g}_k, |\tilde{g}_k| \in W^{1,p}([-1,0],\mathbb{R}^n)$, $|\tilde{g}_k| \leq f$, and that $\lim_{k \to \infty} \|\tilde{g}_k - g\|_\infty = 0$. Moreover,

$$|\Phi \tilde{g}_k| \leq |\Phi(\tilde{g}_k^+)| + |\Phi(\tilde{g}_k^-)|$$
$$\leq \hat{\Phi}(\tilde{g}_k^+) + \hat{\Phi}(\tilde{g}_k^-)$$
$$= \hat{\Phi} |\tilde{g}_k| \leq \hat{\Phi} f.$$

This implies $|\Phi g| \leq \hat{\Phi} f$. $\qquad\qquad\qquad\qquad\qquad\qquad\qquad\qquad\qquad\qquad\square$

We can now state the main result of this subsection.

Theorem 6.27. *Under the main assumptions formulated in the beginning of this subsection, the operator $\tilde{A}$, defined by Equation (6.15), is the generator of the modulus semigroup of $(\mathcal{T}(t))_{t\geq 0}$, i.e., $\mathcal{A}_{\#} = \tilde{A}$.*

Proof. Let $\left(\begin{smallmatrix}x\\f\end{smallmatrix}\right) \in D(\mathcal{A})_+$. By Lemma 6.26 we have

$$\mathcal{A}_{\#}\left(\tfrac{x}{f}\right) = \lim_{t\to 0} \frac{1}{t}\left(\mathcal{T}_{\#}(t)\left(\tfrac{x}{f}\right) - \left(\tfrac{x}{f}\right)\right).$$

Hence, for the sequence (t_k) from Lemma 6.26, we have

$$\pi_1 \mathcal{A}_{\#}\left(\tfrac{x}{f}\right) = \lim_{k\to\infty} \frac{1}{t}\left(\pi_1 \mathcal{T}_{\#}(t_k)\left(\tfrac{x}{f}\right) - x\right)$$

$$= \lim_{k\to\infty} \frac{1}{t_k}\left(\pi_1 \mathcal{T}_{\#}(t_k)\left(\tfrac{x}{0}\right) - x\right) + \lim_{k\to\infty}\frac{1}{t_k}\pi_1 \mathcal{T}_{\#}(t_k)\left(\tfrac{0}{f}\right)$$

$$= B_{\#}x + |\Phi|f.$$

By Equation (6.21) we also have

$$\mathcal{A}\left(\tfrac{x}{f}\right) \leq \mathcal{A}_{\#}\left(\tfrac{x}{f}\right) \leq \tilde{A}\left(\tfrac{x}{f}\right),$$

i.e.,

$$\begin{pmatrix} Bx + \Phi f \\ f' \end{pmatrix} \leq \mathcal{A}_{\#}\left(\tfrac{x}{f}\right) \leq \begin{pmatrix} B_{\#}x + |\Phi|f \\ f' \end{pmatrix}.$$

Therefore, we have

$$\pi_2 \mathcal{A}_{\#}\left(\tfrac{x}{f}\right) = f',$$

where π_2 is the canonical projection from $\mathcal{E}_p = \mathbb{R}^n \times L^p([-1,0],\mathbb{R}^n)$ onto $L^p([-1,0],\mathbb{R}^n)$ defined in Definition 3.8. Thus, we have shown that

$$\mathcal{A}_{\#}\left(\tfrac{x}{f}\right) = \begin{pmatrix} B_{\#}x + |\Phi|f \\ f' \end{pmatrix} = \tilde{A}\left(\tfrac{x}{f}\right)$$

for all $0 \leq \left(\begin{smallmatrix}x\\f\end{smallmatrix}\right) \in D(\mathcal{A})_+$. To finish the proof, we note that $D(\tilde{A}) = D(\tilde{A})_+ - D(\tilde{A})_+$. Therefore, $\mathcal{A}_{\#} \supset \tilde{A}$, and since both operators are generators, $\mathcal{A}_{\#} = \tilde{A}$. $\square$

6.4 Notes and References

In the abstract framework of Banach lattices, a powerful stability theory has been developed for such positive semigroups in Nagel et al. [148] and successfully applied to delay equations as semigroups on the history space $C([-1,0],X)$ (see [148, Section B.IV and C-IV] and [72, Section VI.6]).

The results presented in Section 6.2 are in complete analogy to the ones in Engel and Nagel [72, Section VI.6]). Lemma 6.16 is due to W. Arendt and we quote it from [72, Lemma VI.6.15].

The application of Theorem 6.9 to delay equations as in Theorem 6.19 seems to be new. It was recently applied to a population equation with delayed birth process in [167], where the investigations presented in this chapter are extended to show balanced exponential growth and to consider also boundary delays. Example 6.20 is a modification of Greiner [91].

The existence of the modulus semigroup for the delay semigroup in the history space $C([-1,0],X)$ with $X = \mathbb{R}^n$ finite-dimensional was shown by Becker and Greiner [23].

The results in Section 6.3 are due to S. Boulite, L. Maniar, A. Rhandi, and J. Voigt [25]. These were extended to the case where X is a Banach lattice with order continuous norm by Voigt [207] and Stein, Vogt, and Voigt [197].

Chapter 7

Small Delays

It was observed by Datko [47] (see also [44, 46]) that small delays may destroy stability for a partial differential equation. This phenomenon has important consequences for the applications and will be studied systematically in this chapter.

We start by formulating the problem of small delays in precise mathematical terms. Then we show in several examples how uniform exponential stability or hyperbolicity can be destroyed by arbitrary small delays, a phenomenon which is impossible in the finite-dimensional context (more generally, whenever the semigroup is eventually norm continuous). In applications, for example, if we are stabilizing a vibrating string, it is important to know whether our stabilizing procedure is sensitive to small delays or not, because we cannot make sure that the stabilization acts immediately and not with some microseconds delay. Therefore, we present a strategy that allows us to give conditions on the delay equation ensuring that uniform exponential stability or hyperbolicity is not sensitive to small delays. Because the proofs are at some points rather involved, we present the conditions first in the Hilbert space case where the calculations are much simpler and then afterwards in the general Banach space case. Finally, we show that if we have unbounded operators in the delay term as it was discussed in Section 3.4, then the stability or hyperbolicity cannot be sensitive to small delays.

7.1   The Effect of Small Delays

We start from the equation

$$
(DE)_0 \qquad
\begin{cases}
u'(t) = (B + C)u(t), & t \geq 0, \\
u(0) = x, \\
u_0 = f,
\end{cases}
$$

145

where $C \in \mathcal{L}(X)$ and assume that the solutions are uniformly exponentially stable (or hyperbolic).

The question is, whether stability or hyperbolicity prevails for the solutions of the delayed equation

$$(\mathrm{DE})_\tau \qquad \begin{cases} u'(t) = Bu(t) + Cu(t - \tau), & t \geq 0, \\ u(0) = x, \\ u_0 = f, \end{cases}$$

where $\tau > 0$ is "small."

Example 7.1. If C is a positive operator and $(B, D(B))$ generates a positive semigroup on an L^p-space, then the delay semigroup corresponding to $(\mathrm{DE})_\tau$ is uniformly exponentially stable if and only if the semigroup generated by $(B + C, D(B))$ is uniformly exponentially stable. This is a special case of Theorem 6.19 and Proposition 6.17. Hence, in the positive case, uniform exponential stability is not sensitive to small delays.

It is an open question, however, as to what happens to the hyperbolicity in the positive case. The example due to Montgomery-Smith [147] suggests that this result may not remain true.

Before considering the general problem, we demonstrate by some simple examples that in fact the stability can be destroyed by small delays.

Example 7.2. Let $(B, D(B))$ be the (unbounded) generator of a unitary group in an infinite-dimensional Hilbert space H and let $C := d \cdot Id$ for $d < 0$. Then $(B + C, D(B))$ generates an exponentially stable semigroup. We show that there exists a sequence (τ_k), $\tau_k \searrow 0$, such that the solution semigroup of each of the corresponding equations $(\mathrm{DE})_{\tau_k}$ does not decay exponentially.

To construct this sequence, take $(\mu_k) \subset \mathbb{R}$ such that $i\mu_k \in \sigma(B)$, $|\mu_k| \to \infty$ and $\mu_k \neq -d$. If we define

$$\tau_k := \begin{cases} \frac{3\pi}{2(\mu_k + d)}, & \mu_k + d > 0, \\ \frac{-\pi}{2(\mu_k + d)}, & \mu_k + d < 0, \end{cases}$$

we obtain $\lambda_k \in \sigma(B + \Phi_{\lambda_k})$ for the numbers $\lambda_k := (\mu_k + d)i \in i\mathbb{R}$. By the spectral characterization in Lemma 3.19 it follows that the associated operator $(\mathcal{A}, D(\mathcal{A}))$ cannot generate a uniformly exponentially stable semigroup. If we assume further that $i\mu_k \in P\sigma(B)$, which, e.g., is satisfied if $(B, D(B))$ has compact resolvent, then we even find classical solutions of $(\mathrm{DE})_{\tau_k}$ not decaying exponentially.

This example can be modified in the following way to include the case in which $R(\lambda, B)$ is compact and the essential growth bound satisfies $\omega_{\text{ess}}(B) < 0$.

Example 7.3. Consider the space $X := l^2$, the multiplication operator $B := \text{diag}(\lambda_0, \lambda_1, \lambda_{-1}, \lambda_2, \lambda_{-2}, ...)$ where $\lambda_0 > 0$ large, and $\lambda_{\pm k} := -\rho \pm 2^k i$ for some $\rho > 0$ small. Defining the operator $C := -\mu \cdot Id$ for some $\mu > \lambda_0$ and $\mu > \rho$, we obtain that

$$\omega_{\text{ess}}(B) = -\rho < 0, \qquad \omega_0(B) = \lambda_0 > 0,$$
$$\omega_{\text{ess}}(B + C) = -\rho - \mu < 0, \qquad \omega_0(B + C) = \lambda_0 - \mu < 0.$$

Now let us take, analogously to the previous example, the delay

$$\tau := 2^{-k}\pi$$

for some fixed $k \in \mathbb{N}$. The corresponding characteristic equations (see Proposition 3.19) are

$$z - \lambda_j + \mu e^{-z\tau} = 0, \qquad j \in \mathbb{Z}.$$

We show that for $j = k$ there exist solutions with positive real part, which means, by Lemma 3.24, that the associated semigroup cannot be exponentially stable.

Put $z := 2^k i + \varepsilon$ for $\varepsilon > 0$. Then we have

$$2^k i + \varepsilon + \rho - 2^k i + \mu e^{-2^k i\pi 2^{-k}} \cdot e^{-\varepsilon\pi 2^{-k}} = 0,$$

and hence

$$\varepsilon = \mu e^{-\varepsilon 2^{-k}\pi} - \rho.$$

Since $\mu > \rho$, there exists a positive real solution ε. This means that $s(A_{\tau_k}) \geq \varepsilon > 0$, and the assertion follows.

The essence of these examples can be formulated as follows.

Theorem 7.4. *Let X be a Hilbert space and assume that $(B, D(B))$ generates a strongly continuous semigroup such that the spectrum $\sigma(B)$ is unbounded along an imaginary line. Then there exists $C \in \mathcal{L}(X)$ and $(\tau_k) \subset \mathbb{R}^+$, $\tau_k \searrow 0$, such that $\omega_0(B + C) < 0$ while for all $k \in \mathbb{N}$ there are solutions of $(DE)_{\tau_k}$ not converging to zero.*

Proof. Assume, as in the previous example that $\sigma(B)$ is unbounded along $-\rho + i\mathbb{R}$ and take $C := -\mu \cdot Id$, $\mu > 0$, $\mu > -\rho$, and further define

$$\tau_k := \frac{\pi}{|\mu_k|}$$

for each $k \in \mathbb{N}$. Here $\mu_k \in \sigma(B)$, $\Re\mu_k = -\rho$ is a sequence with the property $|\mu_k| \to +\infty$.

Then the corresponding characteristic equations are again

$$z - \lambda + \mu e^{-zT_k} = 0, \qquad \lambda \in \sigma(B).$$

Take now $\lambda := \rho + i\mu_k$ and put $z := \varepsilon + i\mu_k$ for some $\varepsilon > 0$. Then we obtain

$$i\mu_k + \varepsilon - \rho - i\mu_k + \mu e^{-i\mu_k T_k} \cdot e^{-\varepsilon T_k} = 0,$$

and hence

$$\varepsilon = \mu e^{-\varepsilon T_k} + \rho.$$

Since $\mu > -\rho$, there exists a positive real solution ε. This means that $s(\mathcal{A}_{T_k}) \geq \varepsilon > 0$, and the assertion follows. $\qquad\square$

In order to find conditions under which stability is not sensitive to small delays, we use an idea similar to Hale and Verduyn Lunel [103, Section 5.4 (4.9)] and transform the equation $(DE)_\tau$ into

$$u'(t) = (B + C)u(t) + C\left(u(t - \tau) - u(t)\right).$$

The main task now is to transform $\Phi u_t = C\left(u(t - \tau) - u(t)\right)$ in a form such that it is possible to use the estimates from Corollary 5.7.

Integration on $0 \leq t_1 \leq t_2$, the original form of $(DE)_\tau$, yields the Variation of Parameters Formula

$$u(t_2) - u(t_1) = [S(t_2 - t_1) - Id]\,u(t_1) + \int_{t_1}^{t_2} S(t_2 - s)Cu(s - \tau)ds. \quad (7.1)$$

Here the semigroup $(S(t))_{t\geq 0}$ denotes the semigroup generated by $(B, D(B))$. By taking $t_1 = t - \tau$ and $t_2 = t$, we obtain that

$$u(t) - u(t - \tau) = [S(\tau) - Id]\,u(t - \tau) + \int_{-\tau}^{0} S(-s)Cu(t + s - \tau)ds.$$

Thus, $(DE)_\tau$ can be written in the form

$$u'(t) = (B + C)u(t) \qquad\qquad\qquad\qquad\qquad (7.2)$$
$$- C\left([S(\tau) - Id]\,u(t - \tau) + \int_{-\tau}^{0} S(-s)Cu(t + s - \tau)ds\right).$$

Define now

$$\Phi f := -C\,[S(\tau) - Id]\,\delta_{-\tau}f - \int_{-\tau}^{0} CS(-s)C\delta_{s-\tau}f\,ds, \qquad (7.3)$$

where $\delta_r \in \mathcal{L}\left(W^{1,p}([-1,0], X), X\right)$ is given by $\delta_r(f) := f(r)$ for $r \in [-1,0]$. Then $(\mathrm{DE})_\tau$ has the form

$$u'(t) = (B + C)\, u(t) + \Phi u_t, \tag{7.4}$$

where $(B + C, D(B))$ generates a uniformly exponentially stable or hyperbolic semigroup and Φ is of the form of Equation (3.38).

In order to apply the stability results from Corollary 5.7, or the hyperbolicity results from Corollary 5.18, we have to calculate

$$\Phi_\lambda R(\lambda, B + C)x = -C\left[S(\tau) - Id\right] e^{-\lambda \tau} R(\lambda, B + C)x$$
$$- \int_{-\tau}^0 CS(-s)Ce^{-\lambda(s-\tau)} R(\lambda, B + C)x\, ds. \tag{7.5}$$

As we saw in Example 7.2, the unboundedness of the spectrum of $(B, D(B))$ along imaginary axes may cause trouble. Therefore, we make an additional assumption on the semigroup generated by B implying that $\sigma(B)$ is bounded along the imaginary lines; see Theorem 2.22.

Theorem 7.5. *Assume that $(B, D(B))$ generates an immediately norm continuous semigroup and that the semigroup generated by $(B + C, D(B))$ is uniformly exponentially stable (hyperbolic, respectively) on the Hilbert space X. Then there exists $\kappa > 0$ such that the solution semigroup of $(\mathrm{DE})_\tau$ is uniformly exponentially stable (hyperbolic, respectively) for all $\tau \in (0, \kappa)$. Thus, in this situation, stability and hyperbolicity are not sensitive to small delays.*

Proof. Define

$$I_1^\omega(\tau) := C\left[S(\tau) - Id\right] e^{-i\omega\tau} R(i\omega, B + C) \tag{7.6}$$

and

$$I_2^\omega(\tau)x := \int_{-\tau}^0 CS(-s)Ce^{-i\omega(s-\tau)} R(i\omega, B + C)x\, ds. \tag{7.7}$$

We show that there exists $\kappa > 0$ such that $\sup_{\omega \in \mathbb{R}} \|I_i^\omega(\tau)\| < \frac{1}{2}$ for $i = 1, 2$ and all $\tau \in (0, \kappa)$. Then, using Corollary 5.7 or Corollary 5.18 and Equation (7.5), the assertion follows since $\sup_{\omega \in \mathbb{R}} \|\Phi_{i\omega} R(i\omega, B + C)\| \leq \sup_{\omega \in \mathbb{R}} \|I_1^\omega(\tau)\| + \sup_{\omega \in \mathbb{R}} \|I_2^\omega(\tau)\| < 1$.

The estimate on I_2^ω is

$$\|I_2^\omega(\tau)\| \leq \tau \|C\|^2 K \|R(i\omega, B + C)\|, \tag{7.8}$$

where $K := \sup_{0 \leq t \leq 1} \|S(t)\|$. Since $\|R(i\omega, B + C)\|$ is uniformly bounded for all $\omega \in \mathbb{R}$, there exists $\kappa_2 > 0$ such that for all $\tau \in (0, \kappa_2)$ the estimate $\sup_{\omega \in \mathbb{R}} \|I_2^\omega(\tau)\| < \frac{1}{2}$ holds.

The estimate on I_1^ω is

$$\|I_1^\omega(\tau)\| \leq \|C\| \cdot \|(S(\tau) - Id)R(i\omega, B + C)\|$$
$$\leq \|C\| \cdot \|(S(\tau) - Id)R(\lambda, B)\| \cdot \|(\lambda - B)R(i\omega, B + C)\|, \quad (7.9)$$

where $\lambda > \max\{\omega_0(B), 0\}$ is fixed.

Since $\|(\lambda - B)R(i\omega, B + C)\|$ is independent of τ, we only have to consider $(S(\tau) - Id)R(\lambda, B)$.

But then it follows from

$$\|(S(\tau) - Id)R(\lambda, B)\| \leq \|S(\tau)\|(1 - e^{\lambda\tau})\|R(\lambda, B)\| + \left\| \int_0^\tau e^{-\lambda s} S(s)ds \right\|$$
$$\leq \tau K \left(\|R(\lambda, B)\| |\lambda| + 1 \right)$$

that

$$\lim_{\tau \to 0} \left\| C\left[S(\tau) - Id \right] e^{i\omega\tau} R(i\omega, B + C) \right\| = 0 \qquad (7.10)$$

for every $\omega \in \mathbb{R}$.

To finish the proof we have to show that this convergence is uniform in ω.

To this end, we use the immediate norm continuity of the semigroup generated by $(B + C, D(B))$; see Proposition 1.47. An important property of these semigroups is that $\lim_{|\omega| \to \infty} \|R(i\omega, B + C)\| = 0$; see Theorem 2.21. Thus, there exists $L > 0$ such that

$$\|R(i\omega, B + C)\| < \frac{1}{2\|C\|(K + 1)} \quad \text{for } |\omega| > L,$$

where $K := \sup_{0 \leq t \leq 1} \|S(t)\|$.

For $\omega \in [-L, \overline{L}]$, we recall that the function

$$(\omega, \tau) \mapsto \left\| C\left[S(\tau) - Id \right] e^{i\omega\tau} R(i\omega, B + C) \right\|$$

is uniformly continuous on $[-L, L] \times [0, 1]$. Thus, there exists $\kappa_1 > 0$ such that for all $\tau \in (0, \kappa_1)$ and for all $\omega \in [-L, L]$

$$\left\| C\left[S(\tau) - Id \right] e^{i\omega\tau} R(i\omega, B + C) \right\| < \frac{1}{2}.$$

Combining these estimates we obtain the desired statement.

The proof can be finished by choosing $\kappa := \min\{\kappa_1, \kappa_2\}$. $\square$

In the previous theorem we gave a condition on the generator $(B, D(B))$ without any restriction on the stabilizing operator C. In the following we provide a condition also involving C.

Proposition 7.6. *Let $(B, D(B))$ be a generator of a strongly continuous semigroup $(S(t))_{t \geq 0}$ in the Hilbert space X; let $C \in \mathcal{L}(X)$ be a compact operator commuting with $S(t)$, $t \geq 0$. Assume that the semigroup generated by $(B+C, D(B))$ is uniformly exponentially stable (hyperbolic, respectively). Then there exists $\kappa > 0$ such that the solution semigroup of (DE)$_\tau$ is uniformly exponentially stable (hyperbolic, respectively) for all $\tau \in (0, \kappa)$. Thus, stability and hyperbolicity are not sensitive to small delays.*

Proof. We repeat the arguments in the proof of Theorem 7.5 and only have to show that the convergence in Equation (7.10) is uniform in ω. Since C commutes with the semigroup, we obtain

$$C\left[S(\tau) - Id\right] e^{i\omega\tau} R(i\omega, B + C) = \left[S(\tau) - Id\right] e^{i\omega\tau} R(i\omega, B + C)C.$$

By our assumptions, the set $C(B(0,1)) \subset X$ is relatively compact in X. The assertion follows since on compact sets the strong and the uniform topology coincide (see Engel and Nagel [72, Proposition A.3]).   □

Remark 7.7. We can actually sharpen the statements of the previous results for the hyperbolicity case. Applying Theorem 5.19, we obtain under the conditions of Theorem 7.5 or Proposition 7.6 that if $\tau > 0$ is sufficiently small, then the dimension of the unstable manifold of the semigroup generated by $(B+C, D(B))$ in X coincides with the dimension of the unstable manifold of the semigroup generated by $\mathcal{A}_\tau$ in $\mathcal{E}_p$ corresponding to the equation (DE)$_\tau$.

The previous two results are based on the stability theorem from Section 5.1 valid for Hilbert spaces only. If we use the theory of Fourier multipliers as in Section 5.3, it is possible to extend them to arbitrary spaces. Since the proofs are much more involved, we chose to treat the Hilbert space case independently. The following result has been obtained in collaboration with B. Farkas; see [15].

Theorem 7.8. *Suppose that either*

(i) *the semigroup $(S(t))_{t \geq 0}$, generated by $(B, D(B))$ commutes with the compact operator C or*

(ii) *$(S(t))_{t \geq 0}$ is an immediately norm continuous semigroup.*

Assume moreover that the semigroup $(T(t))_{t \geq 0}$, generated by $(B+C, D(B))$, is uniformly exponentially stable with $\omega_0(B + C) < \alpha \leq 0$.
    Then the growth bound of the delay semigroup generated by $\mathcal{A}_\tau := \begin{pmatrix} B & C\delta_\tau \\ 0 & \frac{d}{d\sigma} \end{pmatrix}$ satisfies $\omega_0(\mathcal{A}_\tau) < \alpha$ for τ sufficiently small.

In particular, the solutions of (DE)$_\tau$ have uniform exponential growth smaller than α for τ sufficiently small.

Proof. First we fix some constants. Take $\omega \in [\omega_0(B+C), \alpha)$ and $M \geq 1$ such that $T(t) \leq Me^{\omega t}$ for all $t \geq 0$. Furthermore, define $M' = \sup_{[0,1]} \|S(t)\|$ and take $\beta \geq \alpha$. By the Hausdorff-Young Inequality (see Theorem 5.24), we have

$$\|\mathcal{F}^{-1}\Phi_{\beta+i\cdot}R(\beta+i\cdot, B+C)\mathcal{F}f\|_{L^p(\mathbb{R},X)} \leq$$
$$\|\mathcal{F}^{-1}\Phi_{\beta+i\cdot}R(\beta+i\cdot, B+C)\|_{L^1(\mathbb{R},\mathcal{L}(X))}\|f\|_{L^p(\mathbb{R},X)}. \tag{7.11}$$

Therefore, to estimate the Fourier multiplier norm of $\Phi_{\beta+i\cdot}R(\beta+i\cdot, B+C)$ it suffices to give L^1-estimates on the inverse Fourier transform. From Equation (7.5) we have for all $x \in X$ that

$$\Phi_{\beta+i\cdot}R(\beta+i\cdot, B+C)x$$
$$= -C[S(\tau)-Id]e^{-(\beta+i\cdot)\tau}R(\beta+i\cdot, B+C)x \tag{7.12}$$
$$- \int_{-\tau}^0 CS(-s)Ce^{-(\beta+i\cdot)(s-\tau)}R(\beta+i\cdot, B+C)x\,ds.$$

Let $I_1^{\beta+i\cdot}(\tau)$ and $I_2^{\beta+i\cdot}(\tau)$ denote the first and the second part in this expression, respectively. We calculate them separately. To that purpose fix $x \in D(B)$. Then for all $t \in \mathbb{R}$ we obtain

$$-\mathcal{F}^{-1}(I_1^{\beta+i\cdot}(\tau))(t)x = -\frac{1}{2\pi i}\int_{-\infty}^{+\infty} e^{itr}I_1^{\beta+ir}(\tau)x\,dr$$
$$= \frac{1}{2\pi i}\int_{-\infty}^{+\infty} e^{itr}C[S(\tau)-Id]e^{-(\beta+ir)\tau}R(\beta+ir, B+C)x\,dr$$
$$= \frac{1}{2\pi i}C[S(\tau)-Id]e^{-\beta\tau}\int_{-\infty}^{+\infty} e^{ir(t-\tau)}R(\beta+ir, B+C)x\,dr \tag{7.13}$$
$$= \frac{1}{2\pi i}C[S(\tau)-Id]e^{-\beta\tau}\int_{-\infty}^{+\infty} e^{ir(t-\tau)}R(ir, B+C-\beta)x\,dr$$
$$= C[S(\tau)-Id]e^{-\beta\tau}e^{-\beta(t-\tau)}\tilde{T}(t-\tau)x$$
$$= C[S(\tau)-Id]e^{-\beta t}\tilde{T}(t-\tau)x,$$

where

$$\tilde{T}(t) = \begin{cases} T(t) & \text{if } t \geq 0 \\ 0 & \text{otherwise.} \end{cases}$$

Here, we have used that the inverse Fourier transform of the resolvent $R(i\cdot, B + C - \beta)$ is $e^{-\beta \cdot} \tilde{T}(\cdot)$ since $(e^{-\beta t} T(t))_{t \geq 0}$ is uniformly exponentially stable. From Equation (7.13), it follows immediately that

$$\mathcal{F}^{-1}(I_1^{\beta + i\cdot}(\tau))(t) = -C[S(\tau) - Id]e^{-\beta t}\tilde{T}(t - \tau) \quad \text{for all } t \in \mathbb{R}.$$

(i) Suppose now that C is compact and commutes with the semigroup $(S(t))_{t \geq 0}$. Then for each $x \in X$

$$\left\| \mathcal{F}^{-1}(I_1^{\beta + i\cdot}(\tau))(t)x \right\|_X = \left\| C[S(\tau) - Id]e^{-\beta t}\tilde{T}(t - \tau)x \right\|_X$$

$$= \left\| [S(\tau) - Id]Ce^{-\beta t}\tilde{T}(t - \tau)x \right\|_X \to 0 \text{ as } \tau \to 0,$$

and the convergence is uniform for $\beta \in [\alpha, +\infty)$ and x in bounded sets. Hence, it follows that

$$\left\| \mathcal{F}^{-1}(I_1^{\beta + i\cdot}(\tau))(t) \right\|_{\mathcal{L}(X)} \to 0$$

uniformly in $t \in \mathbb{R}$ and $\beta \in [\alpha, +\infty)$. Now, by Lebesgue's Dominated Convergence Theorem, we obtain

$$\left\| \mathcal{F}^{-1}(I_1^{\beta + i\cdot}(\tau)) \right\|_{L^1(\mathbb{R}, \mathcal{L}(X))}$$

$$= \int_{-\infty}^{+\infty} \left\| [S(\tau) - Id]Ce^{-\beta t}\tilde{T}(t - \tau) \right\|_{\mathcal{L}(X)} dt$$

$$= \int_0^{+\infty} \left\| [S(\tau) - Id]Ce^{-\beta(s + \tau)}T(s) \right\|_{\mathcal{L}(X)} ds \to 0$$

uniformly in $\beta \in [\alpha, +\infty)$.

(ii) For the other case, assume that $(S(t))_{t \geq 0}$ is immediately norm continuous. Then by Proposition 1.47, $(T(t))_{t \geq 0}$ is also immediately norm continuous.

Let $t \in \mathbb{R}$, $x \in X$, and $\beta \in [\alpha, +\infty)$. We can estimate the above expression as

$$\left\| \mathcal{F}^{-1}(I_1^{\beta + i\cdot}(\tau))(t)x \right\|_X \leq \|C\| \left\| [S(\tau) - Id]e^{-\beta t}\tilde{T}(t - \tau)x \right\|_{\mathcal{L}(X)}$$

$$= \|C\| \left\| [S(\tau) - T(\tau) + T(\tau) - Id]e^{-\beta t}\tilde{T}(t - \tau)x \right\|_{\mathcal{L}(X)}$$

$$\leq \|C\| \cdot \|S(\tau) - T(\tau)\|_X \cdot \left\| e^{-\beta t}\tilde{T}(t - \tau)x \right\|_X$$

$$+ \|C\| \cdot \left\| [T(\tau) - Id]e^{-\beta t}\tilde{T}(t - \tau)x \right\|_X.$$

Again, the two parts will be dealt with separately. For the first term we obtain

$$\|S(\tau) - T(\tau)\|_X \cdot \left\|e^{-\beta t}\tilde{T}(t-\tau)x\right\|_X \leq \tau K \left\|e^{-\beta t}\tilde{T}(t-\tau)x\right\|_X \to 0$$

uniformly in $\beta \in [\alpha, +\infty)$ and x in bounded sets. Hence,

$$\left\|[S(\tau) - T(\tau)]e^{-\beta \cdot}\tilde{T}(\cdot - \tau)\right\|_{L^1(\mathbb{R}, \mathcal{L}(X))} \to 0,$$

and the convergence is uniform in $\beta \in [\alpha, +\infty)$. For the second term, we have

$$\left\|[T(\tau) - Id]e^{-\beta t}\tilde{T}(t-\tau)x\right\|_X \to 0$$

uniformly for $\beta \in [\alpha, +\infty)$, x in bounded sets, and $t \in \mathbb{R}$. Therefore, again using Lebesgue's theorem, we obtain

$$\left\|[T(\tau) - Id]e^{-\beta \cdot}\tilde{T}(\cdot - \tau)\right\|_{L^1(\mathbb{R}, \mathcal{L}(X))} \to 0$$

uniformly in $\beta \in [\alpha, +\infty)$.

Summarizing our results, we see that under either assumption

$$\left\|\mathcal{F}^{-1}I_1^{\beta+i\cdot}(\tau)\right\|_{L^1(\mathbb{R}, X)} \to 0 \text{ as } \tau \to 0, \tag{7.14}$$

and the convergence is uniform for $\beta \in [\alpha, +\infty)$.

Now we estimate the second term in Equation (7.12). Let $x \in D(B)$ and $t \in \mathbb{R}$ be arbitrary. Then it holds that

$$-\mathcal{F}^{-1}(I_2^{\beta+i\cdot}(\tau))(t)x$$

$$= \frac{1}{2\pi i}\int_{-\infty}^{+\infty} e^{itr}\int_{-\tau}^{0} CS(-s)Ce^{(\beta+ir)(s-\tau)}R(\beta+ir, B+C)x\,ds\,dr$$

$$= \int_{-\tau}^{0} CS(-s)C\frac{1}{2\pi i}\int_{-\infty}^{+\infty} e^{itr}e^{(\beta+ir)(s-\tau)}R(\beta+ir, B+C)x\,dr\,ds$$

$$= \int_{-\tau}^{0} CS(-s)Ce^{\beta(s-\tau)}\frac{1}{2\pi i}\int_{-\infty}^{+\infty} e^{ir(t+s-\tau)}R(ir, B+C-\beta)x\,dr\,ds$$

$$= \int_{-\tau}^{0} CS(-s)Ce^{\beta(s-\tau)}e^{-\beta(t+s-\tau)}\tilde{T}(t+s-\tau)x\,ds$$

$$= \int_{-\tau}^{0} CS(-s)Ce^{-\beta t}\tilde{T}(t+s-\tau)x\,ds.$$

The interchange of the integrals can be justified since

$$e^{-\beta t}\tilde{T}(t)x = \lim_{r \to +\infty} \int_{-r}^{r} e^{ist} R(is, B + C - \beta)\, ds,$$

uniformly for t in compact intervals.

From the above identity we obtain for the operator norm

$$\left\| \mathcal{F}^{-1}(I_2^{\beta+i\cdot}(\tau))(t) \right\|_{\mathcal{L}(X)} \leq \int_{-\tau}^{0} \left\| CS(-s)Ce^{-\beta t}\tilde{T}(t+s-\tau) \right\|_{\mathcal{L}(X)}\, ds.$$

Therefore the L^1-estimate is obtained as follows. By Fubini's Theorem

$$\begin{aligned}
&\left\| \mathcal{F}^{-1}(I_2^{\beta+i\cdot}(\tau)) \right\|_{L^1(\mathbb{R},\mathcal{L}(X))} \\
&= \int_{-\infty}^{+\infty} \int_{-\tau}^{0} \left\| CS(-s)Ce^{-\beta t}\tilde{T}(t+s-\tau) \right\|_{\mathcal{L}(X)}\, ds\, dt \\
&= \int_{-\tau}^{0} \int_{\tau-s}^{+\infty} \left\| CS(-s)Ce^{-\beta t}T(t+\tau-s) \right\|_{\mathcal{L}(X)}\, dt\, ds \\
&\leq \|C\|^2 M' \int_{-\tau}^{0} \int_{\tau-s}^{+\infty} \left\| e^{-\beta t}T(t+\tau-s) \right\|_{\mathcal{L}(X)}\, dt\, ds \\
&\leq \|C\|^2 MM' \int_{-\tau}^{0} \int_{\tau-s}^{+\infty} e^{-\beta t + \omega(t+s-\tau)}\, dt\, ds \\
&= \frac{\|C\|^2 MM'}{\beta - \omega} \int_{-\tau}^{0} e^{-\omega(\tau-s)}\, ds \leq \frac{K'}{\alpha - \omega}\tau \to 0 \text{ as } \tau \to 0,
\end{aligned} \tag{7.15}$$

again the convergence being uniform in $\beta \in [\alpha, +\infty)$.

Combining our results in Equations (7.14) and (7.15), we obtain that

$$\left\| \mathcal{F}^{-1}\Phi_{\beta+i\cdot}R(\beta+i\cdot, B+C)\mathcal{F} \right\|_{\mathcal{M}_p^{\mathcal{L}(X)}} < 1,$$

whenever τ is sufficiently small. This completes the proof by Corollary 5.30.
□

Remark 7.9. To show that the estimates on I_1^ω and I_2^ω are not sharp, take $B = 0$ and $C = d \cdot Id$ for $d < 0$. The direct calculation of Equation (7.5) shows that the solutions of $(DE)_\tau$ are exponentially stable if $|d|\tau < 1$.

However, applying directly the spectral characterization of Lemma 3.19 and using that $\sigma(B + \Phi_\lambda) = \{d \cdot e^{-\lambda \tau}\}$, we obtain (see Hale and Verduyn Lunel [103, page 135]) that the solutions decay exponentially if

$$|d|\tau < \frac{\pi}{2}, \tag{7.16}$$

which is the best possible estimate.

This result can be generalized in the following way.

Corollary 7.10. *Let X be a Hilbert space, $(B, D(B))$ selfadjoint and negative semidefinite, and let $C := d \cdot Id$ for $d < 0$. Then the solutions of* $(\mathrm{DE})_\tau$ *decay exponentially if*

$$|d|\tau < \frac{\pi}{2}. \tag{7.17}$$

If $0 \in \sigma(B)$, then this is the best possible estimate.

Proof. Assume that the Inequality (7.17) holds, but that there exists $\lambda = \alpha + i\beta$ with $\alpha, \beta \in \mathbb{R}$ and $\alpha \geq 0$ such that

$$\alpha + i\beta = \mu + de^{-\alpha \tau}\left(\cos(-\beta \tau) + i\sin(-\beta \tau)\right),$$

where $\mu \in \sigma(B)$ and therefore $\mu \leq 0$.

Taking imaginary parts it follows that $|\beta| \leq |d|e^{-\alpha \tau} \leq |d|$ by our assumptions and thus

$$|\beta \tau| < \frac{\pi}{2}$$

and therefore $\cos(-\beta \tau) > 0$. This is a contradiction since it means that $\alpha = \mu + de^{-\alpha \tau}\cos(-\beta \tau) < 0$.   □

Now we continue with the investigation of the effect of small delays for equations but admit unbounded operators in the delay terms (compare Section 3.4). To do this we make Conditions (3.42) and (3.43) hold on the operator $(B, D(B))$ from page 71. The problem is then the following: assume that the uniform exponential stability is known for the solutions of the equation

$$(\mathrm{DE})_0 \qquad \begin{cases} u'(t) = (B + C)u(t), & t \geq 0, \\ u(0) = x, \\ u_0 = f, \end{cases}$$

where $C \in \mathcal{L}(D(-B + \delta)^\vartheta, X)$ for $0 < \vartheta < 1$ and some $\delta > \omega_0(B)$.

The question is whether the same stability holds for the solutions of the equation

$$\text{(DE)}_\tau \quad \begin{cases} u'(t) = Bu(t) + Cu(t - \tau), & t \geq 0, \\ u(0) = x, \\ u_0 = f, \end{cases}$$

where $\tau > 0$.

Using the same idea as in the case of bounded delay operators, the analogous versions of the Variation of Parameters Formula (Equations (7.1) and (7.2)) yield the following. The equation (DE)_τ has the form

$$u'(t) = (B + C)\, u(t) + \Phi u_t, \tag{7.18}$$

where $\Phi \in \mathcal{L}(W^{1,p}([-1,0], D(-B+\delta)^\vartheta), X)$ is defined by

$$\Phi f := -C\,[S(\tau) - Id]\,\delta_{-\tau} f - \int_{-\tau}^0 CS(-s)C\delta_{s-\tau} f ds. \tag{7.19}$$

The analogous formula to Equation (7.5) is

$$\Phi_\lambda R(\lambda, B + C)x = -C\,[S(\tau) - Id]\,e^{-\lambda \tau} R(\lambda, B + C)x$$
$$- \int_{-\tau}^0 CS(-s)Ce^{-\lambda(s-\tau)} R(\lambda, B + C)x ds. \tag{7.20}$$

Having developed the appropriate techniques, we formulate the results on the stability, the counterpart of Theorem 7.5. The results on the hyperbolicity can be formulated analogously with the same proof.

Theorem 7.11. *Assume that Conditions (3.42) and (3.43) hold, $C \in \mathcal{L}(D(-B+\delta)^\vartheta, X)$ for $0 < \vartheta < 1$, and that the semigroup generated by the operator $(B+C, D(B))$ is uniformly exponentially stable in the Banach space X. Then there exists $\kappa > 0$ such that the solutions of (DE)_τ decay exponentially for $\tau \in (0, \kappa)$. Thus, the stability is not sensitive to small delays.*

Proof. First we show that under these conditions $(B+C, D(B))$ generates an analytic semigroup. From Theorem 1.42 it follows that $(C, D(C))$ is relatively B-bounded with bound 0. Then the analyticity follows from Theorem 1.41.

The other parts of the proof are analogous to the proof of Theorem 7.5. Define

$$I_1^\omega(\tau) := C\,[S(\tau) - Id]\,e^{-i\omega\tau} R(i\omega, B + C) \tag{7.21}$$

and

$$I_2^\omega(\tau)x := \int_{-\tau}^0 CS(-s)Ce^{-i\omega(s-\tau)}R(i\omega, B+C)x\,ds. \qquad (7.22)$$

We show that there exists $\kappa > 0$ such that $\sup_{\omega \in \mathbb{R}} \|I_i^\omega(\tau)\| < \frac{1}{2}$ for $i = 1, 2$ and all $\tau \in (0, \kappa)$. Then, using Corollary 4.17 and Equation (7.20), the assertion follows since

$$\sup_{\omega \in \mathbb{R}} \|\Phi_{i\omega}R(i\omega, B+C)\| \le \sup_{\omega \in \mathbb{R}} \|I_1^\omega(\tau)\| + \sup_{\omega \in \mathbb{R}} \|I_2^\omega(\tau)\| < 1.$$

The estimate on I_2^ω is

$$\|I_2^\omega(\tau)\| \le \frac{K}{1-\vartheta}\tau^{1-\vartheta}\|CR(i\omega, B+C)\|, \qquad (7.23)$$

where we used the fact that $\|CS(t)\| \le \frac{K}{t^\vartheta}$. Since

$$\|CR(i\omega, B+C)\| \le \|C(B+C)^{-1}\| \cdot \|(B+C)R(i\omega, B+C)\|$$

is uniformly bounded for $\omega \in \mathbb{R}$ because of the analyticity of the semigroup generated by $(B+C, D(B))$, there exists $\kappa_2 > 0$ such that for all $\tau \in (0, \kappa_2)$ the estimate $\sup_{\omega \in \mathbb{R}} \|I_2^\omega(\tau)\| < \frac{1}{2}$ holds.

The estimate on I_1^ω is

$$\begin{aligned}
\|I_1^\omega(\tau)\| &\le \|C(S(\tau) - Id)R(i\omega, B+C)\| \\
&\le \|C(S(\tau) - Id)R(\lambda, B)\| \cdot \|(\lambda - B)(B+C)^{-1}\| \qquad (7.24) \\
&\quad \times \|(B+C)R(i\omega, B+C)\|,
\end{aligned}$$

where $\lambda > \max\{\omega_0(B), 0\}$ is fixed.

So we only have to consider $C(S(\tau) - Id)R(\lambda, B)$.

However, it follows from

$$\begin{aligned}
\|C(S(\tau) - Id)R(\lambda, B)\| &\le \|CR(\lambda, B)\| \cdot \|S(\tau)\|(1 - e^{\lambda\tau}) \\
&\quad + \|\int_0^\tau e^{-\lambda s}CS(s)\,ds\| \\
&\le \tau L + \tau^{1-\vartheta}K'
\end{aligned}$$

with suitable constants $L, K' > 0$, that there exists $\kappa_1 > 0$ such that for all $\tau \in (0, \kappa_1)$ the estimate $\|I_1^\omega(\tau)\| < \frac{1}{2}$ holds uniformly in ω.

The proof can be finished by choosing $\kappa := \min\{\kappa_1, \kappa_2\}$. $\qquad \square$

7.2 Notes and References

Section 7.1 is mainly based on the articles [12, 15, 18]. Theorem 7.8 was obtained in collaboration with B. Farkas.

The first to examine the effect of small delays was R. Datko [44–48]. It is known that for finite-dimensional equations, the stability cannot be destroyed by small delays. In addition, there exists an extensive literature on delay-dependent stability conditions; see, e.g., the papers by Győri and coauthors [95, 97]. For similar questions in the nonautonomous parabolic case we refer to Gühring, Räbiger, and Schnaubelt [94] and Schnaubelt [189]. There is a recent exposition of this problem by Hale and Verduyn Lunel [104, 105], where many examples of functional differential and difference equations are considered. A control theoretic investigation using transfer functions is made for compact feedback in Rebarber and Townly [171]. See also the example in Section 10.3 for the small delay problem with unbounded operators in the delay term.

Chapter 8

More Asymptotic Properties

Up to now we were concerned with stability and hyperbolicity of delay equations. In this chapter, we investigate other asymptotic properties such as strong stability, asymptotic almost periodicity, and ergodicity. The results of this chapter were obtained in collaboration with V. Casarino [29].

In Section 8.1, we study the asymptotic behavior of perturbed semigroups. In particular, we investigate under which conditions the perturbed semigroup inherits the same asymptotic behavior as the unperturbed one.

In Section 8.2, we apply the abstract results of Section 8.1 to delay semigroups.

8.1   Asymptotic Properties of Perturbed Semigroups

Besides stability, there are other properties that describe the asymptotic behavior of semigroups. Such properties are defined as subspaces of the space of bounded continuous Banach space valued functions as follows.

Let X be a Banach space and $\mathbb{R}_+$ denote the interval $[0, +\infty)$. For a function $f : \mathbb{R}_+ \longrightarrow X$, we denote by $H(f)$ the set of all translates $\{f(\cdot + \omega) : \omega \in \mathbb{R}_+\}$. Let $C_b(\mathbb{R}_+, X)$ be the Banach space of all bounded continuous functions from $\mathbb{R}_+$ to X, endowed with the uniform norm. A function $f \in C_b(\mathbb{R}_+, X)$ is called *asymptotically almost periodic* (abbreviated as a.a.p.) if $H(f)$ is relatively compact in $C_b(\mathbb{R}_+, X)$.

Then the following decomposition theorem holds.

Theorem 8.1. *A function $f \in C_b(\mathbb{R}_+, X)$ is asymptotically almost periodic if and only if one of the following two equivalent conditions is satisfied.*

(i) *There exists a unique almost periodic function $g \in C_b(\mathbb{R}, X)$ and a unique $h \in C_b(\mathbb{R}_+, X)$, vanishing at infinity, such that $f = h + g_{|\mathbb{R}_+}$.*

(ii) For every $\varepsilon > 0$, there exists $\Lambda > 0$ and $K \geq 0$ such that every interval of length Λ contains some τ for which the inequality

$$\|f(t + \tau) - f(t)\| \leq \varepsilon$$

holds whenever $t, t + \tau \geq K$.

We recall that $C_b(\mathbb{R}, X)$ is the space of all bounded continuous functions from $\mathbb{R}$ to X and that a function $g \in C_b(\mathbb{R}, X)$ is said to be *almost periodic* if the set $\{g(\cdot + \omega) : \omega \in \mathbb{R}\}$ is relatively compact in $C_b(\mathbb{R}, X)$.

A strongly continuous semigroup is called *strongly asymptotically almost periodic* if the function $t \mapsto T(t)x$ from $\mathbb{R}_+$ to X is a.a.p. for every $x \in X$.

A continuous bounded function $f : \mathbb{R}_+ \to X$ is called *weakly asymptotically almost periodic in the sense of Eberlein* if $H(f)$ is weakly relatively compact in $C_b(\mathbb{R}_+, X)$.

A function $f \in C_{ub}(\mathbb{R}_+, X)$ is said to be *uniformly ergodic* if the limit

$$\lim_{\alpha \to 0^+} \alpha \int_0^\infty e^{-\alpha s} f(\cdot + s) \, ds \tag{8.1}$$

exists and defines an element of $C_{ub}(\mathbb{R}_+, X)$.

A function $f \in C_{ub}(\mathbb{R}_+, X)$ is said to be *totally (uniformly) ergodic* if the function $e^{i\theta \cdot} f(\cdot)$ is uniformly ergodic, for all $\theta \in \mathbb{R}$. Since f is uniformly bounded, this is also equivalent to the existence of the Cesáro limit

$$\lim_{t \to +\infty} \frac{1}{t} \int_0^t e^{i\theta s} f(\cdot + s) \, ds \tag{8.2}$$

in $C_{ub}(\mathbb{R}_+, X)$.

We recall that a closed subspace $\mathcal{F}$ of $C_{ub}(\mathbb{R}_+, X)$ is said to be *translation-invariant* if

$$\mathcal{F} = \{f \in C_{ub}(\mathbb{R}_+, X) : \ f(\cdot + t) \in \mathcal{F}\} \text{ for all } t \geq 0.$$

A closed subspace $\mathcal{F}$ of $C_{ub}(\mathbb{R}_+, X)$ is said to be *operator-invariant* if $M \circ f \in \mathcal{F}$ for every $f \in \mathcal{F}$ and $M \in \mathcal{L}(X)$, where $M \circ f$ is defined by $(M \circ f)(t) = M(f(t))$, $t \geq 0$.

The following classes of X-valued functions are closed, translation- and operator-invariant subspaces of $C_{ub}(\mathbb{R}_+, X)$:

- the space $C_0(\mathbb{R}_+, X)$ of all continuous functions vanishing at infinity,

- the class of all asymptotically almost periodic functions from $\mathbb{R}_+$ to X,

- the class of all weakly asymptotically almost periodic functions in the sense of Eberlein,

- the class of uniformly ergodic functions from $\mathbb{R}_+$ to X,

- the class of totally (uniformly) ergodic functions from $\mathbb{R}_+$ to X.

The following lemma will be crucial.

Lemma 8.2. *Let $L : \mathbb{R}_+ \to \mathcal{L}(X)$ be a bounded function continuous in the strong operator topology. Let $\mathcal{F}$ be a closed, translation-invariant subspace of $C_{ub}(\mathbb{R}_+, X)$ such that the functions $\mathbb{R}_+ \ni t \mapsto L(t)x \in X$ belong to $\mathcal{F}$ for every $x \in X$.*
If $g \in L^1(\mathbb{R}, X)$, then $(L * g)_{|\mathbb{R}_+} \in \mathcal{F}$.

Here, the convolution between L and g is defined as

$$(L * g)(t) = \int_0^{+\infty} L(s)g(t - s)ds = \int_{-\infty}^{t} L(t - s)g(s)ds, \, t \geq 0. \quad (8.3)$$

Our purpose now is to find sufficient conditions such that the semigroup inherits the same asymptotic behavior as the unperturbed one. More precisely, assume to have a semigroup $(T(t))_{t \geq 0}$ on a Banach space X with generator $(A, D(A))$ such that the maps $t \mapsto T(t)x$ belong to a closed, translation- and operator-invariant subspace $\mathcal{F}$ of $C_{ub}(\mathbb{R}_+, X)$ for all $x \in X$. Let $C \in \mathcal{L}(D(A), X)$ such that $(A + C, D(A))$ generates a semigroup $(U(t))_{t \geq 0}$ on X. The question is whether $t \mapsto U(t)x$ still belongs to $\mathcal{F}$ for all $x \in X$.

In general the answer is negative, as one can easily see by taking $\mathcal{F} := C_0(\mathbb{R}_+, X)$, however, it is possible to show that under certain assumptions the asymptotic behavior persists.

We can do this by means of the Dyson-Phillips Series introduced in Equation (1.15). First we find conditions that ensure the uniform convergence of the Dyson-Phillips Series on $\mathbb{R}_+$.

Theorem 8.3. *Let $T := (T(t))_{t \geq 0}$ be a bounded, strongly continuous semigroup on the Banach space X with generator $(A, D(A))$. Let $C \in \mathcal{L}(D(A), X)$ and assume that there is $0 < q < 1$ such that*

$$\int_0^t \|CT(s)x\| \, ds \leq q\|x\| \quad (8.4)$$

for all $t \geq 0$ and $x \in D(A)$. Then, $(A + C, D(A))$ generates a strongly continuous semigroup $\mathcal{U} := (U(t))_{t \geq 0}$, and the Dyson-Phillips Series (1.15) converges uniformly on $\mathbb{R}_+$.

Proof. The sum $(A + C, D(A))$ generates a strongly continuous semigroup $(U(t))_{t \geq 0}$ with

$$U(t)x = \sum_{0}^{\infty} (V^n \mathfrak{I})(t)x \quad \text{for all} \quad x \in X, \, t \geq 0. \qquad (1.15)$$

This follows from Theorem 1.37.

We only show the uniform convergence of Equation (1.15) on $\mathbb{R}_+$.

Let $C_b(\mathbb{R}_+, \mathcal{L}_s(X))$ be the space of all bounded, strongly continuous functions from $\mathbb{R}_+$ to $\mathcal{L}(X)$. Let $F \in C_b(\mathbb{R}_+, \mathcal{L}_s(X))$ and $t \geq 0$. It follows from Equation (8.4) that

$$\|(VF)(t)x\| \leq \|F\|_\infty \int_0^t \|CT(s)x\| \, ds \leq \|F\|_\infty \, q \, \|x\|$$

for every $x \in D(A)$, where V is the abstract Volterra operator defined in Equations (1.13) and (1.14). Therefore, we obtain

$$\|(VF)(t)\| \leq q \|F\|_\infty \qquad (8.5)$$

for every $t \geq 0$, so that the operator $V : C_b(\mathbb{R}_+, \mathcal{L}_s(X)) \to C_b(\mathbb{R}_+, \mathcal{L}_s(X))$ has norm

$$\|V\| \leq q < 1.$$

This implies that the Neumann series

$$\sum_{n=0}^{\infty} V^n$$

converges to a bounded operator on $C_b(\mathbb{R}_+, \mathcal{L}_s(X))$. In particular, the series

$$\sum_{n=0}^{\infty} V^n F$$

converges in the norm of $C_b(\mathbb{R}_+, \mathcal{L}_s(X))$, hence uniformly on $\mathbb{R}_+$ for all $F \in C_b(\mathbb{R}_+, \mathcal{L}_s(X))$. □

We can now state the main result of this section.

Theorem 8.4. *Assume that the conditions of Theorem 8.3 are satisfied. If $\mathfrak{F}$ is a translation-invariant closed subspace of $C_{ub}(\mathbb{R}_+, X)$ and if $t \mapsto T(t)x$ belongs to $\mathfrak{F}$ for every $x \in X$, then $t \mapsto U(t)x$ belongs to $\mathfrak{F}$ for every $x \in X$.*

Proof. By Theorem 8.3, we have

$$\mathcal{U} = \sum_{n=0}^{+\infty} V^n \mathcal{T},$$

the convergence being in the norm of $C_b(\mathbb{R}_+, \mathcal{L}_s(X))$, hence uniformly on $\mathbb{R}_+$.

Now we show that the functions

$$t \mapsto \left(V^n \mathcal{T}\right)(t) x$$

belong to $\mathcal{F}$ for every $n \in \mathbb{N}$ and $x \in X$.

We consider first the case $n = 1$. Take $x \in D(A)$ and define

$$g(t) := \begin{cases} CT(t)x & \text{if } t \geq 0 \\ 0 & \text{if } t < 0 . \end{cases}$$

Since

$$\int_{-\infty}^{+\infty} \|g(t)\| \, dt = \int_{0}^{+\infty} \|CT(t)x\| \, dt \leq q \, \|x\| < \|x\|,$$

it follows that g belongs to $L^1(\mathbb{R}, X)$. Since $\mathcal{T}$ is bounded, strongly continuous and, by hypothesis, $t \mapsto T(t)x$ belongs to $\mathcal{F}$ for every $x \in X$, Lemma 8.2 entails that $(\mathcal{T} * g)_{|\mathbb{R}_+} \in \mathcal{F}$.

Observe now that

$$(\mathcal{T} * g)_{|\mathbb{R}_+}(t) = \int_{-\infty}^{t} T(t - s)g(s)ds = \int_{0}^{t} T(t - s)CT(s)x \, ds.$$

Since $x \in D(A)$, it follows that

$$(V\mathcal{T})(t)x = \int_{0}^{t} T(t - s)CT(s)x \, ds,$$

so that the function $t \mapsto \left(V\mathcal{T}\right)(t) x$ belongs to $\mathcal{F}$ for every $x \in D(A)$.

Take now $x \in X$ and a sequence $(x_j) \subset D(A)$ such that $x_j \to x$.

Since $V\mathcal{T}$ belongs to $C_b(\mathbb{R}_+, \mathcal{L}_s(X))$, we conclude that $\{(V\mathcal{T})(t) : t \geq 0\}$ is uniformly bounded on $\mathbb{R}_+$, and therefore the sequence $\{(V\mathcal{T})(t)x_j : j \in \mathbb{N}\}$ converges uniformly on $\mathbb{R}_+$ to $(V\mathcal{T})(t)x$. Since $\mathcal{F}$ is a closed subspace of $C_{ub}(\mathbb{R}_+, X)$, we conclude that the limit function belongs to $\mathcal{F}$ as well, proving the assertion.

Suppose now that the maps $t \mapsto \left(V^{n-1}\mathcal{T}\right)(t)x$ belong to $\mathcal{F}$ for all $x \in X$. Then, if $x \in D(A)$ and $t \in \mathbb{R}_+$, we have

$$(V^n \mathcal{T})(t)x = V\left(V^{n-1}\mathcal{T}\right)(t)x = \int_{0}^{t} \left(V^{n-1}\mathcal{T}\right)(t - s)CT(s)x \, ds.$$

Since $V^{n-1}\mathcal{T}$ belongs to $C_b(\mathbb{R}_+, \mathcal{L}_s(X))$, the operator-valued function $V^{n-1}\mathcal{T} : \mathbb{R}_+ \to \mathcal{L}(X)$ is strongly continuous. Moreover, by hypothesis, $(V^{n-1}\mathcal{T})(\cdot)x$ belongs to $\mathcal{F}$ for all $x \in X$.

Thus, we can apply Lemma 8.2, taking as L the operator $V^{n-1}\mathcal{T}$ and choosing g as above. Then

$$t \mapsto (L * g)_{|\mathbb{R}_+} (t) = \int_0^t (V^{n-1}\mathcal{T}) (t - s)CT(s)x \, ds$$

belongs to $\mathcal{F}$ for every $x \in D(A)$.

Therefore, every function $t \mapsto (V^n\mathcal{T})(t)x$ belongs to $\mathcal{F}$ for all $x \in D(A)$.

As in the case $n = 1$, we conclude that $(V^n\mathcal{T})(\cdot)x$ belongs to $\mathcal{F}$ for all $x \in X$.

To conclude the proof, we observe that every term of the series

$$U(\cdot)x = \sum_{n=0}^{+\infty}(V^n\mathcal{T})(\cdot)x \, , \; x \in X \, , \, t \geq 0$$

belongs to $\mathcal{F}$. Since $\mathcal{F}$ is closed in $C_{ub}(\mathbb{R}_+, X)$ and the above series converges uniformly on $\mathbb{R}_+$ by Theorem 8.3, every map $t \mapsto U(t)x$ belongs to $\mathcal{F}$. $\quad\square$

8.2 Asymptotic Properties of the Delay Semigroup

In this section, we investigate qualitative properties of the delay semigroup such as asymptotic almost periodicity and ergodicity.

Let $(B, D(B))$ be the generator of a strongly continuous semigroup $(T(t))_{t\geq 0}$ on a Banach space X. On $\mathcal{E}_p := X \times L^p([-1, 0], X)$, $1 \leq p < \infty$, we consider the operator $(\mathcal{A}_0, D(\mathcal{A}_0))$ defined in Equations (3.30) and (3.31) by

$$\mathcal{A}_0 := \begin{pmatrix} B & 0 \\ 0 & \frac{d}{d\sigma} \end{pmatrix}$$

with domain

$$D(\mathcal{A}_0) := \left\{ \left(\begin{smallmatrix} x \\ f \end{smallmatrix}\right) \in D(B) \times W^{1,p}([-1, 0], X) \; : \; f(0) = x \right\}.$$

As we already observed in Section 3.3.3, this operator generates the strongly continuous semigroup $(\mathcal{T}_0(t))_{t\geq 0}$ given by

$$\mathcal{T}_0(t) := \begin{pmatrix} S(t) & 0 \\ S_t & T_0(t) \end{pmatrix},$$

where $(T_0(t))_{t\geq 0}$ is the nilpotent left shift semigroup on $L^p([-1, 0], X)$ and $S_t : X \to L^p([-1, 0], X)$ is defined as

$$(S_t x)(\sigma) := \begin{cases} S(t + \sigma)x, & \sigma \geq -t, \\ 0, & \text{otherwise.} \end{cases}$$

Therefore, the semigroup $(\mathcal{T}_0(t))_{t\geq 0}$ on $\mathcal{E}_p$ is essentially given by the semi-group $(S(t))_{t\geq 0}$ on X. Thus it is reasonable to hope that $(\mathcal{T}_0(t))_{t\geq 0}$ shows the same asymptotic behavior as $(S(t))_{t\geq 0}$. Actually, this can be shown for most of the properties introduced in the previous section.

Lemma 8.5. *Assume that for each $x \in X$ the map $\mathbb{R}_+ \ni t \mapsto S(t)x \in X$ is*

(i) in $C_0(\mathbb{R}_+, X)$, or

(ii) asymptotically almost periodic, or

(iii) uniformly ergodic, or

(iv) totally uniformly ergodic;

then $\mathbb{R}_+ \ni t \mapsto \mathcal{T}_0(t)\left(\begin{smallmatrix}x\\f\end{smallmatrix}\right) \in \mathcal{E}_p$ has the same property for all $\left(\begin{smallmatrix}x\\f\end{smallmatrix}\right) \in \mathcal{E}_p$ and all $1 \leq p < \infty$.

Proof. Since all these classes are translation-invariant, it suffices to show the assertions for the map

$$\mathbb{R}_+ \ni t \mapsto \mathcal{T}_0(t+1)\left(\begin{smallmatrix}x\\f\end{smallmatrix}\right) = \left(\begin{smallmatrix}S(t+1) & 0\\S_{t+1} & 0\end{smallmatrix}\right)\left(\begin{smallmatrix}x\\f\end{smallmatrix}\right) = \left(\begin{smallmatrix}S(t+1)x\\S_{t+1}x\end{smallmatrix}\right) \in \mathcal{E}_p.$$

Let $x \in X$.

(i) If $(t \mapsto S(t)x) \in C_0(\mathbb{R}_+, X)$, then for every $\varepsilon > 0$ there exists a $T > 0$ such that $\|S(t)x\| < \varepsilon$ for all $t > T$. Then we have

$$\|S_{t+1}x\|_p < \varepsilon,$$

and hence

$$\left\|\mathcal{T}_0(t+1)\left(\begin{smallmatrix}x\\f\end{smallmatrix}\right)\right\| < 2\varepsilon$$

for all $t > T$.

(ii) If $t \mapsto S(t)x$ is asymptotically almost periodic, it follows from Theorem 8.1 that for every $x \in X$ and for every $\varepsilon > 0$ there exist $\Lambda > 0$ and $K \geq 0$ such that each interval of length Λ contains some τ for which

$$\|S(t+\tau)x - S(t)x\| \leq \varepsilon$$

holds whenever $t, t+\tau \geq K$.

Hence

$$\int_{-1}^{0} \|(S_{t+1+\tau}x)(\sigma) - (S_{t+1}x)(\sigma)\|^p d\sigma$$

$$= \int_{-1}^{0} \|S(t+1+\tau+\sigma)x - S(t+1+\sigma)x\|^p d\sigma$$

$$= \int_{0}^{1} \|S(t+\tau+\sigma)x - S(t+\sigma)x\|^p d\sigma \leq \varepsilon^p$$

whenever $t, t + \tau \geq K$. This shows that the map $t \mapsto S_{t+1}x$ from $\mathbb{R}_+$ to $L^p([-1,0], X)$ is asymptotically almost periodic for every $x \in X$.

(iii) The limit

$$F(\cdot) := \lim_{\alpha \searrow 0} \alpha \int_0^\infty e^{-\alpha t} S(\cdot + t) x \, dt$$

exists in $C_{ub}(\mathbb{R}_+, X)$. For $s \geq 0$ we define $F_{s+1} : [-1, 0] \longrightarrow X$ by $F_{s+1}(\sigma) := F(s + 1 + \sigma)$. It is easy to see that the map $\mathbb{R}_+ \ni t \mapsto F_{s+1}$ belongs to $C_{ub}(\mathbb{R}_+, L^p([-1,0], X))$. Then

$$\lim_{\alpha \searrow 0} \sup_{s \in \mathbb{R}_+} \|(\alpha \int_0^\infty e^{-\alpha t} S_{\cdot+1+t} \, x \, dt)(s) - F_{s+1}\|_p^p$$

$$= \lim_{\alpha \searrow 0} \sup_{s \in \mathbb{R}_+} \int_{-1}^0 \|\alpha \int_0^\infty e^{-\alpha t} S(s + 1 + t + \sigma) x \, dt - F(s + 1 + \sigma)\|^p \, d\sigma$$

$$= 0$$

since $\|\alpha \int_0^\infty e^{-\alpha t} S(s+1+t+\sigma)x \, dt - F(s+1+\sigma)\| \to 0$ as $\alpha \searrow 0$ uniformly for $s \in \mathbb{R}_+$.

(iv) The proof is analogous to (iii). $\square$

Consider now the perturbing operator $(\mathcal{B}, D(\mathcal{B}))$ on $\mathcal{E}_p$ defined in Equation (3.32) by

$$\mathcal{B} := \begin{pmatrix} 0 & \Phi \\ 0 & 0 \end{pmatrix}, \qquad D(\mathcal{B}) := D(\mathcal{A}_0),$$

where Φ is a bounded linear operator from $W^{1,p}([-1,0], X)$ to X. It follows that

$$\int_0^t \|\mathcal{B}\mathcal{T}_0(s)\left(\begin{smallmatrix} x \\ f \end{smallmatrix}\right)\| \, ds = \int_0^t \|\Phi(S_s x + T_0(s)f)\| \, ds$$

for all $t \geq 0$ and we can state the following theorem.

Theorem 8.6. *Assume that for each $x \in X$ the map $\mathbb{R}_+ \ni t \mapsto S(t)x$ is in one of the classes (i)–(iv) of Lemma 8.5. Moreover, let $1 \leq p < \infty$ and assume that there exists a constant $0 \leq q < 1$ such that*

$$\int_0^t \|\Phi(S_s x + T_0(s)f)\| \, ds \leq q \left\|\left(\begin{smallmatrix} x \\ f \end{smallmatrix}\right)\right\|$$

for all $t \geq 0$ and all $\left(\begin{smallmatrix} x \\ f \end{smallmatrix}\right) \in D(\mathcal{A}_0) \subset \mathcal{E}_p$. Then the operator $\mathcal{A} := \mathcal{A}_0 + \mathcal{B}$ with domain $D(\mathcal{A}) := D(\mathcal{A}_0)$ generates a strongly continuous semigroup $(\mathcal{T}(t))_{t \geq 0}$, and for each $\left(\begin{smallmatrix} x \\ f \end{smallmatrix}\right) \in \mathcal{E}_p$ the map $t \mapsto \mathcal{T}(t)\left(\begin{smallmatrix} x \\ f \end{smallmatrix}\right)$ is in the same class as $t \mapsto S(t)x$.

Proof. From Lemma 8.5 we know that $t \mapsto \mathcal{T}_0(t)\left(\begin{smallmatrix} x \\ f \end{smallmatrix}\right)$ is in the same class as $t \mapsto S(t)x$ for all $\left(\begin{smallmatrix} x \\ f \end{smallmatrix}\right) \in \mathcal{E}_p$. Moreover, $(\mathcal{T}_0(t))_{t \geq 0}$ and $\mathcal{B}$ satisfy Equation (8.4) of Theorem 8.3. So, by Theorem 8.4 we have that the operator $(\mathcal{A}, D(\mathcal{A}))$ generates a strongly continuous semigroup $(\mathcal{T}(t))_{t \geq 0}$ on $\mathcal{E}_p$ and $t \mapsto \mathcal{T}(t)\left(\begin{smallmatrix} x \\ f \end{smallmatrix}\right)$ is in the same class as $t \mapsto \mathcal{T}_0(t)\left(\begin{smallmatrix} x \\ f \end{smallmatrix}\right)$ for all $\left(\begin{smallmatrix} x \\ f \end{smallmatrix}\right) \in \mathcal{E}_p$. $\square$

We can now state a criterion for, e.g., strong stability of the solution of the delay equation (DE).

Corollary 8.7. *Let $(B, D(B))$ be the generator of a strongly stable semigroup $(T(t))_{t \geq 0}$ on a Banach space X. Let $1 \leq p < \infty$ and $\Phi : W^{1,p}([-1,0], X) \longrightarrow X$ be a bounded linear operator such that there exists a constant $0 \leq q < 1$ with*

$$\int_0^t \|\Phi(S_s x + T_0(s)f)\| \, ds \leq q \left\| \left(\begin{smallmatrix} x \\ f \end{smallmatrix}\right) \right\|$$

for all $t \geq 0$ and $\left(\begin{smallmatrix} x \\ f \end{smallmatrix}\right) \in D(\mathcal{A}_0)$. Then the solutions of the delay equation $(DE)_p$ vanish at infinity.

We can also state a criterion for asymptotic almost periodicity of the solutions of $(DE)_p$.

Corollary 8.8. *Let $(B, D(B))$ be the generator of a strongly asymptotic almost periodic semigroup $(T(t))_{t \geq 0}$ on a Banach space X. Let $1 \leq p < \infty$ and $\Phi : W^{1,p}([-1,0], X) \longrightarrow X$ be a bounded linear operator such that there exists a constant $0 \leq q < 1$ with*

$$\int_0^t \|\Phi(S_s x + T_0(s)f)\| \, ds \leq q \left\| \left(\begin{smallmatrix} x \\ f \end{smallmatrix}\right) \right\|$$

for all $t \geq 0$ and $\left(\begin{smallmatrix} x \\ f \end{smallmatrix}\right) \in D(\mathcal{A}_0)$. Then the solutions of the partial differential equation with delay $(DE)_p$ are asymptotically almost periodic.

Similarly, we obtain a criterion for (total) uniform ergodicity.

Corollary 8.9. *Let $(B, D(B))$ be the generator of a strongly (totally) uniformly ergodic semigroup $(T(t))_{t \geq 0}$ on a Banach space X. Let $1 \leq p < \infty$ and $\Phi : W^{1,p}([-1,0], X) \longrightarrow X$ be a bounded linear operator such that there exists a constant $0 \leq q < 1$ with*

$$\int_0^t \|\Phi(S_s x + T_0(s)f)\| \, ds \leq q \left\| \left(\begin{smallmatrix} x \\ f \end{smallmatrix}\right) \right\|$$

for all $t \geq 0$ and $\left(\begin{smallmatrix} x \\ f \end{smallmatrix}\right) \in D(\mathcal{A}_0)$. Then the solutions of the partial differential equation with delay $(DE)_p$ are (totally) uniformly ergodic.

Now we give an application of Theorem 8.6.

Example 8.10. Let $p > 1$ and $\Phi f := Cf(-1)$, where $C \in \mathcal{L}(X)$ with $\|C\| < 1$. Moreover, assume that there exists $0 < c < 1$ such that

$$\int_0^t \|CS(s)x\| \, ds \leq c\|x\|$$

for all $t \geq 0$ and $x \in D(B)$. Then

$$\int_0^t \|\Phi(S_s x + T_0(s)f)\| \, ds = \begin{cases} \int_0^t \|Cf(s-1)\| \, ds, & 0 \leq t \leq 1, \\ \int_0^1 \|Cf(s-1)\| \, ds + \int_1^t \|CS(s-1)x\| \, ds, \\ & t > 1. \end{cases}$$

For $t \in [0,1]$, we have

$$\int_0^t \|\Phi(S_s x + T_0(s)f)\| \, ds \leq \|C\| \, t^{1/p'} \, \|f\|_p$$

for $\frac{1}{p} + \frac{1}{p'} = 1$.

If, for every $x \in X$, the map $t \mapsto S(t)x$ is in one of the classes (i)–(iv) in Lemma 8.5, then the assumptions of Theorem 8.6 are satisfied and the solutions of the delay equation

$$\begin{cases} u'(t) = Bu(t) + Cu(t-1), & t \geq 0, \\ u(0) = x, \\ u_0 = f \end{cases} \tag{8.6}$$

are in the same class as $t \mapsto S(t)x$ for every $\left(\begin{smallmatrix} x \\ f \end{smallmatrix}\right) \in \mathcal{E}_p$.

For example, if $(B, D(B))$ generates a strongly asymptotically almost periodic semigroup $(S(t))_{t \geq 0}$ (see Appendix 8.1) and there exists $0 \leq c < 1$ such that

$$\int_0^t \|CS(s)x\| \, ds \leq c\|x\|$$

for all $t \geq 0$ and $x \in D(B)$, then the solutions of Equation (8.6) are asymptotically almost periodic for every initial value $\left(\begin{smallmatrix} x \\ f \end{smallmatrix}\right) \in \mathcal{E}_p$.

8.3   Notes and References

The results of this chapter were obtained in collaboration with V. Casarino and are taken from [29].

Theorem 8.1 is due to Ruess and Summers [179]. Equation 8.2 is taken from Arendt and Batty [5]. Lemma 8.2 was proved by Batty and Chill [22, Lemma 7.2].

For other results on the asymptotic properties of delay equations, see also Ruess [176] and Ruess and Summers [180], and for Volterra integro-differential equations, see Chill and Prüß [33].

Part IV
More Delay Equations

Chapter 9

Second-Order Cauchy Problems with Delay

After having developed the necessary tools to study the well-posedness and asymptotic behavior of first-order delay differential equations, we now turn our attention to second-order problems. We will treat second-order problems by rewriting them as a system of first-order equations and then applying the results obtained previously.

In Section 9.1 and 9.2, we recall recent results on well-posedness and uniform exponential stability for wave equations. These results were obtained in collaboration with K.-J. Engel [14].

In Section 9.3 and 9.4, we study well-posedness and uniform exponential stability of wave equations with delay.

9.1 Dissipative Wave Equations in a Hilbert Space

The aim of this section is to collect some results on dissipative wave equations in Hilbert spaces that we will need later on for the treatment of second-order Cauchy problems with delay.

We consider the complete second-order abstract Cauchy problem

$$\begin{cases} u''(t) = Du'(t) - C^*Cu(t), & t \geq 0, \\ u(0) = x, \quad u'(0) = y, \end{cases} \tag{ACP$_2$}$$

on a Hilbert space H with linear operators C and D.

Here we assume C to be densely defined, closed, and invertible, hence $A := C^*C$ is selfadjoint and positive definite; see Theorem 1.29.

Using the standard reduction $\mathcal{U} := \left(\begin{smallmatrix} u \\ u' \end{smallmatrix} \right)$ as in Fattorini [77, Section VIII], we can transform (ACP$_2$) into a first-order system

$$(\text{ACP}) \qquad \begin{cases} \mathcal{U}'(t) = \hat{\mathcal{B}}_0\,\mathcal{U}(t), & t \geq 0 \\ \mathcal{U}(0) = \left(\begin{smallmatrix} x \\ y \end{smallmatrix} \right) \end{cases}$$

for the matrix operator

$$\hat{\mathcal{B}}_0 := \left(\begin{smallmatrix} 0 & Id \\ -A & D \end{smallmatrix} \right) \tag{9.1}$$

with domain

$$D(\hat{\mathcal{B}}_0) := D(A) \times (D(D) \cap D(A^{\frac{1}{2}})) \tag{9.2}$$

in the product space

$$\mathcal{F} := H_{\frac{1}{2}} \times H, \quad \text{with} \quad H_{\frac{1}{2}} := \left(D(A^{\frac{1}{2}}), \| \cdot \|_{\frac{1}{2}} \right), \tag{9.3}$$

where $\| \cdot \|_{\frac{1}{2}}$ is defined by $\|x\|_{\frac{1}{2}} := \|A^{\frac{1}{2}}x\|$.

In many applications the square of this norm in $\mathcal{F}$ is proportional to the energy of the solutions. This is why, for this norm, this particular phase space is usually called the *energy space* associated to (ACP$_2$).

We remark that $D(C) = D(A^{\frac{1}{2}})$ and $C : H_{\frac{1}{2}} \to H$ is a unitary operator by Theorem 1.29.

This allows us to consider the unitary operator $\mathcal{S} := \left(\begin{smallmatrix} C & 0 \\ 0 & Id \end{smallmatrix} \right) \in \mathcal{L}(\mathcal{F}, \mathcal{W})$ for $\mathcal{W} := H \times H$. Defining the operator matrix

$$\mathcal{B}_0 := \left(\begin{smallmatrix} 0 & C \\ -C^* & D \end{smallmatrix} \right) \tag{9.4}$$

with domain

$$D(\mathcal{B}_0) := D(C^*) \times (D(C) \cap D(D)) \tag{9.5}$$

in the Hilbert space $\mathcal{W} = H \times H$, we obtain that $\mathcal{S}$ defines a unitary equivalence between $\mathcal{B}_0$ and $\hat{\mathcal{B}}_0$, i.e.,

$$\mathcal{B}_0 = \mathcal{S}\hat{\mathcal{B}}_0\mathcal{S}^{-1}. \tag{9.6}$$

To proceed, we assume that D is dissipative and that $D(C) \cap D(D)$ is dense in H. Then one easily verifies that $\mathcal{B}_0$ becomes densely defined and dissipative, hence closable (see Lemma 1.15) and we denote by $\mathcal{B} := \overline{\mathcal{B}_0}$ its closure. Using unpublished ideas of S.-Z. Huang, we are able to prove some conditions for the generator property of $\mathcal{B}$.

Theorem 9.1. *Let $C : D(C) \to H$ be a densely defined, invertible operator and $D : D(D) \to H$ a densely defined, dissipative operator on the Hilbert space H. Moreover, define $A := C^*C$ and assume further that $D(A) \cap D(D)$ is dense in H. Then the following assertions are equivalent:*

(i) The operator $\mathcal{B}$ generates a strongly continuous semigroup in the product space $\mathcal{W}$.

(ii) There exists $\lambda > 0$ such that the operator

$$P_\lambda := -\left(\overline{\lambda^2 - \lambda D + A}\right)$$

is invertible.

(iii) There exists $\lambda > 0$ such that the operator $G_\lambda := \overline{\lambda D - A}$ is a generator.

We call the family $(P_\lambda)_{\lambda > 0}$ the operator pencil or the characteristic operator polynomial associated to (ACP_2).

Proof. The equivalence of (ii) and (iii) can be verified easily using the Lumer-Phillips Theorem; see Theorem 1.16. We show that (i) and (ii) are equivalent.

Again, by the Lumer-Phillips Theorem, $\mathcal{B}$ is a generator if and only if there exists $\lambda > 0$ such that $\lambda - \mathcal{B}_0$ has dense range. Take $\left(\begin{smallmatrix}x_0\\y_0\end{smallmatrix}\right) \in \mathcal{W}$ such that

$$\langle \left(\begin{smallmatrix}x_0\\y_0\end{smallmatrix}\right), (\lambda - \mathcal{B}_0)\left(\begin{smallmatrix}x\\y\end{smallmatrix}\right)\rangle = 0 \quad \text{for all } \left(\begin{smallmatrix}x\\y\end{smallmatrix}\right) \in D(\mathcal{B}_0).$$

Then simple calculations show that this is equivalent to

$$\langle \lambda x_0, x \rangle + \langle y_0, C^* x \rangle = 0 \quad \text{for all} \quad x \in D(C) \tag{9.7}$$

and

$$\langle x_0, Cy \rangle + \langle y_0, Dy - \lambda y \rangle = 0 \quad \text{for all} \quad y \in D(C) \cap D(D). \tag{9.8}$$

By Equation (9.7) we see that $y_0 \in D((C^*)^*) = D(C)$ and $\lambda x_0 = -Cy_0$. Substituting this into Equation (9.8), we obtain

$$\langle y_0, \lambda^2 y - \lambda D y + A y \rangle = 0 \quad \text{for all} \quad y \in D(A) \cap D(D). \tag{9.9}$$

If P_λ is surjective, then $y_0 = 0$ and the surjectivity of $(\lambda - \mathcal{B})$ follows.

The other implication can be proved by a similar argument. $\square$

Let us remark that the operator G_λ in condition (iii) can be interpreted as the sum of the two operators $-A$ and λD. Thus, for the generator property of G_λ, we can use well-known perturbation arguments. Since this result is marginal for our further investigations, we omit the proof.

Corollary 9.2. Suppose that the assumptions of Theorem 9.1 are satisfied. Then the operator $(\mathcal{B}, D(\mathcal{B}))$ generates a contraction semigroup if one of the following conditions holds:

(i) D *is A-bounded.*

(ii) $\bar{D}$ *is m-dissipative and A is D-bounded.*

(iii) $\bar{D}$ *is m-dissipative and the "angle conditions"*

$$\Re\langle Dx, (Id - A)^{-1}x\rangle \leq 0 \quad \text{for all} \quad x \in D(D)$$

and

$$\Re\langle D^*x, (Id - A)^{-1}x\rangle \leq 0 \quad \text{for all} \quad x \in D(D^*)$$

hold.

(iv) $\bar{D}$ *is m-dissipative and the resolvents of A and $\bar{D}$ commute.*

To obtain other criteria for semigroup generation, we observe that by the Lumer-Phillips Theorem it is sufficient to find criteria ensuring the invertibility of $\mathcal{B}$.

If $\mathcal{B}$ is invertible, we can represent its inverse in matrix form

$$\mathcal{B}^{-1} = \begin{pmatrix} U & V \\ W & S \end{pmatrix} \in \mathcal{L}(\mathcal{W}),$$

where U, V, W, and S are bounded linear operators in H. We formally conclude from $\mathcal{B}^{-1}\mathcal{B}_0 = Id|_{D(\mathcal{B}_0)}$ that its entries satisfy

$$V = -C^{*-1},\; S = 0,\; W = C^{-1}, \text{ and} \tag{9.10}$$

$$Uz = C^{*-1}DC^{-1}z \quad \text{for all } z \in D(C^{*-1}DC^{-1}). \tag{9.11}$$

These operators give rise to a bounded inverse of $\mathcal{B}$ if and only if

$$C(D(D) \cap D(C)) \text{ is dense in } H \text{ and } \quad Q := \overline{(C^*)^{-1}DC^{-1}} \in \mathcal{L}(H). \tag{9.12}$$

In particular, these conditions are satisfied if D is C-bounded, i.e.,

$$D(C) \subseteq D(D), \tag{9.13}$$

or if

$$D = dA^{\alpha}, \quad \alpha \in [0,1],\; \Re d \leq 0. \tag{9.14}$$

Proposition 9.3. *Let C be a densely defined, invertible operator on the Hilbert space H. Moreover, assume that D is a densely defined, dissipative operator such that Equation (9.12) is satisfied. Then the operator $\mathcal{B} = \overline{\mathcal{B}_0}$ generates a contraction semigroup $(\mathcal{T}(t))_{t\geq 0}$ on $\mathcal{W}$.*

Proof. We define the operator

$$\widetilde{\mathcal{B}} := \begin{pmatrix} Id & 0 \\ 0 & C^* \end{pmatrix} \begin{pmatrix} 0 & Id \\ -Id & Q \end{pmatrix} \begin{pmatrix} Id & 0 \\ 0 & C \end{pmatrix} \tag{9.15}$$

with domain

$$D(\widetilde{\mathcal{B}}) := \left\{ \left(\begin{smallmatrix} x \\ y \end{smallmatrix} \right) \in H \times D(C) : x - QCy \in D(C^*) \right\}. \tag{9.16}$$

Then it is easy to check that $D(\mathcal{B}_0) \subseteq D(\widetilde{\mathcal{B}})$. Further, for $\left(\begin{smallmatrix} x \\ y \end{smallmatrix} \right) \in D(\mathcal{B}_0)$,

$$\widetilde{\mathcal{B}} \left(\begin{smallmatrix} x \\ y \end{smallmatrix} \right) = \begin{pmatrix} Cy \\ C^*(QCy - x) \end{pmatrix} = \begin{pmatrix} Cy \\ Dy - C^* x \end{pmatrix} = \mathcal{B}_0 \left(\begin{smallmatrix} x \\ y \end{smallmatrix} \right),$$

i.e., $\mathcal{B}_0 \subseteq \widetilde{\mathcal{B}}$.

Furthermore,

$$\widetilde{\mathcal{B}} D(\mathcal{B}_0) = \left\{ \begin{pmatrix} Cy \\ Dy - C^* x \end{pmatrix} \mid \left(\begin{smallmatrix} x \\ y \end{smallmatrix} \right) \in D(\mathcal{B}_0) \right\}$$

contains

$$C\left(D(D) \cap D(C) \right) \times H,$$

which is dense in $\mathcal{W}$ by Equation (9.12). Here we used again that

$$D(\mathcal{B}_0) = D(C^*) \times \left(D(C) \cap D(D) \right).$$

Thus $\widetilde{\mathcal{B}} D(\mathcal{B}_0)$ is dense in $\mathcal{W}$. Moreover, $\widetilde{\mathcal{B}}$ is invertible and hence $D(\mathcal{B}_0)$ is a core for $\widetilde{\mathcal{B}}$, i.e., $\mathcal{B} = \overline{\mathcal{B}_0} = \widetilde{\mathcal{B}}$. Since the assumptions on D and C imply that $\mathcal{B}_0$ is dissipative, we conclude by the Lumer-Phillips Theorem (see Theorem 1.16) that $\mathcal{B}$ generates a contraction semigroup. □

Let now the assumptions in Proposition 9.3 hold and consider a solution $\mathcal{U} = \left(\begin{smallmatrix} u \\ v \end{smallmatrix} \right)$ of (ACP). Then $v(t) = Cu'(t)$ and hence u is a solution of the second-order abstract Cauchy problem

$$u''(t) = C^*(QCu'(t) - Cu(t)), \quad t \geq 0, \tag{CP$_2$}$$

with appropriate initial conditions. If Equation (9.13) or (9.14) is satisfied, then this is equivalent to (ACP$_2$), but not in general. This means that solutions of (ACP$_2$) are always solutions of (CP$_2$), but there are solutions of (CP$_2$) not solving (ACP$_2$). Thus, the solutions of (CP$_2$) can be considered as generalized solutions of the second-order abstract Cauchy problem.

In order to apply the theory developed for unbounded delay operators in Section 3.4, we have to consider operator matrices arising from second-order Cauchy problems and generating analytic semigroups.

For this purpose, we recall a rather general result from Xiao and Liang [224, Section 6.4] without proof.

Assume that $\sigma(A) \subset [\sigma_0, \infty)$ and take a measurable function

$$f : [\sigma_0, \infty) \to (\sigma_1, \infty)$$

for some $\sigma_1 > 0$ with the properties

$$\lim_{s \to +\infty} f(s) = +\infty, \tag{9.17}$$

the function

$$\left(s \mapsto \frac{f(s)}{s+1} \right) \in L^\infty [\sigma_0, \infty), \tag{9.18}$$

and there exist $C_1, L_1 > 0$ such that

$$\left| f^2(s) - 4s \right| \geq C_1 \max \left\{ f^2(s), s \right\} \quad \text{for all} \quad s > L_1. \tag{9.19}$$

Assume further that $D(A) \subset D(D)$, $D = -D_1 + iD_2$, where D_1 is a positive definite, selfadjoint operator and D_2 is a selfadjoint operator in H satisfying

$$\langle f(A)x, x \rangle \leq \langle D_1 x, x \rangle \leq b\langle f(A)x, x \rangle, \tag{9.20}$$

$$|\langle D_2 x, x \rangle| \leq a\langle D_1 x, x \rangle \tag{9.21}$$

for some constants $b \geq 1$, $a > 0$, and for all $x \in D(A)$.

Then by Corollary 9.2 (i), the operator $(\mathcal{B}, D(\mathcal{B}))$ generates a contraction semigroup $(\mathcal{S}(t))_{t \geq 0}$ on the product space $\mathcal{W}$ with the following properties depending on the function f.

Theorem 9.4. *Assume Equations (9.17), (9.18), (9.19), (9.20), and (9.21). Then the following assertions hold:*

(i) *The semigroup $(\mathcal{S}(t))_{t \geq 0}$ is norm continuous for $t > 0$ and exponentially stable.*

(ii) *If*

$$\lim_{s \to +\infty} \frac{\ln s}{f(s)} = 0, \tag{9.22}$$

then the semigroup $(\mathcal{S}(t))_{t \geq 0}$ is differentiable for $t > 0$.

(iii) *If*

$$K := \sup_{s \geq \sigma_0} \left\{ \frac{s^{\frac{1}{2}}}{f(s)} \right\} < +\infty, \tag{9.23}$$

then $(\mathcal{S}(t))_{t \geq 0}$ is an analytic semigroup.

(iv) If

$$\lim_{s \to +\infty} \frac{s^{\frac{1}{2}}}{f(s)} = 0, \tag{9.24}$$

then $(\mathcal{S}(t))_{t \geq 0}$ *is an analytic semigroup of angle* $\frac{\pi}{2}$.

9.2 Uniform Exponential Stability

After having studied the well-posedness, now we consider the exponential stability of the complete second-order abstract Cauchy problem

$$\begin{cases} u''(t) = Du'(t) - C^*Cu(t), & t \geq 0, \\ u(0) = x, \quad u'(0) = y, \end{cases} \tag{ACP$_2$}$$

with linear operators $(D, D(D))$ and $(C, D(C))$ on a Hilbert space H. We assume C to be invertible. Our aim is to develop the techniques needed for the robust stability of damped wave equations with delay.

First, we develop some results on the exponential energy decay of second-order Cauchy problems. Let the general assumptions of Proposition 9.3 hold in this section. The important point here is that we not only show uniform exponential stability, but also obtain estimates on the decay rate of the associated semigroup and a bound on the resolvent.

Now we look for criterions implying the uniform exponential stability of the semigroup generated by $\mathcal{B}$. Our main aim will be to show that for uniform exponential stability, it is enough that D is sectorial and strictly dissipative, i.e., that there exist constants $\gamma \geq 0$, $\delta > 0$ such that

$$|\Im\langle Dy, y \rangle| \leq \gamma \, \Re\langle -Dy, y \rangle \quad \text{and} \quad \delta \|y\|^2 \leq \Re\langle -Dy, y \rangle \quad \forall y \in D(D). \tag{9.25}$$

To be able to use the stability estimates of Corollary 5.8, we have to estimate the resolvent of $\mathcal{B}$. To this purpose, we need first the following technical lemma.

Lemma 9.5. *Assume that* $(C, D(C))$ *is invertible and that Equation (9.25) holds, and let* $0 < \varepsilon < \frac{\delta}{2}$ *and* $\alpha \in (-\frac{\delta}{2} + \varepsilon, 0]$. *If*

$$\inf_{\mathcal{X} \in D(\mathcal{B}), \|\mathcal{X}\|=1} \|(\alpha + i\omega - \mathcal{B})\mathcal{X}\| < \varepsilon,$$

then

$$|\omega| < \frac{(\varepsilon - \alpha)\gamma + 3\varepsilon}{\delta + 2(\alpha - \varepsilon)} \cdot \delta.$$

Proof. Let $\mathcal{X} := \begin{pmatrix} x \\ y \end{pmatrix} \in D(\mathcal{B})$ satisfy $\|\mathcal{X}\| = 1$ and $\|(\alpha + i\beta - \mathcal{B})\mathcal{X}\| < \varepsilon$. Since $D(\mathcal{B}_0)$ is a core for $\mathcal{B}$, we may assume that $\mathcal{X} \in D(\mathcal{B}_0)$ and therefore obtain

$$\|(\alpha + i\beta - \mathcal{B})\mathcal{X}\| = \left\| \begin{pmatrix} \alpha x + i\beta x - Cy \\ C^* x + \alpha y + i\beta y - Dy \end{pmatrix} \right\| < \varepsilon. \tag{9.26}$$

Thus $|\langle (\alpha + i\beta - \mathcal{B})\mathcal{X}, \mathcal{X} \rangle| < \varepsilon$, i.e.,

$$\left| \alpha + i\beta + 2i\Im\langle x, Cy \rangle - \langle Dy, y \rangle \right| < \varepsilon \tag{9.27}$$

and hence

$$\left| \Re\langle Dy, y \rangle - \alpha \right| < \varepsilon. \tag{9.28}$$

This estimate combined with Equation (9.25) implies

$$\delta \min\left\{ \|x\|^2, \|y\|^2 \right\} \leq -\Re\langle Dy, y \rangle < \varepsilon - \alpha. \tag{9.29}$$

Thus,

$$\|y\|^2 < \frac{\varepsilon - \alpha}{\delta}.$$

Then

$$\|x\|^2 = 1 - \|y\|^2 > 1 - \frac{\varepsilon - \alpha}{\delta},$$

hence the assumption on α implies

$$1 - 2\|x\|^2 < 2\frac{\varepsilon - \alpha}{\delta} - 1 < 0,$$

and therefore

$$\left| 1 - 2\|x\|^2 \right| > 1 - 2\frac{\varepsilon - \alpha}{\delta}. \tag{9.30}$$

On the other hand, we obtain from Equation (9.26) the estimate

$$\| -Cy + (\alpha + i\beta)x \| < \varepsilon. \tag{9.31}$$

By taking the inner product with x, this implies

$$\left| \langle -Cy + (\alpha + i\beta)x, x \rangle \right| < \varepsilon,$$

hence, by taking imaginary parts,

$$\left| \Im\langle x, Cy \rangle + \beta\|x\|^2 \right| < \varepsilon. \tag{9.32}$$

Next, we combine Equation (9.32) with Equation (9.27) and conclude

$$\begin{aligned} 3\varepsilon &> \left| \beta \left(1 - 2\|x\|^2 \right) - \Im\langle Dy, y \rangle \right| \\ &\geq |\beta| \cdot \left| 1 - 2\|x\|^2 \right| - \left| \Im\langle Dy, y \rangle \right|. \end{aligned} \tag{9.33}$$

Together with Equations (9.30), (9.25), and (9.29), this implies

$$|\beta|\left(1 - 2\frac{\varepsilon - \alpha}{\delta}\right) \leq 3\varepsilon + \gamma\left(-\Re\langle Dy, y\rangle\right) < 3\varepsilon + \gamma(\varepsilon - \alpha), \qquad (9.34)$$

and the assertion follows. $\square$

This result allows us to prove a stability theorem for second-order abstract Cauchy problems.

Theorem 9.6. *Assume that Equations (9.25) and (9.12) hold. Then the growth bound $\omega_0(\mathcal{B})$ satisfies*

$$\omega_0(\mathcal{B}) \leq \max\left\{s(\mathcal{B}), -\frac{\delta}{2}\right\} < 0. \qquad (9.35)$$

Proof. Due to the Gearhart's Theorem (see Theorem 2.26), it suffices to show that

$$\alpha + i\mathbb{R} \subset \rho(\mathcal{B}) \quad \text{and} \quad \sup_{\beta \in \mathbb{R}} \|R(\alpha + i\beta, \mathcal{B})\| < +\infty \quad \text{for all} \quad \alpha > w, \qquad (9.36)$$

where $w := \max\left\{s(\mathcal{B}), -\frac{\delta}{2}\right\}$. Since $\mathcal{B}$ generates a contraction semigroup, we know that $\{z \in \mathbb{C} : \Re z > 0\} \subset \rho(\mathcal{B})$. We will also make use of the fact that the boundary $\partial\sigma(\mathcal{B})$ is always contained in the approximate point spectrum $A\sigma(\mathcal{B})$; see Theorem 2.5.

First we show by contradiction that $\sigma(\mathcal{B}) \cap i\mathbb{R} = \emptyset$. Assume that there exists $\beta \in \mathbb{R}$ such that $i\beta \in \sigma(\mathcal{B})$. Then, by assumption, $\beta \neq 0$, and by Lemma 9.5, $i\beta \notin A\sigma(\mathcal{B})$. But this is a contradiction, since $i\beta$ is in the boundary of the spectrum. Hence, $i\mathbb{R} \in \rho(\mathcal{B})$.

Furthermore, if $|\beta| > \frac{\varepsilon\delta(\gamma+3)}{\delta - 2\varepsilon}$ for $\varepsilon \in (0, \frac{\delta}{2})$, then Lemma 9.5 gives

$$\|R(i\beta, \mathcal{B})\| \leq \frac{1}{\varepsilon}.$$

Hence it follows from our invertibility assumption on $\mathcal{B}$ that the resolvent is bounded on $i\mathbb{R}$ and therefore $s(\mathcal{B}) < 0$.

Consider now $\alpha > w := \max\left\{s(\mathcal{B}), -\frac{\delta}{2}\right\}$. Since $\alpha + i\mathbb{R} \subset \rho(\mathcal{B})$, we only have to show the boundedness of the resolvent. But this follows by an analogous argument as above. $\square$

In applications it is usually difficult to determine $s(\mathcal{B})$ and therefore it is not easy to apply Theorem 9.6 directly. However, it is possible to estimate $\omega_0(\mathcal{B})$ by a constant depending on δ and the norm of $\mathcal{B}^{-1}$.

Corollary 9.7. *Let the conditions of Theorem 9.6 hold.*

(i) If $\gamma \neq 0$, then

$$\omega_0(\mathcal{B}) \leq w,$$

where $w \in (-\frac{\delta}{2}, 0)$ is the unique solution of the equation

$$w^2 + \frac{w^2 \gamma^2 \delta^2}{(\delta + 2w)^2} = \left\| \mathcal{B}^{-1} \right\|^{-2}. \tag{9.37}$$

(ii) If $\gamma = 0$, then

$$\omega_0(\mathcal{B}) \leq w := \max \left\{ -\tfrac{\delta}{2}, - \left\| \mathcal{B}^{-1} \right\|^{-1} \right\}.$$

Proof. We visualize the situation in (i) by Figure 9.1.

We know from Lemma 9.5 that if $\alpha \in (-\frac{\delta}{2}, 0]$ and $|\beta| > \frac{-\alpha \gamma \delta}{\delta + 2\alpha}$, then $\alpha + i\beta \in \rho(\mathcal{B})$. From elementary spectral theory, it follows that the disk with radius $\left\| \mathcal{B}^{-1} \right\|^{-1}$ centered at zero is also contained in the resolvent set.

We obtain the desired estimate on the spectral bound by intersecting the two curves.

Statement (ii) can be obtained by choosing first $\gamma > 0$ and then taking the limit as $\gamma \searrow 0$.   □

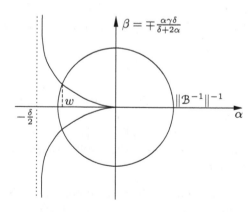

Figure 9.1. The situation (i).

By a simple density argument, we obtain the following result.

Corollary 9.8. *Let $\mathcal{B}$ be maximal dissipative and assume that $\gamma \neq 0$. If there exists $\kappa > 0$ such that*

$$\|\mathcal{B}_0 \mathcal{X}\| \geq \kappa \|\mathcal{X}\| \quad \text{for all } \mathcal{X} \in D(\mathcal{B}_0), \tag{9.38}$$

then

$$\omega_0(\mathcal{B}) \leq w,$$

where $w \in (-\frac{\delta}{2}, 0)$ is the unique solution of the equation

$$w^2 + \frac{w^2 \gamma^2 \delta^2}{(\delta + 2w)^2} = \kappa^2. \tag{9.39}$$

Proof. Using that $\mathcal{B}$ is the closure of $\mathcal{B}_0$, we obtain that $\frac{1}{\kappa} \geq \|\mathcal{B}^{-1}\|$, and the desired estimate follows. $\qquad\square$

Corollary 9.9. *Assume that $\mathcal{B}$ is maximal dissipative and take*

$$\kappa^{-1} := \|Q\| + 2\|C^{-1}\|, \tag{9.40}$$

then the following estimates for the growth bound $\omega_0(\mathcal{B})$ of $\mathcal{B}$ hold:

  (i) *If $\gamma \neq 0$, then*
$$\omega_0(\mathcal{B}) \leq w,$$
  where $w \in (-\frac{\delta}{2}, 0)$ is the unique solution of Equation (9.39).

  (ii) *If $\gamma = 0$, then*
$$\omega_0(\mathcal{B}) \leq \max\left\{-\kappa, -\tfrac{\delta}{2}\right\}.$$

Proof. The assertions follow from Corollaries 9.7 and 9.8 and Theorem 9.6, respectively, and the estimate (see Equations (9.10)–(9.12))

$$\begin{aligned}
\left\|\mathcal{B}^{-1}\left(\begin{smallmatrix} x \\ y \end{smallmatrix}\right)\right\| &= \left(\|Qx - C^{*-1}y\|^2 + \|C^{-1}x\|^2\right)^{\frac{1}{2}} \\
&\leq \|Qx - C^{*-1}y\| + \|C^{-1}x\| \\
&\leq \|Q\|\|x\| + \|C^{*-1}\|\|y\| + \|C^{-1}\|\|x\| \\
&\leq \left((\|Q\| + \|C^{-1}\|)^2 + \|C^{*-1}\|^2\right)^{\frac{1}{2}} \cdot \left(\|x\|^2 + \|y\|^2\right)^{\frac{1}{2}} \\
&\leq (\|Q\| + 2\|C^{-1}\|) \cdot \left(\|x\|^2 + \|y\|^2\right)^{\frac{1}{2}}
\end{aligned}$$

for $\left(\begin{smallmatrix} x \\ y \end{smallmatrix}\right) \in D(\mathcal{B}_0)$. $\qquad\square$

Lemma 9.5 also allows us to estimate the resolvent $R(\alpha + i\omega, \mathcal{B})$, which will become important for second-order Cauchy problems with delay. Recall that in Corollary 5.7 the estimate for the resolvent along imaginary lines was a central tool.

We consider here only the case $\alpha = 0$. Other cases can be treated similarly.

Lemma 9.10. *Let $(C, D(C))$ be a densely defined invertible and $(D, D(D))$ be a densely defined and closed operator such that Equations (9.12) and (9.25) hold. Then for every $0 < c < 1$, we have*

$$\|R(i\omega, \mathcal{B})\| \leq \begin{cases} \frac{\|\mathcal{B}^{-1}\|}{1-c} & \text{for } |\omega| \leq \frac{c}{\|\mathcal{B}^{-1}\|}, \\ \frac{(3+\gamma)\delta\|\mathcal{B}^{-1}\|+2c}{\delta c} & \text{for } |\omega| > \frac{c}{\|\mathcal{B}^{-1}\|}. \end{cases}$$

Proof. Since $\mathcal{B}$ is invertible, the resolvent of $\mathcal{B}$ is given by the power series

$$R(\lambda, \mathcal{B}) = \sum_{n=0}^{\infty} \lambda^n \mathcal{B}^{-(n+1)}$$

for all $|\lambda| < \frac{1}{\|\mathcal{B}^{-1}\|}$. Moreover, for $0 < c < 1$, we have that

$$\|R(\lambda, \mathcal{B})\| \leq \|\mathcal{B}^{-1}\| \sum_{n=0}^{\infty} |\lambda|^n \|\mathcal{B}^{-1}\|^n$$

$$\leq \|\mathcal{B}^{-1}\| \sum_{n=0}^{\infty} c^n$$

$$= \frac{\|\mathcal{B}^{-1}\|}{1-c}$$

for all $|\lambda| \leq \frac{c}{\|\mathcal{B}^{-1}\|}$.

It now remains to estimate $R(i\omega, \mathcal{B})$ for $|\omega| > \frac{c}{\|\mathcal{B}^{-1}\|}$. To this purpose, we apply Lemma 9.5. Let $0 < \varepsilon < \frac{\delta}{2}$, then for

$$|\omega| > \frac{(3\varepsilon + \gamma)\delta}{\delta - 2\varepsilon}$$

we have

$$\|R(i\omega, \mathcal{B})\| \leq \frac{1}{\varepsilon}.$$

Now we have to look for one $0 < \varepsilon < \frac{\delta}{2}$ such that

$$\frac{(3\varepsilon + \gamma)\delta}{\delta - 2\varepsilon} = \frac{c}{\|\mathcal{B}^{-1}\|}.$$

This is true for

$$\varepsilon := \frac{\delta c}{(3+\gamma)\delta\|\mathcal{B}^{-1}\| + 2c},$$

which concludes the proof. □

Since $\mathcal{B}$ is invertible, we can consider lower bounds $\kappa > 0$ of $\mathcal{B}$, that is constants that satisfy the condition

$$\|\mathcal{B}x\| \geq \kappa\|x\| \qquad \text{for all } x \in D(\mathcal{B}). \tag{9.41}$$

For example, by Corollary 9.9,

$$\kappa := \left(\|Q\| + 2\|C^{-1}\|\right)^{-1} \tag{9.42}$$

is such a bound.

Using this concept, we obtain the following uniform estimate for the resolvent on the imaginary axis $\alpha = 0$.

Corollary 9.11. *Let Equations (9.12) and (9.25) hold. Then*

$$\|R(i\omega, \mathcal{B})\| \leq \frac{2(3+\gamma)\delta\kappa^{-1} + 2}{\delta}$$

for all $\omega \in \mathbb{R}$ and every lower bound κ of $\mathcal{B}$.

Proof. We choose $c = \frac{1}{2}$ in Lemma 9.10 and obtain

$$\|R(i\omega, \mathcal{B})\| \leq \frac{2(3+\gamma)\delta\|\mathcal{B}^{-1}\| + 2}{\delta}$$

for all $\omega \in \mathbb{R}$. Since $\|\mathcal{B}^{-1}\| \leq \kappa^{-1}$, the assertion holds. □

Now we show that in an important special case our result is optimal. This is due to the validity of the Weak Spectral Mapping Theorem for matrix multiplication operators (see Holderrieth [112]).

Corollary 9.12. *Assume the conditions in Theorem 9.6 to be satisfied and further that C and D are commuting normal operators. Then*

$$\omega_0(\mathcal{B}) = s(\mathcal{B}).$$

Proof. Using the Spectral Theorem for normal operators, we can transform the operator $\mathcal{B}$ into a matrix multiplication on an L^2–space. Using the spectral characterization as in Holderrieth [112] and the fact that for a scalar matrix $\left(\begin{smallmatrix} 0 & c \\ -\bar{c} & d \end{smallmatrix}\right)$ one has $s\left(\left(\begin{smallmatrix} 0 & c \\ -\bar{c} & d \end{smallmatrix}\right)\right) \geq \frac{\Re d}{2}$, one easily shows that $s(\mathcal{B}) \geq -\frac{\delta}{2}$. Hence, $\omega_0(\mathcal{B}) = s(\mathcal{B})$ by Theorem 9.6. □

Example 9.13. (Internally Damped Wave Equation.) Let Ω be a bounded, connected domain in $\mathbb{R}^n$ and let Γ be its boundary. We suppose Γ to be piecewise smooth and to consist of two closed parts Γ^0 and Γ^1 such that $\Gamma = \Gamma^0 \cup \Gamma^1$ and $\Gamma^0 \cap \Gamma^1 = \emptyset$. Moreover, denote by ν the outer unit normal of Γ.

Next we consider an internally damped wave equation

$$\begin{cases} \partial_t^2 y(x,t) = \mu \Delta \partial_t y(x,t) + \Delta y(x,t), & (x,t) \in \Omega \times (0,\infty), \\ y(x,0) = y_0(x), \quad \partial_t y(x,0) = y_1(x), & x \in \Omega, \end{cases} \quad \text{(IDW)}$$

with mixed boundary conditions

$$y(x,t) = 0 \quad \text{on} \quad \Gamma^0 \times (0,\infty), \tag{9.43}$$

$$\partial_\nu y(x,t) = 0 \quad \text{on} \quad \Gamma^1 \times (0,\infty), \tag{9.44}$$

where $\mu < 0$ is a constant.

Then the energy of a solution y of (IDW) is given by

$$E(t) = \frac{1}{2} \int_\Omega \left(|\partial_t y(x,t)|^2 + |\nabla y(x,t)|^2 \right) dx.$$

In order to apply our abstract results, we have to reformulate this problem as a second-order abstract Cauchy problem.

To this end, we define the following operators on the Hilbert space $H := L^2(\Omega)$:

$$A := -\Delta, \ \ D := \mu \Delta, \ \ D(D) := D(A) := \left\{ y \in H^2 : y|_{\Gamma^0} = 0, \ \partial_\nu y|_{\Gamma^1} = 0 \right\},$$

and take $C := A^{\frac{1}{2}}$.

By Equation (9.14) and Proposition 9.3 we have that $\mathcal{B} = \begin{pmatrix} 0 & C \\ -C^* & D \end{pmatrix}$ generates a contraction semigroup in the product space $\mathcal{W}$.

As A and D are commuting, normal operators, so are $A^{\frac{1}{2}}$ and D, where we used that $D(A^{\frac{1}{2}}) = H^1_{\Gamma^0} := \left\{ f \in H^1 : f|_{\Gamma^0} = 0 \right\}$; see Triggiani [202, Formula (2.11)].

Hence, we can calculate the exact decay rate by Corollary 9.12 as

$$\omega_0(\mathcal{B}) = s(\mathcal{B}) = \max \left\{ -\frac{\mu \lambda_1}{2}, -\frac{1}{\mu} \right\}, \tag{9.45}$$

where $\lambda_1 > 0$ denotes the first eigenvalue of A.

Actually, in this case, we also know that the generated contraction semigroup is analytic; see Theorem 9.4 (iv) with $f(s) = s$.

This is the exact decay rate, i.e., an optimal result, and therefore better than the (not optimal) decay rate derived in Gorain [90].

Example 9.14. (Damped Wave Equation.) Let Ω be a bounded domain in $\mathbb{R}^n$. Consider the damped wave equation

$$\begin{cases} \partial_t^2 u(t,x) - q(x)\partial_t u(t,x) - \Delta u(t,x) = 0, & (t,x) \in \mathbb{R}_+ \times \Omega, \\ u(0,x) = u_0(x), \quad \partial_t u(0,x) = u_1(x), & x \in \Omega, \qquad \text{(DWE)} \\ u(t,x) = 0, & (t,x) \in \mathbb{R}_+ \times \partial\Omega \end{cases}$$

for initial values $u_0, u_1 \in L^2(\Omega)$. Here, we assume that $q : \Omega \to \mathbb{C}$ is measurable and satisfies

$$|\Im q(x)| \leq -\gamma \Re q(x) \quad \text{and} \quad \delta \leq -\Re q(x) \text{ a.e.} \qquad (9.46)$$

for some constants $\gamma, \delta > 0$.

In order to reformulate (DWE) as a second-order abstract Cauchy problem we take $H := L^2(\Omega)$, $A := -\Delta$ the Dirichlet-Laplacian with domain $D(A) := \{u \in H_0^1(\Omega) : \Delta u \in L^2(\Omega)\}$ and $D = M_q$ the multiplication operator with the function q and maximal domain; see Engel and Nagel [72, Proposition I.4.10] or Weidmann [214, 4.1. Example 1].

This ensures that D satisfies Equation (9.25). Hence, in order to apply our results we have to look for conditions implying Equation (9.13).

We use Sobolev embedding theorems as in Adams [1, Theorem 5.4.C], that is,

$$H_0^1(\Omega) \hookrightarrow C_b(\Omega) \text{ in the one-dimensional case,}$$
$$H_0^1(\Omega) \hookrightarrow L^p(\Omega) \quad (p \geq 2) \text{ in two dimensions, and}$$
$$H_0^1(\Omega) \hookrightarrow L^p(\Omega) \quad (p \geq 2n/(n-2)) \text{ in higher dimensions.}$$

We assume that $q \in L^r(\Omega)$

$$\text{with } r \geq 2 \text{ if } n = 1,$$

$$\text{with } r > 2 \text{ if } n = 2, \text{ and}$$

$$\text{with } r \geq n \text{ if } n \geq 3.$$

These embeddings combined with Hölder's inequality imply that D is $A^{\frac{1}{2}}$-bounded, i.e., $D(A^{\frac{1}{2}}) = H_0^1(\Omega) \subseteq D(D)$.

Let us consider more in detail the case $n \geq 3$. For $f \in H_0^1(\Omega)$, we have

$$\int_\Omega |fq|^2 \leq \left(\int_\Omega |f|^p\right)^{\frac{1}{p}} \left(\int_\Omega |q|^r\right)^{\frac{1}{r}} = \|f\|_{L^p}\|q\|_{L^r} \leq K\|f\|_{H_0^1}\|q\|_{L^r},$$

where K denotes the appropriate Sobolev embedding constant. Hence, by Theorem 9.6 we obtain exponential decay of the energy

$$E(t) = \frac{1}{2} \int_\Omega \left(|\partial_t u(t,x)|^2 + |\nabla u(t,x)|^2 \right) dx$$

of the solution u of (DWE).

To obtain an explicit estimate on the decay rate, we can use Equation (9.40) in Corollary 9.9 for

$$\kappa^{-1} = \frac{\sqrt{-\lambda_1}}{2K\|q\|_{L^r} + 2}, \tag{9.47}$$

where λ_1 is the first eigenvalue of the Laplacian.

9.3   Wave Equations with Bounded Delay Operators

In this section we consider the complete second-order abstract Cauchy problem with delay

$$\begin{cases} u''(t) = Du'(t) - C^*Cu(t) + \phi u_t + \psi u_t', & t \geq 0, \\ u(0) = x, \quad u'(0) = y, \\ u_0 = f, \quad u_0' = g, \end{cases} \tag{DACP$_2$}$$

with linear operators $(D, D(D))$ and $(C, D(C))$ on a Hilbert space H. We consider $H_{\frac{1}{2}} := (D(C), \|\cdot\|_C)$ and assume that the operators

$$\phi \in \mathcal{L}(W^{1,2}([-1,0], H_{\frac{1}{2}}), H_{\frac{1}{2}}) \text{ and } \psi \in \mathcal{L}(W^{1,2}([-1,0], H), H)$$

are linear and bounded.   The initial functions f and g belong to $L^2([-1,0], H_{\frac{1}{2}})$ and $L^2([-1,0], H)$, respectively, while for the initial values we assume $x \in D(C^*C)$, $y \in D(D) \cap D(C)$.

Our strategy here is to apply the techniques developed previously. Therefore, as a first step, let us rewrite this problem as a first-order delay equation.

Using the standard reduction $\mathcal{U} := \left(\begin{smallmatrix} u \\ u' \end{smallmatrix}\right)$, we can transform (DACP$_2$) into a first-order system

$$
\begin{cases}
\mathcal{U}'(t) = \hat{\mathcal{B}}\,\mathcal{U}(t) + \Phi\,\mathcal{U}_t, & t \geq 0, \\
\mathcal{U}(0) = \left(\begin{smallmatrix} x \\ y \end{smallmatrix}\right), \\
\mathcal{U}_0 = \left(\begin{smallmatrix} f \\ g \end{smallmatrix}\right),
\end{cases}
\tag{DACP}
$$

for the operator matrix

$$
\hat{\mathcal{B}} := \begin{pmatrix} 0 & Id \\ -C^*C & D \end{pmatrix}, \quad D(\hat{\mathcal{B}}) := D(C^*C) \times D(D) \cap D(C)
$$

on the product space $\mathcal{F} := H_{\frac{1}{2}} \times H$. The delay operator becomes then

$$
\Phi := \begin{pmatrix} 0 & 0 \\ \phi & \psi \end{pmatrix} : W^{1,2}([-1,0], H_{\frac{1}{2}}) \times W^{1,2}([-1,0], H) \to \mathcal{F}.
$$

Definition 9.15. We say that (DACP$_2$) is *well-posed* if

(i) for every $\left(\begin{smallmatrix} x \\ y \end{smallmatrix}\right) \in D(\hat{\mathcal{B}})$, $f \in W^{1,2}([-1,0], H_{\frac{1}{2}})$ and $g \in W^{1,2}([-1,0], H)$ with $f(0) = x$, $g(0) = y$, there is a unique (classical) solution $u(x, y, f, g, \cdot)$, and

(ii) the solutions depend continuously on the initial values, i.e., if a sequence

$$
\begin{pmatrix} x_n \\ y_n \\ f_n \\ g_n \end{pmatrix} \subset D(\hat{\mathcal{B}}) \times W^{1,2}([-1,0], H_{\frac{1}{2}}) \times W^{1,2}([-1,0], H)
$$

with $f_n(0) = x_n$ and $g_n(0) = y_n$ converges to

$$
\begin{pmatrix} x \\ y \\ f \\ g \end{pmatrix} \in D(\hat{\mathcal{B}}) \times W^{1,2}([-1,0], H_{\frac{1}{2}}) \times W^{1,2}([-1,0], H)
$$

with $f(0) = x$ and $g(0) = y$, then

$$
(u(x_n, y_n, f_n, g_n, t), u'(x_n, y_n, f_n, g_n, t))
$$

converges to $(u(x, y, f, g, t), u'(x, y, f, g, t))$ in the space $\mathcal{F}$ uniformly for t in compact intervals.

In the sequel, we impose the well-posedness assumptions of Section 9.1 for equations without delay, i.e., we assume that C is densely defined, invertible and D is densely defined, dissipative, and the compatibility condition of Equation (9.12) between them is satisfied.

Furthermore we assume that there exist $\mu : [-1,0] \longrightarrow \mathcal{L}(H_{\frac{1}{2}}, H)$ and $\nu : [-1,0] \longrightarrow \mathcal{L}(H)$ of bounded variation such that

$$\phi(f) := \int_{-1}^{0} d\mu f \quad \text{and} \quad \psi(g) := \int_{-1}^{0} d\nu g. \tag{9.48}$$

Defining $\eta : [-1,0] \longrightarrow \mathcal{F}$, $\eta := \left(\begin{smallmatrix} 0 & 0 \\ \mu & \nu \end{smallmatrix}\right)$, and $\Phi := \left(\begin{smallmatrix} 0 & 0 \\ \phi & \psi \end{smallmatrix}\right)$, we obtain that Φ is of the form of Equation (3.38), i.e.,

$$\Phi\left(\begin{smallmatrix} f \\ g \end{smallmatrix}\right) = \int_{-1}^{0} d\eta\left(\begin{smallmatrix} f \\ g \end{smallmatrix}\right). \tag{9.49}$$

We are now able to formulate a well-posedness result for equation (DACP$_2$).

Theorem 9.16. *The complete second-order abstract Cauchy problem with delay* (DACP$_2$) *is well-posed if C is densely defined and invertible; D is densely defined, dissipative, and invertible; D and C satisfy Equation (9.13) or (9.14); and ϕ and ψ are given by Equation (9.48).*

Proof. After the preparations in Sections 9.1 and 3.3, the proof is a straightforward combination of the well-posedness results collected there. It follows from $D(C) \subseteq D(D)$ or from $D = dA^\alpha$ that $\hat{\mathcal{B}}$ is closed and generates a C_0-semigroup on the space $\mathcal{W}$. So by Theorem 3.29 and Theorem 3.12, we have that (DACP$_2$) is well-posed. $\qquad\square$

Similarly, we can reformulate Corollary 5.8 for the new situation. The proof follows by the same arguments. Recall that $\Phi_{\alpha+i\omega} \in \mathcal{L}(\mathcal{F})$ is defined by $\Phi_{\alpha+i\omega}\left(\begin{smallmatrix} x \\ y \end{smallmatrix}\right) = \Phi\left(e^{(\alpha+i\omega)\cdot}\left(\begin{smallmatrix} x \\ y \end{smallmatrix}\right)\right)$.

Theorem 9.17. *Under the assumptions of Theorem 9.16, the energy of the solutions decays exponentially if D satisfies Equation (9.25) and the inequality*

$$\sup_{\omega \in \mathbb{R}} \|\Phi_{\alpha+i\omega}\| < \frac{1}{\sup_{\omega \in \mathbb{R}} \left\| R(\alpha + i\omega, \hat{\mathcal{B}}) \right\|} \tag{9.50}$$

holds for some $\omega_0(\hat{\mathcal{B}}) < \alpha \leq 0$.

We remind the reader that the energy of a solution is usually defined by $\mathcal{E}(u(t)) := \frac{1}{2} \left\| \left(\begin{smallmatrix} u(t) \\ u'(t) \end{smallmatrix}\right) \right\|_{\mathcal{F}}^{2}$.

Assume now that D and C satisfy the hypothesis of Theorem 9.17. We look for a simple criterion on ϕ and ψ such that Φ satisfies Equation (9.50).

We denote by $\text{Var}(g)^0_{-1}$ the total variation of the function g on the interval $[-1, 0]$. For Φ given by Equation (3.38), we can then estimate

$$\|\Phi_{\alpha+i\omega}x\| \leq \left\| \int_{-1}^0 e^{\alpha+i\omega t} x \, d\eta(t) \right\| \leq \text{Var}(\eta)^0_{-1}\|x\|. \qquad (9.51)$$

Using this we obtain a criterion for exponential stability of the energy of the solutions of (DACP$_2$), which is quite useful in concrete applications.

Corollary 9.18. *Under the assumptions of Theorem 9.17, the energy of the solutions of the considered complete second-order abstract Cauchy problem with delay (DACP$_2$) decays exponentially if*

$$\sup_{\omega \in \mathbb{R}} \left\{ \|\phi_{i\omega}\|_{H_{\frac{1}{2}} \to H} + \|\psi_{i\omega}\|_H \right\} < \frac{\kappa\delta}{2(3+\gamma)\delta + 2\kappa}, \qquad (9.52)$$

where γ, δ, and κ are defined as in Equations (9.25) and (9.41), respectively.

Proof. Use Theorem 9.17, Corollary 9.11, and the estimate of Equation (9.51) to obtain the assertion. $\square$

Now we illustrate our results on a damped wave and on a structurally damped plate equation with delay.

Example 9.19. Consider the damped wave equation with delay

$$\begin{cases} \partial_t^2 u(x,t) = q(x)\partial_t u(x,t) + \Delta u(t,x) + c_1\sum_{j=1}^n \partial_j u(x, t-h_1) \\ \qquad\qquad + c_2\partial_t u(x, t-h_2), & (x,t) \in \Omega \times \mathbb{R}_+, \\ u(x,s) = f(x,s), \quad \partial_t u(x,s) = g(x,s), & (x,s) \in \Omega \times [-1,0], \\ u(x,t) = 0, & (x,t) \in \partial\Omega \times \mathbb{R}_+. \end{cases}$$
$$\text{(DDWE)}$$

Here $\Omega \subset \mathbb{R}^n$ is bounded, $q \in L^r(\Omega)$, where

- $r \geq 2$ if $n = 1$,

- $r > 2$ if $n = 2$, and

- $r \geq n$ if $n \geq 3$.

Moreover, we assume that there exists $\delta > 0$ such that $\Re q(x) < -\delta$ a.e., $|\Im q(x)| \leq -\Re q(x)$ a.e., and $h_1, h_2 \in (0,1]$.

To write this problem in abstract form, consider the space

$$\mathcal{F} := H_0^1(\Omega) \times L^2(\Omega),$$

where we equip the Sobolev space $H_0^1(\Omega)$ with the norm

$$\|u\|_1 := \|\sum_{j=1}^{n} \partial_j u\|_{L^2} = \||\nabla u|\|_{L^2},$$

which is equivalent to the usual Sobolev norm. Next we define the operators

$$A := -\Delta, \quad D(A) := \left\{u \in H_0^1 : \Delta u \in L^2(\Omega)\right\},$$

$$C := A^{\frac{1}{2}}, \quad D(C) := H_0^1(\Omega),$$

$$Du := qu, \quad D(D) := \left\{u \in L^2 : qu \in L^2\right\},$$

$$\phi := c_1 \sum_{j=1}^{n} \partial_j \delta_{-h_1},$$

$$\psi := c_2 Id\, \delta_{-h_2},$$

where δ_{-h_1} and δ_{-h_2} are the point evaluations in $-h_1$ and $-h_2$, respectively. We note that $\|Ch\|_{L^2} = \||\nabla h|\|_{L^2}$ for $h \in H_0^1(\Omega)$; see Triggiani [202, Formula (2.9)]. We make the remark here that if Ω has a sufficiently smooth boundary, then $\left\{u \in H_0^1 : \Delta u \in L^2(\Omega)\right\} = H_0^2(\Omega)$.

We make the following assumptions on the initial functions in order to obtain classical solutions:

- $f(\cdot, 0) \in \left\{u \in H_0^1 : \Delta u \in L^2(\Omega)\right\}$.

- $f(\cdot, s) \in H_0^1(\Omega)$.

- The map $[-1, 0] \ni s \mapsto f(\cdot, s) \in H_0^1(\Omega)$ belongs to the Sobolev space $W^{1,2}([-1, 0], H_0^1(\Omega))$.

- $g(\cdot, 0) \in H_0^1(\Omega) \cap (D(D))$ for the operator D defined above.

- $g(\cdot, s) \in L^2(\Omega)$.

- The map $[-1, 0] \ni s \mapsto g(\cdot, s) \in L^2(\Omega)$ belongs to the Sobolev space $W^{1,2}([-1, 0], L^2(\Omega))$.

The calculations above and the estimate in Equation (9.51) (replacing Φ with ϕ and ψ) show that

$$\sup_{\omega \in \mathbb{R}} \|\phi_{i\omega}\|_{H_{\frac{1}{2}} \to H} \leq |c_1|$$

and

$$\sup_{\omega \in \mathbb{R}} \|\psi_{i\omega}\|_H \leq |c_2|.$$

By Example 9.14, where we considered damped wave equation without delay, we can choose

$$\kappa^{-1} := \frac{\sqrt{-\lambda_1}}{2K\|q\|_{L^r} + 2},$$

where λ_1 is the first eigenvalue of the Laplacian and K is the appropriate Sobolev embedding constant. After these preparations we arrive at the following result.

Corollary 9.20. *The problem* (DDWE) *is well-posed. The solutions decay exponentially if*

$$(|c_1| + |c_2|) \leq \frac{\sqrt{-\lambda_1}\delta}{12\delta K\|q\|_{L^r} + 12\delta + 2\sqrt{-\lambda_1}}.$$

The proof is a direct application of Corollary 9.18 with the constants calculated above.

Example 9.21. Consider the structurally damped plate equation with delay

$$\begin{cases} \partial_t^2 u(x,t) = \partial_t \Delta u(x,t) + \Delta^2 u(t,x) + \int_{-1}^0 \Delta u(x, t+\tau) dh(\tau) \\ \qquad + c_1 \partial_t u(x, t-1), & (x,t) \in \Omega \times \mathbb{R}_+, \\ u(x,s) = u_0(x,s), \ \frac{\partial}{\partial t} u(x,s) = u_1(x,s), & (x,s) \in \Omega \times [-1, 0], \\ u(x,t) = 0, \ \Delta u(x,t) = 0, & (x,t) \in \partial\Omega \times \mathbb{R}_+. \end{cases}$$
$$\tag{DPE}$$

Here, $\Omega \subset \mathbb{R}^n$ satisfies condition (O), i.e., is contained in a strip, and h is a scalar function of bounded variation.

To write this problem in abstract form, consider the space

$$\mathcal{F} = H_{\frac{1}{2}} \times L^2(\Omega),$$

where we take $H_{\frac{1}{2}} := \left\{ u \in H_0^1 : \Delta u \in L^2(\Omega) \right\}$ with the norm $\|u\|_2 := \|\Delta u\|_{L^2}$, which is equivalent to the usual Sobolev norm.

Moreover, we define the operators

$$A := \Delta^2, \quad D(A) := \left\{ u \in H_0^1(\Omega) : \Delta u \in H_{\frac{1}{2}}, \Delta^2 u \in L^2(\Omega) \right\}$$

$$C := A^{\frac{1}{2}},$$

$$D := -\Delta = A^{\frac{1}{2}},$$

$$\eta := \Delta \cdot h,$$

$$\phi := \int_{-1}^0 d\eta,$$

$$\psi := c_1 Id\delta_{-1}.$$

The conditions on the initial functions yielding to classical solutions are the following:

- $u_0(\cdot, 0) \in D(A)$.

- $f(\cdot, s) \in \{u \in H_0^1 : \Delta u \in L^2(\Omega)\}$.

- The map defined by $[-1, 0] \ni s \mapsto u_0(\cdot, s) \in H_0^1(\Omega) \cap H^2(\Omega)$ belongs to the function space $W^{1,2}([-1, 0], H_{\frac{1}{2}})$.

- $u_1(\cdot, 0) \in H_{\frac{1}{2}}$.

- $u_1(\cdot, s) \in L^2(\Omega)$.

- The map defined by $[-1, 0] \ni s \mapsto g(\cdot, s) \in L^2(\Omega)$ belongs to $W^{1,2}([-1, 0], L^2(\Omega))$.

The calculations above and the estimate in Equation (9.51) show that

$$\sup_{\omega \in \mathbb{R}} \|\phi_{i\omega}\|_{H_{\frac{1}{2}} \to H} \leq \mathrm{Var}(h)_{-1}^0$$

and

$$\sup_{\omega \in \mathbb{R}} \|\psi_{i\omega}\|_H \leq |c_1|.$$

By Corollary 9.9, we can choose

$$\kappa^{-1} := \left(\|C^{*-1} D C^{-1}\| + 2\|C^{-1}\| \right) = 3\|C^{-1}\| = \frac{3}{|\lambda_1|},$$

where $\lambda_1 < 0$ is the first eigenvalue of the Dirichlet-Laplacian.

Corollary 9.22. *The problem* (DPE) *is well-posed. The solutions decay exponentially if*

$$(|c_1| + \mathrm{Var}(h)_{-1}^0) \leq \frac{-\lambda_1}{2(\lambda_1^2 + 1)}.$$

The proof of this result is a direct application of Corollary 9.18 with the constants calculated above and $\gamma = 0$, $\delta = -\lambda_1$.

9.4 Wave Equations with Unbounded Delay Operators

In this section, we apply the theory developed for unbounded operators in the delay term from Section 3.3.4 to second-order abstract Cauchy problems. We use the standard reduction developed previously for second-order equations.

The first restriction we need is the analyticity of the semigroup associated to the second-order equation. We consider here a special case, and apply a characterization of the domains of fractional powers of the corresponding operator matrices; see Theorem 9.27 below. Since this result is only a technical tool for our investigations, we present it only at the end of this section.

Consider the complete second-order abstract Cauchy problem with delay

$$\begin{cases} u''(t) = -f(A)u'(t) - Au(t) + \phi u_t + \psi u'_t, & t \geq 0, \\ u(0) = x, \quad u'(0) = y, \\ u_0 = f, \quad u'_0 = g, \end{cases} \tag{DACP$_2$}$$

in the Hilbert space H, where A is a positive definite, invertible selfadjoint operator and f satisfies the conditions of Theorem 9.4 (iii). The operators ψ and ϕ will be specified later.

Using the standard reduction $\mathcal{U} := \left(\begin{smallmatrix} u \\ u' \end{smallmatrix}\right)$, we can transform (DACP$_2$) into a first-order system

$$\begin{cases} \mathcal{U}'(t) = \hat{\mathcal{B}}\,\mathcal{U}(t) + \Phi\,\mathcal{U}_t, & t \geq 0 \\ \mathcal{U}(0) = \left(\begin{smallmatrix} x \\ y \end{smallmatrix}\right) \\ \mathcal{U}_0 = \left(\begin{smallmatrix} f \\ g \end{smallmatrix}\right) \end{cases} \tag{DACP}$$

for the operator matrices

$$\hat{\mathcal{B}} := \left(\begin{smallmatrix} 0 & Id \\ -A & -f(A) \end{smallmatrix}\right) \text{ with } D(\hat{\mathcal{B}}) := D(A) \times D(f(A))$$

and $\Phi := \left(\begin{smallmatrix} 0 & 0 \\ \phi & \psi \end{smallmatrix}\right)$ on the product space $\mathcal{F} := H_{\frac{1}{2}} \times H$, where $H_{\frac{1}{2}} := D(A^{\frac{1}{2}})$ with the norm $\|\cdot\|_{\frac{1}{2}} := \|A^{\frac{1}{2}} \cdot \|$.

We refer, for example, to Theorem 9.4 for conditions implying the analyticity of the semigroup generated by $\hat{\mathcal{B}}$.

To fulfill the conditions in Section 3.3.4, we assume the operator $\Phi := \left(\begin{smallmatrix} 0 & 0 \\ \phi & \psi \end{smallmatrix}\right)$ to be bounded from $D(\mathcal{A}_0)$ into $\mathcal{E}$ where $\mathcal{A}_0$ is defined as in Equations (3.30) and (3.31) with B replaced by $\hat{\mathcal{B}}$.

We consider, using the notations of Theorem 9.27 below, the case in which $0 \leq \vartheta \leq \frac{1}{2}$ and $\frac{1}{p} > \vartheta$.

This means that

$$\phi \in \mathcal{L}(W^{1,p}([-1,0], D(A^{\frac{1}{2}+\vartheta}f(A)^{-\vartheta})), H_{\frac{1}{2}}) \tag{9.53}$$

and

$$\psi \in \mathcal{L}(W^{1,p}([-1,0], D(f(A)^{\vartheta})), H). \tag{9.54}$$

Now we are in the position to formulate our main theorem for this class of equations.

Theorem 9.23. *Let A be a positive definite, invertible operator in the Hilbert space H, $\frac{1}{2} \leq \gamma \leq 1$, $0 \leq \vartheta \leq \frac{1}{2}$, $p < \frac{1}{\vartheta}$. Assume further that the operators ϕ and ψ are given by the Riemann-Stieltjes Integrals*

$$\phi h := \int_{-1}^{0} d\mu h \quad and \quad \psi k := \int_{-1}^{0} d\nu k, \tag{9.55}$$

where $\mu : [-1,0] \to \mathcal{L}(D(A^{\frac{1}{2}+\vartheta} f(A)^{-\vartheta}), H)$, $\nu : [-1,0] \to \mathcal{L}(D(f(A)^{\vartheta}), H)$, and

$$h \in C([-1,0], D(A^{\frac{1}{2}+\vartheta(1-\gamma)})) \quad and \quad k \in C([-1,0], D(A^{\gamma\vartheta})).$$

Then the complete second-order abstract Cauchy problem with delay (DACP$_2$) is well-posed.

Proof. Consider the space

$$\mathcal{Z} := D((-\hat{\mathcal{B}})^{\vartheta}) = D(A^{\frac{1}{2}+\vartheta} f(A)^{-\vartheta}) \times D(f(A)^{\vartheta}) \overset{d}{\hookrightarrow} \mathcal{F};$$

see Theorem 9.27.

Defining

$$\eta : [-1,0] \longrightarrow \mathcal{L}(\mathcal{Z}, \mathcal{F}), \quad \eta := \left(\begin{smallmatrix} 0 & 0 \\ \mu & \nu \end{smallmatrix}\right)$$

and using the representation formula of Equation (9.55), it is easy to show that $\Phi = \left(\begin{smallmatrix} 0 & 0 \\ \phi & \psi \end{smallmatrix}\right)$ is of the form of Equation (3.38), i.e.,

$$\Phi\left(\begin{smallmatrix} h \\ k \end{smallmatrix}\right) = \int_{-1}^{0} d\eta \left(\begin{smallmatrix} h \\ k \end{smallmatrix}\right). \tag{9.56}$$

Moreover, the operator $(\hat{\mathcal{B}}, D(\hat{\mathcal{B}}))$ generates a uniformly exponentially stable analytic semigroup; see Theorem 9.4 (iii). Combining Theorem 9.27, and Equations (9.53), (9.54), and (9.56), and using Theorem 3.34, we obtain that the transformed abstract delay equation is well-posed. Using the equivalence of the second-order equation to the abstract first-order equation, we obtain the desired statement. $\qquad\square$

By the same argument we can also reformulate Corollary 4.17 for this situation.

Theorem 9.24. *Under the assumptions of Theorem 9.23, the energy of the solutions decays exponentially if the inequality*

$$\sup_{\omega \in \mathbb{R}} \left\| \Phi_{\alpha + i\omega} R(\alpha + i\omega, \hat{\mathcal{B}}) \right\| < 1 \qquad (9.57)$$

holds for some $\omega_0(\hat{\mathcal{B}}) < \alpha \leq 0$.

Example 9.25. Consider the structurally damped plate equation with delay

$$\begin{cases} \partial_t^2 u(x,t) = \partial_t \Delta u(x,t) - \Delta^2 u(t,x) + \int_{-1}^0 \nabla^3 u(x,t+\tau) dh(\tau) \\ \qquad + c\partial_t \sum_{j=1}^n \partial_j u(x,t-1), & (x,t) \in \Omega \times \mathbb{R}_+, \\ u(x,t) = u_0(x,t), \quad \frac{\partial}{\partial t} u(x,t) = u_1(x,t), & (x,t) \in \Omega \times [-1,0], \\ u(x,t) = 0, \ \Delta u(x,t) = 0, & (x,t) \in \partial\Omega \times \mathbb{R}_+. \end{cases}$$
$$\text{(DPED)}$$

Here $\Omega \subset \mathbb{R}^n$ is bounded, h is a scalar function of bounded variation, and we use the abbreviation $\nabla^3 w(x,t) := \sum_{i,j=1}^n \partial_i \partial_j^2 w(x,t)$.

To write this problem in abstract form, consider the space

$$\mathcal{F} = H_{\frac{1}{2}} \times L^2(\Omega),$$

where we equip the Sobolev space $H_{\frac{1}{2}} := \{u \in H_0^1, \Delta u \in L^2(\Omega)\}$ with the norm $\|u\|_2 := \|\Delta u\|_{L^2}$, which is equivalent to the usual Sobolev norm.

Moreover, we define the operators

$$A := \Delta^2, \quad D(A) := \left\{ u \in H_0^1(\Omega), \Delta u \in H_{\frac{1}{2}}, \Delta^2 u \in L^2(\Omega) \right\},$$

$$D := -\Delta = A^{\frac{1}{2}},$$

$$\eta := \nabla^3 \cdot h,$$

$$\phi := \int_{-1}^0 d\eta,$$

$$\psi := c \sum_{j=1}^n \partial_j \delta_{-1}.$$

Then the following conditions on the initial functions yield to classical solutions of (DDPE):

- The function $u_0(\cdot, 0) \in D(A)$.

- $u_0(\cdot, s) \in H_{\frac{1}{2}}$.

- The map defined by $[-1,0] \ni s \mapsto u_0(\cdot, s) \in H_{\frac{1}{2}}$ belongs to the function space $W^{1,2}([-1,0], H_{\frac{1}{2}})$.

- The function $u_1(\cdot, 0) \in H_{\frac{1}{2}}$.

- $u_1(\cdot, s) \in L^2(\Omega)$.

- The map defined by $[-1, 0] \ni s \mapsto u_1(\cdot, s) \in L^2(\Omega)$ belongs to $W^{1,2}([-1, 0], L^2(\Omega))$.

Analogous calculations to the ones in Examples 4.18 and 9.21 show that

$$\sup_{\omega \in \mathbb{R}} \|\phi_{i\omega}\|_{D(A^{\frac{3}{4}}) \to H} \leq \operatorname{Var}(h)^0_{-1},$$

and

$$\sup_{\omega \in \mathbb{R}} \|\psi_{i\omega}\|_{D(A^{\frac{1}{4}}) \to H} \leq |c|.$$

Thus, using an analogous argument as in Equation (4.9), we obtain that the solutions of the damped plate equation with delay decay exponentially if

$$\operatorname{Var}(h)^0_{-1} + |c| < \frac{3\sqrt{|\lambda_1|}}{2\sqrt{3}},$$

where λ_1 denotes the first eigenvalue of the Dirichlet-Laplacian.

Let us finish this section with the characterization of the domain of fractional powers of operator matrices corresponding to second-order Cauchy problems.

We assume that the function f satisfies Equation (9.23). In the special case $f(s) = s^\alpha$, $\frac{1}{2} \leq \alpha \leq 1$; this characterization has been done by Chen and Triggiani [31]. The ideas presented here are analogous to theirs.

Let us start with some technical results before stating our main theorem in this section. We take $0 < \vartheta < 1$ fixed. It will be useful to define the functions

$$g_1(\mu) := \int_0^\infty \frac{\lambda^{1-\vartheta}}{\lambda^2 + \lambda f(\mu) + \mu} d\lambda, \quad \mu \geq \sigma_0, \tag{9.58}$$

$$g_2(\mu) := \int_0^\infty \frac{\lambda^{-\vartheta}}{\lambda^2 + \lambda f(\mu) + \mu} d\lambda, \quad \mu \geq \sigma_0, \tag{9.59}$$

where $\lambda > 0$ and $\sigma_0 > 0$ are fixed.

Lemma 9.26. *Assume that f fulfills the conditions in Theorem 9.4 and satisfies in addition Equation (9.23). Then for each $0 < \vartheta < 1$, there are constants $c_{1,\vartheta}, C_{1,\vartheta}, c_{2,\vartheta}, C_{2,\vartheta}$ such that*

$$0 < c_{1,\vartheta} \leq \left(\frac{f(\mu)}{\mu^{\frac{1}{2}}}\right)^\vartheta \mu^{\frac{\vartheta}{2}} g_1(\mu) \leq C_{1,\vartheta} < \infty, \quad \forall \mu \geq \sigma_0, \tag{9.60}$$

$$0 < c_{2,\vartheta} \leq \left(\frac{f(\mu)}{\mu^{\frac{1}{2}}}\right)^{1-\vartheta} \mu^{\frac{1+\vartheta}{2}} g_2(\mu) \leq C_{2,\vartheta} < \infty, \qquad \forall \mu \geq \sigma_0. \qquad (9.61)$$

Proof. We start by proving the first inequality in Equation (9.60). Introducing a new variable $\sigma := \frac{\lambda}{\sqrt{\mu}}$, we obtain

$$\mu^{\frac{\vartheta}{2}} g_1(\mu) = \int_0^{\frac{f(\mu)}{\mu^{\frac{1}{2}}}} \frac{\sigma^{1-\vartheta} d\sigma}{\sigma^2 + \sigma \frac{f(\mu)}{\mu^{\frac{1}{2}}} + 1} + \int_{\frac{f(\mu)}{\mu^{\frac{1}{2}}}}^\infty \frac{\sigma^{1-\vartheta} d\sigma}{\sigma^2 + \sigma \frac{f(\mu)}{\mu^{\frac{1}{2}}} + 1}. \qquad (9.62)$$

Now we drop the second integral and replace σ in the denominator of the first integral by $\frac{f(\mu)}{\mu^{\frac{1}{2}}}$. Integrating in σ, this gives

$$\mu^{\frac{\vartheta}{2}} g_1(\mu) \geq \int_0^{\frac{f(\mu)}{\mu^{\frac{1}{2}}}} \frac{\sigma^{1-\vartheta} d\sigma}{\sigma^2 + \sigma \frac{f(\mu)}{\mu^{\frac{1}{2}}} + 1} \geq \frac{\left(\frac{f(\mu)}{\mu^{\frac{1}{2}}}\right)^{2-\vartheta}}{\left(2\left(\frac{f(\mu)}{\mu^{\frac{1}{2}}}\right)^2 + 1\right)(2-\vartheta)}.$$

Thus using our assumptions on f, we obtain

$$\left(\frac{f(\mu)}{\mu^{\frac{1}{2}}}\right)^\vartheta \mu^{\frac{\vartheta}{2}} g_1(\mu) \geq \frac{1}{(2+K^2)(2-\vartheta)}.$$

To show the other inequality in Equation (9.60), we drop the first integral in Equation (9.62) and eliminate $\sigma \frac{f(\mu)}{\mu^{\frac{1}{2}}} + 1$ in the denominator of the second integral. Taking $0 < \varepsilon < \vartheta$, we obtain

$$\mu^{\frac{\vartheta}{2}} g_1(\mu) \leq \int_{\frac{f(\mu)}{\mu^{\frac{1}{2}}}}^\infty \frac{\sigma^{1-\vartheta} d\sigma}{\sigma^2 + \sigma \frac{f(\mu)}{\mu^{\frac{1}{2}}} + 1} \leq \int_{\frac{f(\mu)}{\mu^{\frac{1}{2}}}}^\infty \frac{\sigma^{1-\vartheta} d\sigma}{\sigma^{2-\vartheta+\varepsilon} \sigma^{\vartheta-\varepsilon}}$$

$$\leq \left(\frac{\mu^{\frac{1}{2}}}{f(\mu)}\right)^{\vartheta-\varepsilon} \int_{\frac{f(\mu)}{\mu^{\frac{1}{2}}}}^\infty \sigma^{-1-\varepsilon} d\sigma = \frac{1}{\varepsilon\left(\frac{f(\mu)}{\mu^{\frac{1}{2}}}\right)^\vartheta}.$$

Let us consider now Equation (9.61). Introducing again a new variable $\xi := \lambda \frac{f(\mu)}{\mu}$, we obtain

$$\left(\frac{f(\mu)}{\mu^{\frac{1}{2}}}\right)^{1-\vartheta} \mu^{\frac{1+\vartheta}{2}} g_2(\mu) = \int_0^\infty \frac{\xi^{-\vartheta} d\xi}{\left(\frac{\mu}{f(\mu)^2}\right)\xi^2 + \xi + 1}.$$

We can finish the proof by using the assumptions on f to conclude that

$$0 < \int_0^\infty \frac{\xi^{-\vartheta} d\xi}{K^2\xi^2 + \xi + 1} \le \left(\frac{f(\mu)}{\mu^{\frac{1}{2}}}\right)^{1-\vartheta} \mu^{\frac{1+\vartheta}{2}} g_2(\mu) \le \int_0^\infty \frac{\xi^{-\vartheta} d\xi}{\xi + 1} < \infty.$$

$\square$

In the following, we assume that f satisfies Equation (9.23) and define $D := f(A)$. Recall that in this case

$$R(\lambda, \hat{B}) = \begin{pmatrix} \frac{I - P_\lambda^{-1}A}{\lambda} & P_\lambda^{-1} \\ -AP_\lambda^{-1} & \lambda P_\lambda^{-1} \end{pmatrix}, \tag{9.63}$$

where

$$P_\lambda = \lambda^2 I + \lambda D + A \tag{9.64}$$

and

$$\frac{I - P_\lambda^{-1}A}{\lambda} = (\lambda I + D)P_\lambda^{-1}. \tag{9.65}$$

Our main goal is to identify the space $D((-\hat{B})^\vartheta)$.

Theorem 9.27. *Assume that the operators A and $D = f(A)$ satisfy the conditions of Theorem 9.4 (iii).*

(i) If $0 < \vartheta \le \frac{1}{2}$, then

$$D\left((-\hat{B})^\vartheta\right) = D\left(A^{\frac{1}{2}+\vartheta} f(A)^{-\vartheta}\right) \times D\left(f(A)^\vartheta\right).$$

(ii) If $\frac{1}{2} \le \vartheta < 1$, then

$$D\left((-\hat{B})^\vartheta\right) = \left\{ \begin{pmatrix} x \\ y \end{pmatrix} \in D\left(A^{\frac{1}{2}+\vartheta} f(A)^{-\vartheta}\right) \times D\left(A^{\vartheta-\frac{1}{2}} f(A)^{1-\vartheta}\right) : \right.$$
$$\left. Af(A)^{-1}x + y \in D\left(f(A)^\vartheta\right) \right\}.$$

Proof. We use the fact that $D((-\hat{B})^\vartheta) = (-\hat{B})^{-\vartheta}\mathcal{F}$. Thus, we have $\begin{pmatrix} x \\ y \end{pmatrix} \in D((-\hat{B})^\vartheta)$ if and only if

$$\begin{pmatrix} x \\ y \end{pmatrix} = (-\hat{B})^{-\vartheta}w = \frac{\sin \pi\vartheta}{\pi} \int_0^\infty \lambda^{-\vartheta} R(\lambda, \hat{B})w \, d\lambda$$

for some $w = \begin{pmatrix} w_1 \\ w_2 \end{pmatrix} \in \mathcal{F}$. Setting

$$z_1 := \frac{\sin \pi\vartheta}{\pi} A^{\frac{1}{2}} w_1 \in H, \qquad z_2 := \frac{\sin \pi\vartheta}{\pi} w_2 \in H,$$

and using Equation (9.63) for the resolvent, we obtain that

$$x = \int_0^\infty \lambda^{-\vartheta} \left[(\lambda I + D) A^{-\frac{1}{2}} P_\lambda^{-1} z_1 + P_\lambda^{-1} z_2 \right] d\lambda \qquad (9.66)$$

and

$$y = \int_0^\infty \lambda^{-\vartheta} \left[P_\lambda^{-1} A^{\frac{1}{2}} z_1 + \lambda P_\lambda^{-1} z_2 \right] d\lambda \qquad (9.67)$$

for some $z_1, z_2 \in H$.

Since A is selfadjoint, positive definite, we have the spectral representation

$$A = \int_{\mu_0}^\infty \mu \, dE_\mu, \quad \mu_0 > 0,$$

where $(E_\mu)_{\mu \geq \mu_0}$ is the associated family of spectral projections. Recalling Equation (9.65) and the special form of the operator D, we can rewrite Equations (9.66) and (9.67), respectively, as

$$x = \int_0^\infty \int_{\mu_0}^\infty \frac{\lambda^{-\vartheta} (\lambda + f(\mu)) \mu^{-\frac{1}{2}}}{\lambda^2 + \lambda f(\lambda) + \mu} dE_\mu z_1 d\lambda + \int_0^\infty \int_{\mu_0}^\infty \frac{\lambda^{-\vartheta} dE_\mu z_2 d\lambda}{\lambda^2 + \lambda f(\lambda) + \mu} \qquad (9.68)$$

and

$$y = \int_0^\infty \int_{\mu_0}^\infty \frac{\lambda^{-\vartheta} \mu^{\frac{1}{2}} dE_\mu z_1 d\lambda}{\lambda^2 + \lambda f(\lambda) + \mu} + \int_0^\infty \int_{\mu_0}^\infty \frac{\lambda^{1-\vartheta} dE_\mu z_2 d\lambda}{\lambda^2 + \lambda f(\lambda) + \mu}. \qquad (9.69)$$

Note that dE_μ is a finite measure. Since Fubini's Theorem applies on a smooth set dense on $H \times H$, we may interchange the order of integration. Using the functions g_1 and g_2 defined by Equations (9.58) and (9.59), respectively, we obtain that

$$x = A^{-\frac{1}{2}} g_1(A) z_1 + f(A) A^{-\frac{1}{2}} g_2(A) z_1 + g_2(A) z_2$$

and

$$y = -A^{-\frac{1}{2}} g_2(A) z_1 + g_1(A) z_2$$

hold for some $z_1, z_2 \in H$.

Now Lemma 9.26 can be applied to this situation as follows: The operators

$$T_1 := f(A)^\vartheta g_1(A) \qquad (9.70)$$

and

$$T_2 := f(A)^{1-\vartheta} A^{\vartheta} g_2(A) \tag{9.71}$$

are isomorphisms on H. Further, they are strictly positive, selfadjoint operators that commute with any function of A. Making use of these operators, we can rewrite Equations (9.68) and (9.69), respectively, as

$$x = \left(f(A)^{\vartheta} A^{\frac{1}{2}} \right)^{-1} T_1 z_1 + f(A)^{\vartheta} A^{-\frac{1}{2}-\vartheta} T_2 z_1 + f(A)^{\vartheta-1} A^{-\vartheta} T_2 z_2 \tag{9.72}$$

and

$$y = -f(A)^{\vartheta-1} A^{\frac{1}{2}-\vartheta} T_2 z_1 + f(A)^{-\vartheta} T_1 z_2. \tag{9.73}$$

Introducing the positive selfadjoint operator

$$T_3 := \left(f(A)^{\vartheta} A^{\frac{1}{2}} \right)^{-2} T_1 + T_2,$$

we obtain that

$$x = A^{-\frac{1}{2}-\vartheta} f(A)^{\vartheta} T_3 z_1 + f(A)^{\vartheta-1} A^{-\vartheta} T_2 z_2 \tag{9.74}$$

and

$$y = -A^{\frac{1}{2}-\vartheta} f(A)^{\vartheta-1} T_2 z_1 + f(A)^{-\vartheta} T_1 z_2. \tag{9.75}$$

Summarizing, we have shown that $\binom{x}{y} \in D((-A)^{\vartheta})$ if and only if it can be written as Equations (9.74) and (9.75) for suitable $z_1, z_2 \in H$.

For any $z_1, z_2 \in H$, we have

$$A^{-\frac{1}{2}-\vartheta} f(A)^{\vartheta} T_3 z_1 \in D \left(A^{\frac{1}{2}+\vartheta} f(A)^{-\vartheta} \right)$$

$$f(A)^{\vartheta-1} A^{-\vartheta} T_2 z_2 \in D \left(f(A)^{-\vartheta+1} A^{\vartheta} \right).$$

However,

$$A^{\frac{1}{2}+\vartheta} f(A)^{-\vartheta} \supset \left[A^{\frac{1}{2}} f(A)^{-1} \right] f(A)^{-\vartheta+1} A^{\vartheta},$$

where $A^{\frac{1}{2}} f(A)^{-1}$ is a bounded linear operator by assumption.

Hence, by Equation (9.74),

$$x \in D \left(A^{\frac{1}{2}+\vartheta} f(A)^{-\vartheta} \right). \tag{9.76}$$

Analogously, for y we obtain

$$A^{\frac{1}{2}-\vartheta} f(A)^{\vartheta-1} T_2 z_1 \in D \left(A^{\vartheta-\frac{1}{2}} f(A)^{1-\vartheta} \right)$$

and
$$f(A)^{-\vartheta} T_1 z_2 \in D\left(f(A)^{\vartheta}\right).$$

Now we have to distinguish between two cases.

If $0 < \vartheta \le \frac{1}{2}$, then
$$f(A)^{\vartheta} \supset \left[A^{\frac{1}{2}} f(A)^{-1}\right]^{1-2\vartheta} A^{\vartheta - \frac{1}{2}} f(A)^{1-\vartheta},$$

where $\left[A^{\frac{1}{2}} f(A)^{-1}\right]^{1-2\vartheta}$ is a bounded operator by assumption. Hence, in this case,
$$y \in D\left(f(A)^{\vartheta}\right). \tag{9.77}$$

If $\frac{1}{2} \le \vartheta < 1$, then
$$A^{\vartheta - \frac{1}{2}} f(A)^{1-\vartheta} \supset \left[A^{\frac{1}{2}} f(A)^{-1}\right]^{2\vartheta - 1} f(A)^{\vartheta},$$

and hence
$$y \in D\left(A^{\vartheta - \frac{1}{2}} f(A)^{1-\vartheta}\right). \tag{9.78}$$

To finish the proof, we return to Equations (9.72) and (9.73). First applying $f(A)A^{-1}$ to Equation (9.72) and adding the result to Equation (9.73), we obtain

$$
\begin{aligned}
x + f(A)A^{-1}y = A^{-\frac{1}{2}} f(A)^{-\vartheta} T_1 z_1 & \quad \in D\left(A^{\frac{1}{2}} f(A)^{\vartheta}\right) \\
+ A^{-\vartheta} f(A)^{\vartheta - 1} T_2 z_2 & \quad \in D\left(A^{\vartheta} f(A)^{1-\vartheta}\right) \\
+ f(A)^{1-\vartheta} A^{-1} T_1 z_1 & \quad \in D\left(A f(A)^{\vartheta - 1}\right).
\end{aligned}
$$

Since
$$A f(A)^{\vartheta - 1} \supset \left[A^{\frac{1}{2}} f(A)^{-1}\right] A^{\frac{1}{2}} f(A)^{\vartheta},$$
$$A f(A)^{\vartheta - 1} \supset \left[A^{\frac{1}{2}} f(A)^{-1}\right]^{2-2\vartheta} A^{\vartheta} f(A)^{1-\vartheta},$$

it follows that
$$x + f(A)A^{-1}y \in D\left(A f(A)^{\vartheta - 1}\right),$$

or
$$A f(A)^{-1} x + y \in D\left(f(A)^{\vartheta}\right). \tag{9.79}$$

Note that for $0 < \vartheta \le \frac{1}{2}$, this condition is satisfied since by Equation (9.76)
$$A f(A)^{-1} x \in D\left(f(A)^{1-\vartheta} A^{\vartheta - \frac{1}{2}}\right) \subset D\left(f(A)^{\vartheta}\right)$$

for $0 < \vartheta \le \frac{1}{2}$ and by Equation (9.77), $y \in D\left(f(A)^{\vartheta}\right)$.

Thus, we have shown the following inclusions:

- If $0 < \vartheta \leq \frac{1}{2}$, then

$$D\left((-\hat{\mathcal{B}})^{\vartheta}\right) \subset D\left(A^{\frac{1}{2}+\vartheta}f(A)^{-\vartheta}\right) \times D\left(f(A)^{\vartheta}\right).$$

- If $\frac{1}{2} \leq \vartheta < 1$, then

$$D\left((-\hat{\mathcal{B}})^{\vartheta}\right) \subset \left\{ \left(\begin{smallmatrix} x \\ f \end{smallmatrix}\right) \in D\left(A^{\frac{1}{2}+\vartheta}f(A)^{-\vartheta}\right) \times D\left(A^{\vartheta-\frac{1}{2}}f(A)^{1-\vartheta}\right) : \right.$$
$$\left. Af(A)^{-1}x + y \in D\left(f(A)^{\vartheta}\right)\right\}.$$

The other inclusion can be shown in exactly the same way as in Chen and Triggiani [31, Lemma 2.3, Lemma 2.4]. This completes the proof. □

9.5   Notes and References

In Section 9.1, we collect some basic facts on dissipative wave equations in Hilbert spaces.

There is exhaustive literature concerning second-order abstract Cauchy problems. J. Lions [133] and K. Yosida [225] were the first to use the idea of reduction. For more recent literature on well-posedness, see the papers of Engel and Nagel [71], Fattorini [77], Goldstein [87–89], Krein and Langer [125,126], Neubrander [161,162], Obrecht [163], Sandefur [182,183], Xiao and Liang [224], and Zheng [227]. For further references, see [72, Section VI.3]. In Corollary 9.2, statement (iii) follows from a perturbation result of Sohr [195, Satz 2.1]; for a simpler proof see Engel [67, Proposition A.1]. The implication (iv) is a consequence of Clement and Prüß [37, Remark 1]; see also Engel [67, Remark (3)].

The results of Section 9.2 are mainly taken from [14] and were obtained in collaboration with K.-J. Engel. Parts of these results also appeared in [72, Section VI.3.c]. See also the related results of S.-Z. Huang [115]. Lemma 9.10 was obtained in [17].

For internally damped wave equations, see Gorain [90] and Chen and Triggiani [30], especially for the spectral characterization of this matrix operator and for the analyticity of the semigroup generated by it.

The presented damped wave equation and related works are studied in Cox and Zuazua [41,42], Komornik [122, Section 1.3], Lopez-Gomez [134], Tcheugoue-Tebou [200], and Wyler [223].

This estimate on the stability is not optimal in general. In the one-dimensional case, Cox and Zuazua derived the exact decay rates for bounded damping; see [41] and [42]. Similar results were obtained by Shubov for the radially symmetric, two-dimensional equation; see [191] and [192]. These results can be understood in view of a general result of Renardy [172], who gave conditions for the permanence of the principle of linear stability under bounded perturbations. A. Wyler [223] used the theorem of Gearhart to derive stability results under different well-posedness conditions. Unfortunately, we could not use his results to estimate the decay rate and the norm of the resolvent. Refinements of our results can be found, for example, in Veselić [206].

Note that already for two-dimensional equations there are examples for which $s(\mathcal{B}) \neq \omega_0(\mathcal{B})$; see Lebeau [131] and Renardy [173].

For unbounded damping, the only work known to us is Tcheugoué Tébou [200] and Lebeau and Zuazua [132], where other types of conditions ensure polynomial decay rates for not necessarily strict dissipative damping, i.e., where $q(x) = 0$ can occur on a "small" subset of Ω.

Section 9.3 is an attempt to treat wave equations with delay unifying the semigroup treatment of wave equations and the semigroup treatment of first-order delay equations. The results of this section are mainly taken from [17].

For other approaches to wave equations with delay, we refer to the papers by Maniar and Rhandi [138], Prüß [169], Sakawa [181], Sinestrari [194], and Vagabov [203]. Oscillation results for the one-dimensional wave equations with delay were proved in Györi and Krisztin [96].

Section 9.4 is the analogue of the previous section for relatively bounded operators in the delay term and is partly from [18].

For other treatments of second-order equations with more unbounded operators in the delay term, see the works of Fašanga and Prüß [75] and Prüß [169].

Chapter 10

Delays in the Highest-Order Derivatives

In this chapter, we show how the perturbation argument presented in Section 3.3.3 and Section 3.4 can be extended to parabolic problems with delays in the highest-order derivatives. The idea is to use the same additive perturbation method as before. We have to consider this type of equation separately from the rest of the book because of the additional technical difficulties caused by the unboundedness of the delay operator.

In Section 10.1, we present the perturbation theorem of Weiss and Staffans which generalizes the Desch-Schappacher and the Miyadera-Voigt Perturbation Theorems. This theorem will be used in Section 10.2 to show the generation property of the associated matrix operator and hence the existence of the delay semigroup. Section 10.3 is devoted to the spectral properties of the generator and in Section 10.4 we analyze the asymptotic behavior of the solutions of the delay equation. We apply the same method, namely the application of semigroup theoretic results on the delay semigroup. It is important to note that, although the equation without delay is of parabolic type, generally the Spectral Mapping Theorem does not hold for the delay semigroup.

The results of this chapter were obtained in collaboration with R. Schnaubelt and are taken from [20].

10.1   The Perturbation Theorem of Weiss-Staffans

Let Z and U be Banach spaces, $1 < q < \infty$, and $(T(t))_{t\geq 0}$ be a C_0-semigroup on Z generated by G. An *output system* for $(T(t))_{t\geq 0}$ is a family of bounded linear operators $\mathbb{W}_t : Z \to L^q([0,t], U)$, $t \geq 0$, satisfying

$$[\mathbb{W}_{t+s}z](\tau) = [\mathbb{W}_t T(s)z](\tau - s) \qquad (10.1)$$

209

for $\tau \in [s, s + t]$, $z \in Z$, and $t, s \geq 0$. Given $\mathbb{W}_t$, there is a bounded *output operator* $C : D(G) \to U$ such that $(\mathbb{W}_t z)(\tau) = CT(\tau)z$ for $z \in D(G)$ and $0 \leq \tau \leq t$. Moreover, there exists the so-called *Yosida extension* $\tilde{C}$ of C defined by

$$\tilde{C}z = \lim_{s \to \infty} sCR(s, G)z \qquad \text{for} \quad z \in D(\tilde{C}) = \{z \in Z : \text{this limit exists in } U\}.$$

It can be shown that $S(\tau)z \in D(\tilde{C})$ and

$$(\mathbb{W}_t z)(\tau) = \tilde{C}T(\tau)z, \qquad z \in Z, \quad \text{a.e. } \tau \in [0, t]. \tag{10.2}$$

An *input system* for $(T(t))_{t \geq 0}$ is a family of bounded linear operators $\mathbb{V}_t : L^q([0, t], U) \to Z$, $t \geq 0$, such that

$$\mathbb{V}_{t+s} u = \mathbb{V}_t(u(\cdot + s)|_{[0,t]}) + T(t)\mathbb{V}_s(u|_{[0,s]}) \tag{10.3}$$

for $u \in L^q([0, s+t], U)$ and $t, s \geq 0$. To represent $\mathbb{V}_t$, we need the *extrapolation space* Z_{-1} of Z for G, which is the completion of Z with respect to the norm $\|R(w, G)z\|$ for some fixed $w \in \rho(G)$; see Engel and Nagel [72, Section II.5] for the definitions and properties of these spaces. One can uniquely extend G to a bounded operator $G_{-1} : Z \to Z_{-1}$, which generates a C_0-semigroup T_{-1} in Z_{-1} extending the family $(T(t))_{t \geq 0}$. Every input system can be written as

$$\mathbb{V}_t u = \int_0^t T_{-1}(t - s)Du(s)\, ds, \tag{10.4}$$

where the integral exists in Z_{-1} and the *input operator* $D \in \mathcal{L}(U, Z_{-1})$ is given by

$$Du_0 = \lim_{t \to 0} \tfrac{1}{t} \mathbb{V}_t u_0 \qquad \text{(in } Z_{-1}\text{)} \tag{10.5}$$

for $u_0 \in U$ (also denoting the corresponding constant function).

Given $T(t)$, $\mathbb{W}_t$, and $\mathbb{V}_t$, we call bounded operators $\mathbb{F}_t$ on $L^q([0, t], U)$, $t \geq 0$, *input-output operators* if

$$[\mathbb{F}_{t+s} u](\tau) = [\mathbb{F}_t(u(\cdot + s)|_{[0,t]})](\tau - s) + [\mathbb{W}_t \mathbb{V}_s(u|_{[0,s]})](\tau - s) \tag{10.6}$$

for $\tau \in [s, s + t]$, $t, s \geq 0$, and $u \in L^q([0, s+t], U)$. A special continuity property of $\mathbb{F}$ will be important, therefore we introduce a notion for it. The system $\Sigma = (T(t), \mathbb{V}_t, \mathbb{W}_t, \mathbb{F}_t; t \geq 0)$ is called *regular* if

$$\lim_{t \to 0} \frac{1}{t} \int_0^t [\mathbb{F}_t u_0](\tau)\, d\tau = 0. \tag{10.7}$$

It can be shown that $\mathbb{F}_t u(\tau) = \tilde{C} \int_0^\tau T_{-1}(\tau - \sigma)Du(\sigma)\, d\sigma$ for $u \in L^q([0, t], U)$ and almost every $\tau \in [0, t]$, but we will not use this fact.

A bounded operator K on U is called an *admissible feedback* for a regular system Σ, if $I - \mathbb{F}_t K$ is invertible on $L^q([0,t], U)$ on some (and hence all) $t > 0$. Then

$$G_K = G_{-1} + DK\tilde{C} \qquad \text{with} \quad D(G_K) = \{z \in D(\tilde{C}) : G_K z \in Z\}$$

(the sum is defined in Z_{-1}) generates the unique C_0-semigroup S_K on Z such that

$$T_K(t)z = T(t)z + \int_0^t T_{-1}(t-s)DK\tilde{C}T_K(s)z\, ds$$

for $z \in Z$ and $t \geq 0$. Here one has $\tilde{C}T_K(\cdot)z = (I - \mathbb{F}_t K)^{-1}\mathbb{W}_t z$ for $z \in Z$. Observe that the perturbation $DK\tilde{C}$ acts from an intermediate space between $D(G)$ and Z into a space larger than Z.

10.2   Well-Posedness

In this section, we study the general problem

$$\text{(DEU)} \qquad \begin{cases} u'(t) = Bu(t) + \Phi u_t, & t \geq 0, \\ u(0) = x, \\ u_0 = f, \end{cases}$$

where Φ is given by

$$\Phi f := \int_{-1}^0 d\eta\, f, \qquad f \in W^{1,p}([-1,0], D(B))$$

for a map $\eta : [-1,0] \to \mathcal{L}(D(B), X)$ of bounded variation.

Our main hypothesis reads as follows.

Hypothesis 10.1. *The operator B is sectorial with dense domain $X_1 = D(B)$ in X of type (ϕ, K, d), where $\phi \in (\pi/2, \pi]$, $d \in \mathbb{R}$, $K > 0$, B has maximal regularity of type L^p, and the operator-valued function $\eta : [-1,0] \to \mathcal{L}(X_1, X)$ has bounded variation $|\eta|$ with $d|\eta|([-t,0]) \to 0$ as $t \searrow 0$.*

We want to reformulate (DEU) as a perturbation problem for a semigroup on the product space

$$\mathcal{X} = (X, X_1)_{1-\frac{1}{p}, p} \times L^p([-1,0], X_1) = Y \times L^p([-1,0], X_1)$$

endowed with the norm $\|\binom{x}{f}\|_X = \|x\|_Y + \|f\|_{L^p([-1,0],X_1)}$. Recall that the interpolation space Y was defined in Section 1.3.1 by Equation (1.4). Throughout this chapter, we suppose that Hypothesis 10.1 holds for some fixed $p \in (1,\infty)$. Let us denote by $(S(t))_{t\geq 0}$ the analytic semigroup generated by $(B, D(B))$ and define

$$(S_t x)(\theta) = \begin{cases} S(t+\theta)x, & \theta \in [-1,0],\ t+\theta \geq 0, \\ 0, & \theta \in [-1,0],\ t+\theta < 0, \end{cases}$$

and

$$(T_0(t)f)(\theta) = \begin{cases} f(t+\theta), & \theta \in [-1,0],\ t+\theta < 0, \\ 0, & \theta \in [-1,0],\ t+\theta \geq 0, \end{cases}$$

for $x \in X$, $f \in L^p([-1,0],X_1)$, and $t \geq 0$. Observe that $T_0(t)$ yields the left shift semigroup $(T_0(t))_{t\geq 0}$ on $L^p([-1,0],X_1)$ generated by $A_0 f = \frac{d}{d\sigma}f$ with domain $D(A_0) = \{f \in W^{1,p}([-1,0],X_1) : f(0) = 0\}$. Clearly, one has $\rho(A_0) = \mathbb{C}$ and

$$(R(\lambda,A_0)f)(\theta) = \int_\theta^0 e^{\lambda(\theta-s)}f(s)\,ds = \int_\theta^0 e^{\lambda s}f(\theta-s)\,ds, \qquad \theta \in [-1,0].$$

$$(10.8)$$

After these notations, we introduce the matrix operators

$$\mathcal{T}_0(t) = \begin{pmatrix} S(t) & 0 \\ S_t & T_0(t) \end{pmatrix} \qquad \text{and} \qquad \mathcal{A}_0 = \begin{pmatrix} B & 0 \\ 0 & \frac{d}{d\sigma} \end{pmatrix}$$

on $\mathcal{X}$ with $D(\mathcal{A}_0) = \{\binom{x}{f} \in X_1 \times W^{1,p}([-1,0],X_1) : f(0) = x\}$. Finally, we set $(\varepsilon_\lambda x)(\theta) = e^{\lambda\theta}x$ for $x \in X$, $\lambda \in \mathbb{C}$, and $\theta \in [-1,0]$.

Lemma 10.2. *The matrix operator $\mathcal{A}_0$ generates the C_0-semigroup $\mathcal{T}_0 = (\mathcal{T}_0(t))_{t\geq 0}$ on $\mathcal{X}$. Moreover, $\omega_0(\mathcal{A}_0) = \omega_0(B)$, $\sigma(\mathcal{A}_0) = \sigma(B)$, and*

$$R(\lambda,\mathcal{A}_0) = \begin{pmatrix} R(\lambda,B) & 0 \\ \varepsilon_\lambda R(\lambda,B) & R(\lambda,A_0) \end{pmatrix} =: \mathcal{R}_\lambda, \qquad \lambda \in \rho(B).$$

Proof. Recall that we assumed in Hypothesis 10.1 that $(B, D(B))$ generates an analytic semigroup. Due to Equation (1.4), $\mathcal{T}_0(t)$ is a bounded operator on $\mathfrak{X}$ and $\mathcal{T}_0(t)\binom{x}{f} \to \binom{x}{f}$ in $\mathfrak{X}$ as $t \to 0$. It is easy to verify that $\mathcal{T}_0(\cdot)$ is a semigroup with growth bound $\omega_0(B)$; see Section 3.3.2, especially Theorem 3.25. We only have to make sure that the operators $\mathcal{T}_0(t)$ define a bounded operator on $\mathfrak{X}$. This statement follows from the characterization of Y (see Equation (1.4)), meaning that $S_t : Y \to L^p([-1,0], X_1)$ is a bounded operator. Thus $\mathcal{T}_0$ is a C_0-semigroup whose generator is denoted by $\tilde{A}_0$. Taking the Laplace transform of $\mathcal{T}_0$, we see that $R(\lambda, \tilde{A}_0) = \mathcal{R}_\lambda$ for $\Re\lambda > \omega_0(B)$.

Let $\lambda \in \rho(B)$. It follows with the same proof as in Section 3.2 that $\mathcal{R}_\lambda \mathfrak{X} \subseteq D(\mathcal{A}_0)$, $(\lambda - \mathcal{A}_0)\mathcal{R}_\lambda = I$, and $\mathcal{R}_\lambda(\lambda - \mathcal{A}_0)\binom{x}{f} = \binom{x}{f}$ for $\binom{x}{f} \in D(\mathcal{A}_0)$. Hence, $\lambda \in \rho(\mathcal{A}_0)$ and $R(\lambda, \mathcal{A}_0) = \mathcal{R}_\lambda$ for $\lambda \in \rho(B)$. This fact also shows that $\mathcal{A}_0 = \tilde{\mathcal{A}}_0$.

It remains to show that $\sigma(B) \subset \sigma(\mathcal{A}_0)$. Assume that $\lambda \in \rho(\mathcal{A}_0)$. If $(\lambda - B)x = 0$ for some $x \in X_1$, then $\binom{x}{\varepsilon_\lambda x} \in D(\mathcal{A}_0)$ and $(\lambda - \mathcal{A}_0)\binom{x}{\varepsilon_\lambda x} = 0$, so that $x = 0$. If $y \in Y$ is given, then there exists $\binom{x}{f} \in D(\mathcal{A})$ such that $(\lambda - \mathcal{A})\binom{x}{f} = \binom{y}{0}$. In particular, $x \in X_1$ and $(\lambda - B)x = y$, and hence $Bx \in Y$. Therefore, λ belongs to the resolvent set of the part of B in Y, hence $\lambda \in \rho(B)$. $\qquad\square$

Using the matrix $\mathcal{A}_0$ and the substitution $v(t) = u_t$, one can rewrite (DEU) as the evolution equation

$$\binom{u'(t)}{v'(t)} = \mathcal{A}_0\binom{u(t)}{v(t)} + \mathcal{B}\binom{u(t)}{v(t)}, \quad t \geq 0, \qquad \binom{u(0)}{v(0)} = \binom{f(0)}{f} \qquad (10.9)$$

on $\mathfrak{X}$, where $f \in W^{1,p}([-1,0], X_1) \hookrightarrow C([-1,0], X_1)$,

$$\Phi f = \int_{-1}^{0} d\eta(\theta) f(\theta), \quad \text{and} \quad \mathcal{B} = \begin{pmatrix} 0 & \Phi \\ 0 & 0 \end{pmatrix};$$

compare with the proof of Theorem 10.7. Observe that the perturbation $\mathcal{B}$ maps $Y \times C([-1,0], X_1)$ into $X \times L^p([-1,0], X_1)$, that is, it acts from an intermediate space between $D(\mathcal{A}_0)$ and $\mathfrak{X}$ into a space larger than $\mathfrak{X}$. In order to deal with this difficulty, we use the perturbation theorem of Weiss-Staffans introduced in the previous section. The basic idea is to factorize $\mathcal{B} = \mathcal{DC}$ via the space $\mathcal{U} = X_1 \times L^p([-1,0], X_1)$ and the continuous operators

$$\mathcal{D} = \begin{pmatrix} w - B & 0 \\ 0 & 0 \end{pmatrix} : \mathcal{U} \to X \times L^p([-1,0], X_1),$$

$$\mathcal{C} = \begin{pmatrix} 0 & R(w, B)\Phi \\ 0 & 0 \end{pmatrix} : X \times C([-1,0], X_1) \to \mathcal{U},$$

where $w \in \rho(B)$ is fixed. Observe that $\mathcal{C} \in \mathcal{L}(D(\mathcal{A}_0), \mathcal{U})$ if we endow $D(\mathcal{A}_0)$ with its graph norm. We work primarily with the corresponding input and output maps defined by

$$\mathbb{V}_t \begin{pmatrix} u \\ f \end{pmatrix} := \begin{pmatrix} \int_0^t S(t-s)(w-B)u(s)\, ds \\ \int_0^t S_{t-s}(w-B)u(s)\, ds \end{pmatrix}, \tag{10.10}$$

$$\begin{pmatrix} u \\ f \end{pmatrix} \in L^p([0,t], \mathcal{U}), \ t \geq 0,$$

$$\left[\mathbb{W}_t \begin{pmatrix} x \\ f \end{pmatrix} \right](s) := \mathcal{C}\mathcal{T}_0(s) \begin{pmatrix} x \\ f \end{pmatrix} = \begin{pmatrix} R(w,B)\Phi g_{f,x}(s) \\ 0 \end{pmatrix}, \tag{10.11}$$

$$\begin{pmatrix} x \\ f \end{pmatrix} \in D(\mathcal{A}_0), \ 0 \leq s \leq t,$$

where $[g_{f,x}(s)](\theta) = S(s+\theta)x$ if $s+\theta \geq 0$ and $[g_{f,x}(s)](\theta) = f(s+\theta)$ if $s+\theta < 0$, for $\theta \in [-1,0]$ and $s \in [0,t]$. Observe that the second component of $\mathbb{V}_t \begin{pmatrix} u \\ f \end{pmatrix}$ is equal to

$$\int_0^{(t+\theta)^+} S((t+\theta)^+ - s)(w-B)u(s)\, ds, \qquad \theta \in [-1,0],$$

where we used the notation $s^+ := s \vee 0 = \max\{s, 0\}$. In the following we do not distinguish in notation between a function defined on an interval and its restrictions to subintervals.

Lemma 10.3. *Assume that Hypothesis 10.1 holds. Then* $\mathbb{W}_t$ *can be extended to a bounded operator from* X *into* $L^p([0,t], \mathcal{U})$ *such that* $\|\mathbb{W}_t \begin{pmatrix} x \\ f \end{pmatrix}\|_{L^p([0,t],\mathcal{U})} \leq C_1 \|\eta\|_{BV} \|\begin{pmatrix} x \\ f \end{pmatrix}\|_X$ *for* $\begin{pmatrix} x \\ f \end{pmatrix} \in X$, $t \in [0,r]$, *and a constant* C_1 *only depending on* p, w, *and the type of* B. *Moreover* $\mathbb{V}_t$ *and* $\mathbb{W}_t$, $t \geq 0$, *are input and output systems for* $\mathcal{T}_0$.

Proof. (1) By maximal regularity and the embedding in Equation (1.5), $\mathbb{V}_t$ maps continuously into X. Let $\begin{pmatrix} u \\ f \end{pmatrix} \in L^p_{loc}(\mathbb{R}_+, \mathcal{U})$ and $t, s \geq 0$. Then

$$\mathcal{T}_0(t)\mathbb{V}_s \begin{pmatrix} u \\ f \end{pmatrix} = \begin{pmatrix} \int_0^s S(t+s-\tau)(w-B)u(\tau)\, d\tau \\ I(t,s) \end{pmatrix}, \qquad \text{where}$$

$$I(t,s)(\theta) = \begin{cases} \int_0^s S(t+\theta+s-\tau)(w-B)u(\tau)\, d\tau, \\ \qquad\qquad t+\theta \geq 0, \ \theta \in [-1,0], \\ \int_0^{(t+\theta+s)^+} S((t+\theta+s)^+ - \tau)(w-B)u(\tau)\, d\tau, \\ \qquad\qquad t+\theta < 0, \ \theta \in [-1,0]. \end{cases}$$

On the other hand,

$$\mathbb{V}_t \begin{pmatrix} u(\cdot + s) \\ f(\cdot + s) \end{pmatrix} = \begin{pmatrix} \int_s^{s+t} S(t + s - \tau)(w - B)u(\tau)\, d\tau \\ \int_s^{s+(t+\cdot)^+} S((t + \cdot)^+ + s - \tau)(w - B)u(\tau)\, d\tau \end{pmatrix}.$$

Adding these two expressions, Equation (10.3) is verified.

(2) Equation (10.1) is clear on $D(\mathcal{A}_0)$. Let $\binom{x}{f} \in D(\mathcal{A}_0)$, $0 \le t \le r$, and $1/p + 1/p' = 1$. Using Hölder's Inequality, Fubini's Theorem, and Equation (1.4), we estimate

$$\|\mathbb{W}_t \binom{x}{f}\|_{L^p([0,t],\mathfrak{U})} = \left[\int_0^t \|R(w,B)\Phi(S_s x + T_0(s)f)\|_{X_1}^p\, ds \right]^{\frac{1}{p}}$$

$$\le \left[\int_0^t \left[\int_{-s}^0 \|S(s + \theta)x\|_{X_1} d|\eta|(\theta) \right]^p ds \right]^{\frac{1}{p}}$$

$$+ \left[\int_0^t \left[\int_{-1}^{-s\vee -1} \|f(s + \theta)\|_{X_1} d|\eta|(\theta) \right]^p ds \right]^{\frac{1}{p}}$$

$$\le \|\eta\|_{\tilde{B}V}^{\frac{1}{p'}} \left(\left[\int_0^t \int_{-s}^0 \|S(s + \theta)x\|_{X_1}^p d|\eta|(\theta)\, ds \right]^{\frac{1}{p}} \right.$$

$$+ \left. \left[\int_0^t \int_{-1}^{-s\vee -1} \|f(s + \theta)\|_{X_1}^p d|\eta|(\theta)\, ds \right]^{\frac{1}{p}} \right)$$

$$\le \|\eta\|_{\tilde{B}V}^{\frac{1}{p'}} \left(\left[\int_{-t}^0 \int_{-\theta}^t \|S(s + \theta)x\|_{X_1}^p\, ds\, d|\eta|(\theta) \right]^{\frac{1}{p}} \right.$$

$$+ \left. \left[\int_{-1}^0 \int_0^{-\theta \wedge t} \|f(s + \theta)\|_{X_1}^p\, ds\, d|\eta|(\theta) \right]^{\frac{1}{p}} \right)$$

$$\le \|\eta\|_{\tilde{B}V}^{\frac{1}{p'}} \left(\left[\int_{-t}^0 c\, \|x\|_Y^p\, d|\eta|(\theta) \right]^{\frac{1}{p}} + \left[\int_{-1}^0 \|f\|_{L^p([-1,0],X_1)}^p\, d|\eta|(\theta) \right]^{\frac{1}{p}} \right)$$

$$\le c\, \|\eta\|_{BV} \|\binom{x}{f}\|_X.$$

Here, as before, $\|\eta\|_{BV} = |\eta|([-1,0])$. Hence, $\mathbb{W}_t$ can be continuously extended to the whole X with the same property as Equation (10.1). $\square$

In the next lemma we compute the Yosida extension $\tilde{\mathcal{C}}$ of $\mathcal{C}$ for a class of vectors that is large enough for our purposes, namely

$$\mathcal{D}_C = \{ \binom{x}{f} \in X_1 \times C([-1,0], X_1) : f(0) = x \}.$$

Lemma 10.4. *Assume that Hypothesis 10.1 holds. If $\binom{x}{f} \in \mathcal{D}_C$, then $\binom{x}{f} \in D(\tilde{\mathcal{C}})$ and $\tilde{\mathcal{C}}\binom{x}{f} = \mathcal{C}\binom{x}{f} = \binom{R(w,B)\Phi f}{0}$. Moreover, $\mathcal{T}_0(t)\mathcal{D}_C \subseteq \mathcal{D}_C$ for $t \ge 0$.*

Proof. By approximation, we see that $\mathbb{W}_t\binom{x}{f} = \binom{R(w,B)\Phi g_{x,f}}{0}$ for $\binom{x}{f} \in \mathcal{D}_C$; see Equation (10.11). Moreover, the function $s \mapsto g_{x,f}(s) \in C([-1,0], X_1)$ is continuous so that $s \mapsto \mathbb{W}_t\binom{x}{f}(s) \in \mathcal{U}$ is a continuous function. This implies the first assertion thanks to Weiss [220, Proposition 4.7]. The invariance of D_C under $\mathfrak{T}_0(t)$ is clear from the definition. $\qquad\square$

To define the corresponding input-output operator, we use the space $X_2 = D(B^2)$ and introduce

$$\left[\mathbb{F}_t\binom{u}{f}\right](s) = \mathcal{C}\mathbb{V}_s\binom{u}{f}$$

for

$$\binom{u}{f} \in L_{loc}^p(\mathbb{R}_+, X_2 \times L^p([-1,0], X_1))$$

and

$$0 \le s \le t.$$

Observe that for these inputs we have $\mathbb{V}_s\binom{u}{f} \in \mathcal{D}_C$.

Lemma 10.5. *Assume that Hypothesis 10.1 holds. Then $\mathbb{F}_t$, $t \ge 0$, can be extended to bounded operators $\mathbb{F}_t : L^p([0,t],\mathcal{U}) \to L^p([0,t],\mathcal{U})$, where $\|\mathbb{F}_t\| \le C_2 \, d|\eta|([-t,0])$ for $t \le r$ and a constant C_2 only depending on B, p, w. These extensions are regular input-output operators for $\mathfrak{T}_0(t)$, $\mathbb{V}_t$, and $\mathbb{W}_t$.*

Proof. Equation (10.6) is a straightforward consequence of Equation (10.2) and Lemmas 10.3 and 10.4 if $\binom{u}{f} \in L_{loc}^p(\mathbb{R}_+, X_2 \times L^p([-1,0], X_1))$. For $0 \le t \le 1$ and using Hölder's inequality, Fubini's Theorem, and the maximal regularity of B, we further estimate

$$\left\|\mathbb{F}_t\binom{u}{f}\right\|_{L^p([0,t],\mathcal{U})}^p = \int_0^t \left\|\Phi \int_0^s S_{s-\tau}(w-B)u(\tau)\,d\tau\right\|_X^p ds$$

$$\le \int_0^t \left[\int_{-s}^0 \left\|\int_0^{s+\theta} S(s+\theta-\tau)(w-B)u(\tau)\,d\tau\right\|_{X_1} d|\eta|(\theta)\right]^p ds$$

$$\le \int_0^t |\eta|([-s,0])^{p-1} \int_{-s}^0 \left\|\int_0^{s+\theta} S(s+\theta-\tau)(w-B)u(\tau)\,d\tau\right\|_{X_1}^p d|\eta|(\theta)\,ds$$

$$\le |\eta|([-t,0])^{p-1} \int_{-t}^0 \int_0^{t+\theta} \left\|\int_0^{\sigma} S(\sigma-\tau)(w-B)u(\tau)\,d\tau\right\|_{X_1}^p d\sigma\,d|\eta|(\theta)$$

$$\le c\,|\eta|([-t,0])^{p-1} \int_{-t}^0 \int_0^{t+\theta} \|u(\tau)\|_{X_1}^p\,d\tau\,d|\eta|(\theta)$$

$$\le c'\,|\eta|([-t,0])^p \|u\|_{L^p([0,t],X_1)}^p.$$

Thus $\mathbb{F}_t$ can be continuously extended to a bounded operator in $L^p([0,t], \mathcal{U})$ if $t \leq 1$. Due to Equation (10.6) and Lemma 10.3, this extension can be carried out iteratively for all $t \geq 0$ and Equation (10.6) remains valid. To check the regularity property defined in Equation (10.7), first we compute

$$\mathbb{V}_s \begin{pmatrix} x \\ f \end{pmatrix} = \begin{pmatrix} (w - B) \int_0^s S(\tau) x \, d\tau \\ (w - B) \int_0^{(s+\cdot)^+} S(\tau) x \, d\tau \end{pmatrix}$$
$$= w \mathbb{V}_s \begin{pmatrix} R(w, B) x \\ f \end{pmatrix} + \begin{pmatrix} x - S(s) x \\ x - S((s+\cdot)^+) x \end{pmatrix}$$

for $\begin{pmatrix} x \\ f \end{pmatrix} \in \mathcal{U}$ (also denoting the corresponding constant function). Approximating x in X_1 by $x_n \in X_2$, we see that

$$\mathbb{F}_t \begin{pmatrix} x \\ f \end{pmatrix}(s) = \begin{pmatrix} w R(w, B) \Phi \int_0^{(s+\cdot)^+} S(\tau) x \, d\tau \\ 0 \end{pmatrix}$$
$$+ \begin{pmatrix} R(w, B) \Phi (x - S((s+\cdot)^+) x) \\ 0 \end{pmatrix}$$

for $0 \leq s \leq t$. Therefore $\mathbb{F}_t \begin{pmatrix} x \\ f \end{pmatrix}(s) \to 0$ in $\mathcal{U}$ as $s \to 0$, and as a consequence we obtain that $\mathbb{F}_t$ is regular. $\qquad\square$

So we have established that $(\mathcal{T}_0(t), \mathbb{V}_t, \mathbb{W}_t, \mathbb{F}_t)$ is a regular system. But we still have to identify the operator defined in Equation (10.5), which we temporarily denote by $\tilde{B} : \mathcal{U} \to X_{-1}$. Besides the extrapolation space X_{-1} for $\mathcal{A}_0$, we need the space $\mathcal{Z} = X \times L^p([-1, 0], X_1)$ endowed with its usual norm $\left\| \begin{pmatrix} x \\ f \end{pmatrix} \right\|_{\mathcal{Z}} = \|x\|_X + \|f\|_{L^p([-1,0],X_1)}$. Note that B maps $\mathcal{U}$ into $\mathcal{Z}$.

Lemma 10.6. *Assume that Hypothesis 10.1 holds. Then* $X \hookrightarrow \mathcal{Z} \hookrightarrow X_{-1}$ *and* $B = \tilde{B}$.

Proof. One can extend $R(w, \mathcal{A}_0)$ to a continuous and injective operator $\mathcal{R} : \mathcal{Z} \to X$ (having the same representation) due to Lemma 10.2. Let $v \in \mathcal{Z}$. Then there are $v_n \in X$ converging to v in $\mathcal{Z}$. Since

$$\|v_n - v_m\|_{X_{-1}} = \|\mathcal{R}(v_n - v_m)\|_X \leq c \|v_n - v_m\|_{\mathcal{Z}},$$

(v_n) is a Cauchy sequence in X_{-1}. We set $Jv = (v_n) + \mathcal{N}$, where $\mathcal{N}$ is the space of null sequences in $(X, \|\cdot\|_{X_{-1}})$. Observe that $J : \mathcal{Z} \to X_{-1}$ is well defined, linear, and bounded. If $Jv = 0$ for some $v \in \mathcal{Z}$, then $v_n \to 0$ in X_{-1}, i.e., $\mathcal{R}v_n \to 0$ in X. Hence, $\mathcal{R}v = 0$, which shows that $v = 0$. So we have embedded $\mathcal{Z}$ into X_{-1}.

It remains to compute the limit defining $\tilde{\mathcal{B}}$. For $t \in (0,1]$ and $\binom{x}{f} \in \mathcal{U}$, one has

$$R(w, \mathcal{A}_0)\mathbb{V}_t\binom{x}{f}$$

$$= \begin{pmatrix} R(w,B) & 0 \\ \varepsilon_w R(w,B) & R(w,\mathcal{A}_0) \end{pmatrix} \begin{pmatrix} \int_0^t S(t-s)(w-B)x\,ds \\ \int_0^{(t+\cdot)^+} S((t+\cdot)^+ - s)(w-B)x\,ds \end{pmatrix}$$

$$= \begin{pmatrix} \int_0^t S(s)x\,ds \\ \varepsilon_w \int_0^t S(s)x\,ds + I_t \end{pmatrix},$$

where

$$I_t(\theta) = \int_{-t\vee\theta}^0 e^{w(\theta-\tau)}(w-B)\int_0^{t+\tau} S(s)x\,ds\,d\tau$$

$$= \int_{-t\vee\theta}^0 e^{w(\theta-\tau)}\Big[w\int_0^{t+\tau} S(s)x\,ds + x - S(t+\tau)x\Big]\,d\tau,$$

hence

$$\|I_t(\theta)\|_{X_1} \le ct\left(t\|x\|_{X_1} + \sup_{0\le\sigma\le t}\|S(\sigma)x - x\|_{X_1}\right).$$

As a result,

$$R(w,\mathcal{A}_0)\frac{1}{t}\mathbb{V}_t\binom{x}{f} \to \binom{x}{\varepsilon_w x} = R(w,\mathcal{A}_0)\binom{(w-B)x}{0} = R(w,\mathcal{A}_0)\mathcal{B}\binom{x}{f}$$

in $\mathcal{X}$ as $t \to 0$, which means that $\frac{1}{t}\mathbb{V}_t\binom{x}{f} \to \mathcal{B}\binom{x}{f}$ in $\mathcal{X}_{-1}$.   □

Now we come to the main result in this section, where we construct the solution semigroup for (DEU) using the above set-up.

Theorem 10.7. *Assume that Hypothesis 10.1 holds. Then the operator*

$$\mathcal{A} = \begin{pmatrix} B & \Phi \\ 0 & \frac{d}{d\sigma} \end{pmatrix} \qquad \text{with}$$

$$D(\mathcal{A}) = \{\binom{x}{f} \in X_1 \times W^{1,p}([-1,0], X_1) : f(0) = x,\ \Phi f + Bx \in Y\} \tag{10.12}$$

generates the C_0-semigroup $\mathcal{T}$ on $\mathcal{X}$ satisfying

$$\mathcal{T}(t)\binom{x}{f} = \mathcal{T}_0(t)\binom{x}{f} + \int_0^t (\mathcal{T}_0)_{-1}(t-s)\mathcal{D}\mathcal{C}\mathcal{T}(s)\binom{x}{f}\,ds \tag{10.13}$$

$$= \begin{pmatrix} S(t)x \\ S_t x + T_0(t)f \end{pmatrix} + \begin{pmatrix} \int_0^t S(t-s)\Phi v(s)ds \\ \int_0^t S_{t-s}Lv(s)\,ds \end{pmatrix} \tag{10.14}$$

for $v(t) = [\mathcal{T}(t)\binom{x}{f})]_2$, $\binom{x}{f} \in D(\mathcal{A})$, and $t \geq 0$. Moreover,

$$\|\tilde{\mathcal{C}}\mathcal{T}(\cdot)\binom{x}{f}\|_{L^p([0,t],\mathcal{U})} \leq 2C_1 \|\eta\|_{BV} \|\binom{x}{f}\|_X \qquad (10.15)$$

for $t \in [0,t_0]$, $\binom{x}{f} \in \mathcal{X}$, the constant C_1 of Lemma 10.3, and a number $t_0 \in (0,1]$ such that $d|\eta|([-t_0,0]) \leq (2C_2)^{-1}$, where C_2 is given by Lemma 10.5. If $f \in W^{1,p}([-1,0],X_1)$, then $u = [\mathcal{T}(\cdot)\binom{f(0)}{f})]_1$ belongs to $C^1(\mathbb{R}_+, Y) \cap C(\mathbb{R}_+, X_1) \cap W^{1,p}_{loc}(\mathbb{R}_+, X_1)$ and solves (DEU). Moreover, $[v(t)](\theta) = u(t+\theta)$ where we set $u(t) = f(t)$ for $-1 \leq t \leq 0$. The solution of (DEU) is unique in the class $L^p_{loc}(\mathbb{R}_+, X_1) \cap W^{1,p}_{loc}(\mathbb{R}_+, X)$.

Proof. After the preparations that we have done, Lemmas 10.2, 10.3, 10.5, and 10.6 show that $\Sigma = (\mathcal{T}_0(t), \mathbb{V}_t, \mathbb{W}_t, \mathbb{F}_t)$ is a regular well-posed system with input and output operators $\mathcal{B}$ and $\mathcal{C}$. Lemma 10.5 also yields that $I - \mathbb{F}_t$ is invertible on $L^p([0,t],\mathcal{U})$ and $\|(I - \mathbb{F}_t)^{-1}\| \leq 2$ if $0 \leq t \leq t_0$. Therefore we can apply the perturbation theorem of Weiss and Staffans to deduce that the operator

$$\tilde{\mathcal{A}} := (\mathcal{A}_0)_{-1} + \mathcal{B}\tilde{\mathcal{C}} \quad \text{with} \quad D(\tilde{\mathcal{A}}) := \{\binom{x}{f} \in D(\tilde{\mathcal{C}}) : \tilde{\mathcal{A}}\binom{x}{f} \in \mathcal{X}\} \qquad (10.16)$$

(the sum is defined in $\mathcal{X}_{-1}$) generates a C_0-semigroup $\mathcal{T}$ on $\mathcal{X}$ satisfying Equation (10.13) with $\mathcal{C}$ replaced by $\tilde{\mathcal{C}}$ and that

$$\tilde{\mathcal{C}}\mathcal{T}(\cdot)\binom{x}{f} = (I - \mathbb{F}_t)^{-1}\mathbb{W}_t\binom{x}{f} \qquad (10.17)$$

for $\binom{x}{f} \in \mathcal{X}$. Equation (10.15) thus follows from Lemma 10.3. Using the representation of $D(\tilde{\mathcal{A}})$ established below and Lemma 10.4, we also see that we can replace $\tilde{\mathcal{C}}$ by $\mathcal{C}$ in Equation (10.13). This identity further yields Equation (10.14) because of Equation (10.4) and Lemma 10.6.

We have to verify that $\mathcal{A} = \tilde{\mathcal{A}}$. Taking $\binom{x}{f} \in D(\mathcal{A})$ and applying Lemma 10.2 together with Equation (10.8), we obtain

$$\mathcal{A}_0\lambda R(\lambda, \mathcal{A}_0)\binom{x}{f} = \binom{\lambda BR(\lambda, B)x}{\frac{d}{d\sigma}\varepsilon_\lambda \lambda R(\lambda, B)x + \lambda\frac{d}{d\sigma}R(\lambda, A_0)f}$$

$$= \binom{\lambda BR(\lambda, B)x}{\lambda\varepsilon_\lambda \lambda R(\lambda, B)x + \lambda R(\lambda, A_0)\frac{d}{d\sigma}f} \longrightarrow \binom{Bx}{\frac{d}{d\sigma}f}$$

in $\mathcal{Z}$ as $\lambda \to \infty$. This limit also exists in $\mathcal{X}_{-1}$ by Lemma 10.6 so that $(\mathcal{A}_0)_{-1}\binom{x}{f} = \binom{Bx}{Df} \in \mathcal{Z}$ on $D(\mathcal{A})$. Moreover, $D(\mathcal{A}) \subseteq D(\tilde{\mathcal{C}})$ and $\tilde{\mathcal{C}} = \mathcal{C}$ on $D(\mathcal{A})$ by Lemma 10.4. Thus the operator $\tilde{\mathcal{A}}$ defined by Equation (10.16) extends $\mathcal{A}$ given in Equation (10.12). The two operators coincide if $\mu - \mathcal{A}$ is surjective for some $\mu > w_0(\tilde{\mathcal{A}})$. So let $\binom{y}{\psi} \in \mathcal{X}$ and set $f = \varepsilon_\mu x + R(\mu, A_0)\psi$ for $x \in X_1$ and $\mu > \omega_0(\tilde{\mathcal{A}}) \vee d \vee 0$ to be determined. Using $(\mu - \frac{d}{d\sigma})f = \psi$, we see that

$$\binom{x}{f} \in D(\mathcal{A}), \tag{10.18}$$

$$(\mu - \mathcal{A})\binom{x}{f} = \binom{y}{\psi} \iff x - R(\mu, B)\Phi\varepsilon_\mu x = R(\mu, B)(\Phi R(\mu, A_0)\psi + y). \tag{10.19}$$

We have to find μ and x satisfying the right-hand side of this equivalence. Observe that $\|R(\mu, B)\|_{\mathcal{L}(X, X_1)} \leq 1 + 2K$ for $\mu \geq \mu_0 \geq 0$ and a sufficiently large μ_0. We further estimate

$$\begin{aligned}
\|\Phi\varepsilon_\mu x\|_X &\leq \int_{-\varepsilon}^0 e^{\mu\theta} d|\eta|(\theta)\, \|x\|_{X_1} + \int_{-1}^{-\varepsilon} e^{\mu\theta} d|\eta|(\theta)\, \|x\|_{X_1} \\
&\leq (d|\eta|([-\varepsilon, 0]) + \|\eta\|_{BV}\, e^{-\mu\varepsilon})\, \|x\|_{X_1} \\
&\leq \frac{1}{2(1 + 2K)} \|x\|_{X_1}
\end{aligned} \tag{10.20}$$

for sufficiently small $\varepsilon \in (0, 1)$ and large $\mu \geq \mu_0$ (here we use Hypothesis 10.1). Thus we can invert $I - R(\mu, B)\Phi e_\mu$ in $\mathcal{L}(X_1)$ for a fixed μ to obtain $x \in X_1$ satisfying Equation (10.18). So we arrive at the asserted representation for $\tilde{\mathcal{A}}$.

Finally, take $\binom{x}{f} \in D(\mathcal{A})$. In this case the function $\binom{u}{v} = \mathcal{T}(\cdot)\binom{x}{f}$ belongs to $C^1(\mathbb{R}_+, \mathcal{X})$, $\binom{u(t)}{v(t)} \in D(\mathcal{A})$, and we have $\binom{u'(t)}{v'(t)} = \mathcal{A}\binom{u(t)}{v(t)}$. In this particular case $u \in C^1(\mathbb{R}_+, Y)$ and $v \in C(\mathbb{R}_+, W^{1,p}([-1, 0], X_1))$ fulfill the equations $u'(t) = Bu(t) + \Phi v(t)$ (the sum is taken in X) and $v'(t) = \frac{d}{d\sigma}v(t)$. Thus $u \in C(\mathbb{R}_+, X_1)$ and the equation for v yields

$$[v(t)](\theta) = \begin{cases} u(t + \theta), & t + \theta \geq 0,\ \theta \in [-1, 0], \\ f(t + \theta), & t + \theta < 0,\ \theta \in [-1, 0], \end{cases} \tag{10.21}$$

because $[v(t)](0) = u(t)$ and $v(0) = f$. As a result of that, we have solved (DEU) in the required regularity class for the initial history function f. To prove uniqueness, assume that $u \in L_{loc}^p(\mathbb{R}_+, X_1) \cap W_{loc}^{1,p}(\mathbb{R}_+, X)$ solves (DEU) with $f = 0$ and extend u to $[-1, 0]$ by 0. Let $\tau \in (0, r]$. Then we estimate

$$\int_0^\tau \|\Phi u_s\|_X^p \, ds \leq \int_0^\tau \left[\int_{-s}^0 d|\eta|(\theta) \, \|u(s + \theta)\|_{X_1} \right]^p ds$$

$$\leq (d|\eta|[-\tau, 0])^p \int_0^\tau \|u(s)\|_{X_1}^p \, ds$$

using the Hausdorff-Young Inequality. The maximal regularity of B and (DEU) thus implies

$$\int_0^\tau \|u(s)\|_{X_1}^p \, ds \leq c \, (d|\eta|[-\tau, 0])^p \int_0^\tau \|u(s)\|_{X_1}^p \, ds$$

for a constant $c > 0$ only depending on B. As a result of that, $u = 0$ on $[0, \tau]$ for some small $\tau > 0$. Applying the same argument repeatedly, we obtain $u = 0$ on $\mathbb{R}_+$.   □

Using the above setting, we further establish a result on the inhomogeneous Cauchy problem with more general initial data.

Proposition 10.8. *Assume that Hypothesis 10.1 holds. Let $\binom{x}{f} \in X$ and $f \in L^p(J, Y)$ for $J = [0, T]$ and some $T > 0$. Then there is a unique $u \in L^p(J, X_1) \cap W^{1,p}(J, X_1) \cap C(J, Y)$ satisfying*

$$\begin{cases} u'(t) = Bu(t) + f(t) + \int_{-1}^0 d\eta(\theta) u(t + \theta), & a.e. \ t \geq 0, \\ u(t) = f(t), & a.e. \ t \in [-1, 0], \end{cases} \tag{10.22}$$

which is given by

$$\binom{u(t)}{u_t} = \mathcal{T}(t) \binom{x}{f} + \int_0^t \mathcal{T}(t - s) \binom{f(s)}{0} ds, \quad t \geq 0. \tag{10.23}$$

Proof. The uniqueness of solutions can be proved as in Theorem 10.7. Denote by $\binom{u(t)}{v(t)}$ the right-hand side of Equation (10.23) for given $\binom{x}{f} \in X$ and $f \in L^p(J, Y)$, and extend u by setting $u(t) = f(t)$ for $t \in [-1, 0]$.

First we take a regular initial value $\binom{x}{f} \in D(\mathcal{A})$ and $f \in C^1(J, Y)$. In this case, $\binom{u}{v} \in C(\mathbb{R}_+, D(\mathcal{A})) \cap C^1(\mathbb{R}_+, X)$ fulfills

$$\binom{u'(t)}{v'(t)} = \mathcal{A}\binom{u(t)}{v(t)} + \binom{f(t)}{0}, \quad t \geq 0, \qquad \binom{u(0)}{v(0)} = \binom{x}{f},$$

due to standard semigroup theory; see for example, Engel and Nagel [72, Corollary VI.7.6]. Therefore, $v(t) = u_t$ and $u \in C^1(\mathbb{R}_+, Y) \cap C(\mathbb{R}_+, X_1)$ satisfies Equation (10.22). This shows that

$$u(t) = S(t)x + \int_0^t S(t-s)(f(s) + \Phi u_s)\, ds, \qquad t \geq 0. \qquad (10.24)$$

Second, we approximate given $\binom{x}{f} \in \mathfrak{X}$ and $f \in L^p(J, Y)$ by $\binom{x_n}{f_n} \in D(\mathcal{A})$ and $f_n \in C^1(J, Y)$. Using again the above introduced notation, denote by $\binom{u}{v}$ and $\binom{u_n}{v_n}$ the right-hand side of Equation (10.23) for these given and approximating data, respectively. Then Equation (10.23) implies that u_n converges to u in $L^p([-1, T], X_1) \cap C([-1, T], Y)$ and $v(t) = u_t$. Observe that $s \mapsto \Phi u_s$ belongs to $L^p(J, X)$. Thus we can pass to the limit also in Equation (10.24) so that it holds for general data, too. The assertion then follows from the maximal regularity of B. $\qquad \square$

10.3 Spectral Theory

In this section, we compute the spectrum of $\mathcal{A}$. This is important to analyze the asymptotics of the delay semigroup. See also Section 3.2 for analogous results in the case of a more regular delay. We use the notation $\Phi_\lambda = \Phi \varepsilon_\lambda :$ $X_1 \to X$ and $\mathbb{C}_a = \{\lambda \in \mathbb{C} : \Re\lambda \geq a\}$.

Proposition 10.9. *Assume that Hypothesis 10.1 holds. Then*

$$\rho(\mathcal{A}) = \{\lambda \in \mathbb{C} : \exists\, H(\lambda) := (\lambda - B - \Phi_\lambda)^{-1} \in \mathcal{L}(X, X_1)\},$$

$$R(\lambda, \mathcal{A}) = \begin{pmatrix} H(\lambda) & H(\lambda)\Phi R(\lambda, A_0) \\ \varepsilon_\lambda H(\lambda) & (\varepsilon_\lambda H(\lambda)\Phi + I)R(\lambda, A_0) \end{pmatrix} =: \mathcal{R}_\lambda,$$

$$\|R(\lambda, \mathcal{A})\|_{\mathcal{L}(\mathfrak{X})} \leq c_a \|H(\lambda)\|_{\mathcal{L}(X, X_1)},$$

for $\lambda \in \rho(\mathcal{A})$ with $\Re\lambda \geq a$ for some $a \in \mathbb{R}$. Moreover, if $\|H(\lambda)\|_{\mathcal{L}(X, X_1)}$ is unbounded on a closed subset of $\mathbb{C}_a \cap \rho(\mathcal{A})$ for some $a \in \mathbb{R}$, then $\|R(\lambda, \mathcal{A})\|_{\mathcal{L}(\mathfrak{X})}$ is unbounded on the same set.

Note that we introduced the notation $H(\lambda)$ in contrast to Section 3.2 where we plainly used $R(\lambda, B + \Phi_\lambda)$. Our aim was to stress that, though formally the same operators appear here, $R(\lambda, B + \Phi_\lambda) = (\lambda - B - \Phi_\lambda)^{-1}$ is now considered as an operator from X to X_1 and not as an operator in X.

Proof. Assume that $H(\lambda) = (\lambda - B - \Phi_\lambda)^{-1} \in \mathcal{L}(X, X_1)$ exists for some $\lambda \in \mathbb{C}$. Then it is clear that $\mathcal{R}_\lambda$ is a bounded operator on X. More precisely, since $\|\varepsilon_\lambda\|_p$ and the norm of $R(\lambda, A_0) : L^p([-1,0], X_1) \to C([-1,0], X_1)$ are uniformly bounded for $\Re\lambda \geq a$, one has the estimate $\|\mathcal{R}_\lambda\|_{\mathcal{L}(X)} \leq c_a \|H(\lambda)\|_{\mathcal{L}(X,X_1)}$ if $\Re\lambda \geq a$. Take $\binom{x}{f} \in X$. Then $\binom{y}{\psi} = \mathcal{R}_\lambda \binom{x}{f}$ belongs to $X_1 \times W^{1,p}([-1,0], X_1)$, $\psi(0) = y$, and

$$
\begin{aligned}
By + \Phi\psi &= (B + \Phi_\lambda)H(\lambda)x + (B + \Phi_\lambda)H(\lambda)\Phi R(\lambda, A_0)f + \Phi R(\lambda, A_0)f \\
&= \lambda H(\lambda)x - x + \lambda H(\lambda)\Phi R(\lambda, A_0)f \\
&= \lambda y - x \in Y.
\end{aligned}
$$

This shows that $\mathcal{R}_\lambda \binom{x}{f} \in D(A)$ and $(\lambda - A)\mathcal{R}_\lambda = I$ (notice that $(\lambda - \frac{d}{d\sigma})\varepsilon_\lambda = 0$). Thus $\lambda - A$ is surjective. Let $(\lambda - A)\binom{x}{f} = 0$ for some $\binom{x}{f} \in D(A)$. Then $f = \varepsilon_\lambda x$ and hence $(\lambda - B - \Phi_\lambda)x = 0$, so that $x = 0$ and $f = 0$. As a result, $\lambda \in \rho(A)$ and $\mathcal{R}_\lambda = R(\lambda, A)$.

Conversely, let $\lambda \in \rho(A)$. We extend A to an operator on $W = X \times L^p([-1,0], X_1)$ by setting

$$
\tilde{A} = \begin{pmatrix} B & \Phi \\ 0 & \frac{d}{d\sigma} \end{pmatrix}, \quad D(\tilde{A}) = \{\binom{x}{f} \in X_1 \times W^{1,p}([-1,0], X_1) : f(0) = x\}.
$$

Observe that A is the part of $\tilde{A}$ in X. As in Equation (10.20) we can choose a sufficiently large $\mu > \omega_0(A) \vee d$ such that $\|\Phi_\mu R(\mu, B)\|_{\mathcal{L}(X)} \leq 1/2$. Recall that the number d was fixed in Hypothesis 10.1. Hence $\mu - B - \Phi_\mu$ has the inverse

$$
H(\mu) = R(\mu, B)(I - \Phi_\mu R(\mu, B))^{-1} \in \mathcal{L}(X, X_1). \tag{10.25}
$$

As in the first step, one sees that $\mu \in \rho(\tilde{A})$ and $R(\mu, \tilde{A}) = \mathcal{R}_\mu$. Since $\mathcal{R}_\mu : W \to X$ is bounded, we also obtain that the graph norm of $\tilde{A}$ in W dominates the norm of X. Consequently, $\rho(\tilde{A}) = \rho(A) \ni \lambda$ by Engel-Nagel [72, Proposition IV.2.17].

If $(\lambda - B - \Phi_\lambda)x = 0$ for some $x \in X_1$, then $\binom{x}{\varepsilon_\lambda x}$ belongs to the kernel of $\lambda - \tilde{A}$, so that $x = 0$. Given $y \in X$, there exists $\binom{x}{f} \in D(\tilde{A})$ with $(\lambda - \tilde{A})\binom{x}{f} = \binom{y}{0}$. This implies that $f = \varepsilon_\lambda x$ and $(\lambda - B - \Phi_\lambda)x = y$. Summing up, $\lambda - B - \Phi_\lambda : X_1 \to X$ is bijective and continuous, and thus invertible.

Assume that $a_n := \|H(\lambda_n)\|_{\mathcal{L}(X,X_1)} \to \infty$ as $n \to \infty$ for some λ_n in a closed subset of $\mathbb{C}_a \cap \rho(\mathcal{A})$. Then $|\lambda_n| \to \infty$ and thus $\lambda_n \in \Sigma_{\phi,d}$ for $n \geq n_0$ and some $n_0 \geq 1$. Moreover, $H(\lambda) \in \mathcal{L}(X, X_1)$ is uniformly bounded on a half plane $\mathbb{C}_\gamma$ for a sufficiently large $\gamma \geq a$ due to Equation (10.25). Hence, the sequence $\Re\lambda_n$ is uniformly bounded. We take unit vectors $x_n \in X$ such that $\|H(\lambda_n)x_n\|_{X_1} \geq a_n - 1/n$ and define $y_n = R(\lambda_n, B)x_n$ and $b_n = a_n^{p'/2}$ for $1/p + 1/p' = 1$. We can observe that the vectors y_n are uniformly bounded in X_1. Further we take $n_1 \geq n_0$ such that $b_n \geq 1/r$ for $n \geq n_1$ and set

$$f_n(\theta) = b_n\, e^{\lambda_n \theta}\, \chi_{[-b_n^{-1},0]}(\theta), \qquad \theta \in [-1,0],$$

for $n \geq n_1 \geq n_0$. We obtain $\|f_n\|_p \leq e^{ra^-}\, b_n^{\frac{1}{p'}}$ and, using Equation (10.8),

$$R(\lambda_n, A_0)f_n y_n = \varepsilon_{\lambda_n} \psi_n y_n, \quad \text{where } \psi_n(\theta) := \begin{cases} -b_n\theta, & 0 \geq \theta \geq -1/b_n, \\ 1, & -b_n^{-1} \geq \theta \geq -r. \end{cases}$$

This equality yields

$$[R(\lambda_n, \mathcal{A})\begin{pmatrix} 0 \\ f_n y_n \end{pmatrix}]_2 = \varepsilon_{\lambda_n}(H(\lambda_n)\Phi_{\lambda_n} y_n + y_n) + \varepsilon_{\lambda_n} H(\lambda_n)\Phi_{\lambda_n}(\psi_n - 1)y_n$$
$$+ \varepsilon_{\lambda_n}(\psi_n - 1)y_n =: S_1 + S_2 + S_3.$$

First, note that $S_1 = \varepsilon_{\lambda_n} H(\lambda_n)x_n$. Using the inequality $|\psi_n - 1| \leq \chi_{[-b_n^{-1},0]}$, we can estimate

$$\|S_2\|_{L^p([-1,0],X_1)} \leq c_1\, d|\eta|([-b_n^{-1},0])\, a_n \quad \text{and} \quad \|S_3\|_{L^p([-1,0],X_1)} \leq c_2\, b_n^{-\frac{1}{p}}$$

for some constants c_k independent of n. Combining these facts, we arrive at

$$c_3\, b_n^{\frac{1}{p'}} \|R(\lambda_n, \mathcal{A})\|_{\mathcal{L}(X)} \geq \|R(\lambda_n, \mathcal{A})\begin{pmatrix} 0 \\ f_n y_n \end{pmatrix}\|x$$
$$\geq c_4(a_n - 1/n) - c_1\, d|\eta|([-b_n^{-1},0])\, a_n - c_2\, b_n^{-\frac{1}{p}}$$
$$\geq c_5\, a_n$$

for sufficiently large n and constants $c_1, \ldots, c_5 > 0$. As a result, $\|R(\lambda_n, \mathcal{A})\| \geq c\sqrt{a_n}$. $\square$

Before proceeding to theorems on the asymptotic behavior of the solution, we analyze the spectral properties of a simple example.

Example 10.10. Let $\Phi = \beta B \delta_{-r}$ for some $\beta \in \mathbb{R}$.

(i) Let B be sectorial with $\phi > \pi/2$. Then the semigroup generated by $\mathcal{A}$ is not hyperbolic if $|\beta| = 1$ or if $|\beta| > 1$ and $\sigma(B) \cap \mathbb{R}_-$ is unbounded.

(ii) Let $|\beta| > 1$. Then there exists a selfadjoint and negative definite operator B on $X = \ell^2$ such that $i\mathbb{R}$ belongs to $\rho(\mathcal{A})$.

In view of (i), the semigroup $\mathcal{T}$ obtained in (ii) violates the Spectral Mapping Theorem 2.11 and hence cannot be eventually norm continuous. Further, we also see that for this type of equation, the stability and hyperbolicity can be destroyed by small delays (because we did not make any assumption on r). Let us prove (i) and (ii).

(i) If $|\beta| = 1$, then there is $\tau \in \mathbb{R}$ such that $\beta e^{-i\tau r} = -1$. Hence, $i\tau - B - \Phi_{i\tau} = i\tau$ and $i\tau \in \sigma(\mathcal{A})$ by Proposition 10.9. So Theorem 2.26 shows that $\mathcal{T}$ is not hyperbolic. Let $|\beta| > 1$. Then $1 + \beta e^{-i\tau r} \neq 0$ and

$$\frac{i\tau}{1 + \beta e^{-i\tau r}} = \frac{-\tau\beta\sin(\tau r) + i\tau(1 + \beta\cos(\tau r))}{|1 + \beta e^{-i\tau r}|^2}$$

for all $\tau \in \mathbb{R}$. Since $|\beta| > 1$, there are $\tau_n = \tau_0 + 2\pi n/r$ such that $n \in \mathbb{N}_0$, $\tau_0 \in (0, 2\pi/r)$, $\cos\tau_n r = -1/\beta$, and $c_1 = \beta\sin\tau_0 r > 0$. Setting $c_2 = |1 + \beta e^{-i\tau_0 r}|^{-1}$, we obtain

$$\mu_n := \frac{i\tau_n}{1 + \beta e^{-i\tau_n r}} = -c_1 c_2^2 (\tau_0 + 2n\pi). \tag{10.26}$$

If $\mu_n \in \sigma(B)$ for some $n \in \mathbb{N}$, then $i\tau_n \in \sigma(\mathcal{A})$ due to Proposition 10.9, and again $\mathcal{T}$ is not hyperbolic by Theorem 2.26. Otherwise, there exists the operator

$$H(i\tau_n) = (1 + \beta e^{-i\tau_0 r})^{-1} R(\mu_n, B).$$

For sufficiently large n, we obtain

$$\|H(i\tau_n)\|_{\mathcal{L}(X, X_1)} = c_2 \|(w - B)R(\mu_n, B)\|_{\mathcal{L}(X)}$$
$$\geq c_2 \left(\|(w - \mu_n)R(\mu_n, B)\| - 1 \right)$$
$$\geq c_2 \left(\frac{|\mu_n - w|}{d(\mu_n, \sigma(B))} - 1 \right).$$

Now take a sequence $\lambda_k \in \sigma(B) \cap \mathbb{R}$ tending to $-\infty$. One can always find an index n_k such that $|\lambda_k - \mu_{n_k}| \leq 2c_1 c_2^2\pi$ if k is sufficiently large. This shows that $\|H(i\tau_n)\|_{\mathcal{L}(X, X_1)} \geq c\, n_k \to \infty$ as $k \to \infty$. Consequently, (i) holds, owing to Proposition 10.9 and Theorem 2.26.

(ii) Observe that the numbers 0 and μ_n in Equation (10.26) with $n \in \mathbb{Z}$ are the only values of μ on the real axis. Hence we only have to choose negative numbers $a_j \to -\infty$ different from μ_n and to take the diagonal operator $B = (a_j)$.

10.4   Stability and Hyperbolicity

We start with a robustness result for hyperbolicity based on Equation (10.13) and the following lemma. The results obtained are analogous to Theorem 5.4.

Lemma 10.11. *Assume that Hypothesis 10.1 holds and that $\sigma(B) \cap i\mathbb{R} = \emptyset$. Then $\mathcal{T}_0$ is hyperbolic on X with the same exponent, and the dimension of its unstable subspace coincides with that of $(S(t))_{t \geq 0}$.*

Proof. By assumption, $(S(t))_{t \geq 0}$ is hyperbolic on X and Y with projections P and $Q = I - P$ and constants $N, \delta > 0$. Let us denote the inverse of $S(t)$ on QX by $S_Q(-t)$. Then $Q = S(1)S_Q(-1)Q$ so that $Q \in \mathcal{L}(X, X_1)$ and $QX = QY$. We define

$$\mathcal{Q} = \begin{pmatrix} Q & 0 \\ S_Q(\cdot)Q & 0 \end{pmatrix}.$$

Clearly, $\mathcal{Q}$ is a bounded projection on $\mathcal{X}$, $\dim \mathcal{Q}\mathcal{X} = \dim QY = \dim QX$, and $\mathcal{T}_0(t)\mathcal{Q} = \mathcal{Q}\mathcal{T}_0(t)$. The inverse of $\mathcal{T}_0(t)$, $t \geq 0$, on $\mathcal{Q}\mathcal{X}$ is given by

$$\mathcal{T}_Q(-t) = \begin{pmatrix} S_Q(-t) & 0 \\ S_Q(-t + \cdot)Q & 0 \end{pmatrix}.$$

This formula implies that $\|\mathcal{T}_Q(-t)\mathcal{Q}\| \leq N_1 e^{-\delta t}$ for $t \geq 0$. For $t > r$ and $\mathcal{P} = I - \mathcal{Q}$, it holds that

$$\mathcal{T}_0(t)\mathcal{P} = \begin{pmatrix} S(t) & 0 \\ S_t & 0 \end{pmatrix} \begin{pmatrix} P & 0 \\ -S_Q(\cdot)Q & I \end{pmatrix} = \begin{pmatrix} S(t)P & 0 \\ S_t P & 0 \end{pmatrix}$$

so that $\|\mathcal{T}_0(t)\mathcal{P}\| \leq N_2 e^{-\delta t}$ for $t \geq 0$ and a constant N_2. $\square$

     The following results describe the asymptotic behavior of $\mathcal{T}$. Note that the uniform exponential stability of $\mathcal{T}$ implies that

$$\|u(t)\|_Y + \|u_t\|_{L^p([-1,0],X_1)} \leq N e^{-\delta t} \left(\|f(0)\|_Y + \|f\|_{L^p([-1,0],X_1)} \right), \qquad t \geq 0,$$

for the solutions of (DEU). Further, the hyperbolicity of $\mathcal{T}$ leads to a dichotomy on the level of the history functions $u_t \in L^p([-1,0], X_1)$.

Proposition 10.12. *Assume that Hypothesis 10.1 holds and $\sigma(B) \cap i\mathbb{R} = \emptyset$. If $\|\eta\|_{BV}$ is sufficiently small, then $\mathcal{T}$ is hyperbolic on $\mathcal{X}$ and the dimension of its unstable subspace coincides with that of $(S(t))_{t \geq 0}$. Hence, as a special case, if $(S(t))_{t \geq 0}$ is uniformly exponentially stable, then $\mathcal{T}$ is also uniformly exponentially stable.*

Proof. Due to Equations (10.13), (10.15), and (10.4), and Lemma 10.3, we have $\|\mathfrak{T}(t_0) - \mathfrak{T}_0(t_0)\| \leq c\|\eta\|_{BV}$ where c does not depend on B and $t_0 > 0$ is given by Theorem 10.7. Thus Lemma 10.11 and Proposition 2.18 show the assertion provided that $\|\eta\|_{BV}$ is sufficiently small. □

The approach of Proposition 10.12 does not give very sharp conditions and is, of course, restricted to operators B being hyperbolic.

Our following main theorems are immediate consequences of Proposition 10.9 and Gearhart's Theorem.

Theorem 10.13. *Assume that Hypothesis 10.1 holds and that X is Hilbert space. Take $p = 2$.*

(i) *$\mathfrak{T}$ is hyperbolic on $\mathfrak{X}$ if and only if $H(is)$ exists and is uniformly bounded in $\mathcal{L}(X, X_1)$ for $s \in \mathbb{R}$.*

(ii) *$\mathfrak{T}$ is uniformly exponentially stable on $\mathfrak{X}$ if and only if $H(\lambda)$ exists and is uniformly bounded in $\mathcal{L}(X, X_1)$ for $\Re\lambda \geq 0$.*

Proof. Observe that $\mathfrak{X}$ is a Hilbert space if X is a Hilbert space and $p = 2$. Thus Gearhart's Theorem characterizes hyperbolicity and stability of $\mathfrak{T}$ by the uniform boundedness of $R(\lambda, \mathcal{A})$ for $\lambda \in i\mathbb{R}$ and $\Re\lambda \geq 0$, respectively. The assertions hence follow from Proposition 10.9. □

In the following two theorems, we extend part (ii) of Theorem 10.13 to general Banach spaces.

Theorem 10.14. *Assume that Hypothesis 10.1 holds and that X has Fourier type $q \in [1, 2]$. Take $p \in [q, q'] \cap (1, \infty)$. If $H(\lambda)$ exists and is uniformly bounded in $\mathcal{L}(X, X_1)$ for $\Re\lambda \geq 0$, then there are constants $N, \delta > 0$ such that*

$$\left\| \mathfrak{T}(t) \binom{x}{f} \right\|_{\mathfrak{X}} \leq N e^{-\delta t} \left\| (\gamma - \mathcal{A})^{1/q - 1/q'} \binom{x}{f} \right\|$$

for $1/q + 1/q' = 1$ and $\binom{x}{f} \in D((\gamma - \mathcal{A})^{1/q - 1/q'})$ (where $\gamma > \omega_0(\mathcal{A})$ is fixed).

Proof. It is clear that X_1 has Fourier type q. Let us check that the space Y also has Fourier type q. We may assume that $q > 1$ since every Banach space has Fourier type 1. For $1 < q \leq p \leq q'$, the estimate in Equation (1.4) and integral version of Jessen's inequality (see Hardy, Littlewood, and Polya [106, Sections 202, 203]) imply

$$\|\mathcal{F}f\|_{L^{q'}(\mathbb{R},Y)} \leq c \left[\int_{\mathbb{R}} \left[\int_0^1 \|S(\tau)(\mathcal{F}f)(t)\|_{X_1}^p \, d\tau \right]^{\frac{q'}{p}} dt \right]^{\frac{1}{q'}}$$

$$\leq c \left[\int_0^1 \left[\int_{\mathbb{R}} \|S(\tau)(\mathcal{F}f)(t)\|_{X_1}^{q'} \, dt \right]^{\frac{p}{q'}} d\tau \right]^{\frac{1}{p}}$$

$$= c \left[\int_0^1 \left[\int_{\mathbb{R}} \|[\mathcal{F}(S(\tau)f)](t)\|_{X_1}^{q'} \, dt \right]^{\frac{p}{q'}} d\tau \right]^{\frac{1}{p}}$$

$$\leq c' \left[\int_0^1 \left[\int_{\mathbb{R}} \|S(\tau)f(s)\|_{X_1}^{q} \, ds \right]^{\frac{p}{q}} d\tau \right]^{\frac{1}{p}}$$

$$\leq c' \left[\int_{\mathbb{R}} \left[\int_0^1 \|S(\tau)f(s)\|_{X_1}^{p} \, d\tau \right]^{\frac{q}{p}} ds \right]^{\frac{1}{q}} \leq c'' \|f\|_{L^q(\mathbb{R},Y)},$$

where $\mathcal{F}$ denotes the Fourier transform and f is an X_1-valued Schwartz function. Moreover, $L^p([-1,0], X_1)$ and thus $W^{1,p}([-1,0], X_1)$ have Fourier type q by Girardi and Weis [84, Proposition 2.3]. Therefore the space $\mathcal{X}$ has Fourier type q. On the other hand, $s_0(\mathcal{A}) < 0$ by the assumption and Proposition 10.9. The theorem is now a consequence of the Weis-Wrobel Theorem (see Equation (2.11)).   □

Analyzing the previous two theorems, we obtain an analogous version of Theorem 5.5.

Proposition 10.15. *Assume that $s(B) < 0$ and let $\alpha \in (s(B), 0]$ such that*

$$a_{\alpha,n} := \sup_{\omega \in \mathbb{R}} \|(\Phi_{\alpha+i\omega} R(\alpha + i\omega, B))^n\| < \infty. \tag{10.27}$$

If

$$a_\alpha := \sum_{n=0}^{\infty} a_{\alpha,n} < \infty, \tag{10.28}$$

then $s_0(\mathcal{A}) < \alpha \leq 0$. If the Banach space X has Fourier type $q \in [1,2]$, then

$$\omega_{\frac{1}{q} - \frac{1}{q'}}(\mathcal{A}) \leq \alpha.$$

Proof. We have to show the boundedness on the halfplane $\{\Re\lambda > \alpha\}$ of the operator family $H(\lambda)$ given in Proposition 10.9.

Under our assumptions, this is equivalent to the existence and bound-edness of $R(\lambda, B + \Phi_\lambda)$ on the halfplane $\{\Re\lambda > \alpha\}$ as an operator from X to X_1.

Since $s(B) < \alpha$, by Lemma 2.28, we have that $a_{\beta,n} \leq a_{\alpha,n}$ for $\beta > \alpha$.

Defining $M_\alpha := \sup_{\Re\lambda > \alpha} \|R(\lambda, B)\|_{\mathcal{L}(X,X_1)}$, which is bounded by the analyticity of B, we obtain for all $\lambda \in \mathbb{C}$ with $\Re\lambda > \alpha$ that

$$R(\lambda, B) \sum_{n=0}^{\infty} (\Phi_\lambda R(\lambda, B))^n \in \mathcal{L}(X, X_1)$$

and

$$\left\| R(\lambda, B) \sum_{n=0}^{\infty} (\Phi_\lambda R(\lambda, B))^n \right\|_{\mathcal{L}(X,X_1)}$$

$$\leq M_\alpha \sum_{n=0}^{\infty} \| (\Phi_\lambda R(\lambda, B))^n \| \leq M_\alpha a_{\Re\lambda} \leq M_\alpha a_\alpha.$$

This operator is the inverse of $(\lambda - B - \Phi_\lambda)$ and is bounded on the halfplane $\{\Re\lambda > \alpha\}$. Hence, by Lemma 2.25, we have $s_0(\mathcal{A}) < \alpha$.

Finally, the result follows from Theorem 10.14. $\square$

Of course an analogous result can be formulated for hyperbolicity. Now we can partly improve the robustness result in Proposition 10.12.

Corollary 10.16. *Assume that Hypothesis 10.1 holds, that B is invertible, and that X has Fourier type $q \in [1, 2]$. Take $p \in [q, q'] \cap (1, \infty)$ and $w = 0$, i.e., $\|x\|_{X_1} = \|Bx\|_X$.*

(i) If $s(B) < 0$ and

$$\|\eta\|_{BV} < \frac{1}{1 + K},$$

then the assertion of Theorem 10.14 holds. In particular, $\mathcal{T}$ is expo-nentially stable. If X is a Hilbert space, then $\mathcal{T}$ is uniformly exponen-tially stable.

(ii) If X is a Hilbert space (i.e., $q = p = 2$), $i\mathbb{R} \subseteq \rho(B)$, and

$$\|\eta\|_{BV} < \frac{1}{\sup_{s \in \mathbb{R}}(1 + \|sR(is, B)\|_{\mathcal{L}(X)})},$$

then $\mathcal{T}$ is hyperbolic and the unstable subspaces of $(\mathcal{T})_{t \geq 0}$ and $(S(t))_{t \geq 0}$ have the same dimension.

Proof. (i) For $s(B) < 0$ and $\Re\lambda \geq 0$, we have

$$\|\Phi_\lambda R(\lambda, B)\|_{\mathcal{L}(X)} \leq \|\eta\|_{BV} \|R(\lambda, B)\|_{\mathcal{L}(X, X_1)} \leq \|\eta\|_{BV} (1 + K) =: c < 1$$

by the assumption. Thus there exists

$$H(\lambda) = R(\lambda, B)\left[I - \Phi_\lambda R(\lambda, B)\right]^{-1} \in \mathcal{L}(X, X_1),$$

and $\|H(\lambda)\|_{\mathcal{L}(X, X_1)} \leq (1 + K)/(1 - c)$. Therefore assertion (i) follows from Proposition 10.15.

(ii) Let $\sigma(B) \cap i\mathbb{R} = \emptyset$. As above, it follows from the conditions that

$$\|\Phi_{is} R(is, B)\|_{\mathcal{L}(X)} \leq c' < 1 \quad \text{and} \quad \|H(is)\|_{\mathcal{L}(X, X_1)} \leq c''$$

for $s \in \mathbb{R}$. Thus the semigroup generated by $\mathcal{A}$ is hyperbolic by Theorem 10.13. In order to check the assertion concerning the dimensions, we introduce the perturbations $\alpha\mathcal{B}$ for $\alpha \in [0, 1]$. Since the hypotheses hold for $\alpha\mathcal{B}$ with the same constants, we obtain the corresponding exponentially dichotomic semigroups $\mathcal{T}_{\alpha\Phi} =: \mathcal{T}_\alpha$, where $\mathcal{T}_0$ is the unperturbed semigroup as before and $\mathcal{T}_1 = \mathcal{T}$. In view of Proposition 2.18 and Lemma 10.11, we have to show that the projections $\mathcal{Q}_\alpha$ for $\mathcal{T}_\alpha$ depend continuously on α in operator norm. This is the case if $\alpha \mapsto \mathcal{T}_\alpha(t_0) \in \mathcal{L}(\mathcal{X})$ is continuous for some $t_0 > 0$, because of the well-known formula

$$\mathcal{P}_\alpha = \frac{1}{2\pi i} \oint_\Gamma R(\lambda, \mathcal{T}_\alpha(t)) \, d\lambda, \qquad t > 0. \tag{10.29}$$

Now we employ more results from feedback theory in order to establish the continuity of this map. We fix the input operator $\mathcal{B}$ and the output operator $\mathcal{C}$. Then we obtain the semigroup $\mathcal{T}_\alpha$ from $\mathcal{T}_0$ by the admissible feedback $\alpha I : \mathcal{U} \to \mathcal{U}$, see Theorem 10.7. In fact, there exists another regular system $\Sigma^\alpha = (\mathcal{T}_\alpha(t), \Phi_t^\alpha, \mathbb{W}_t^\alpha, \mathbb{F}_t^\alpha)$ with $\mathbb{W}_t^\alpha = (I - \alpha\mathbb{F}_t)^{-1}\mathbb{W}_t$; see Staffans [196, Theorem 7.1.2]. Moreover, $(\beta - \alpha)I : \mathcal{U} \to \mathcal{U}$, $\beta \in [0, 1]$, is an admissible feedback for Σ^α producing Σ^β due to Staffans [196, Lemma 7.1.7]. In particular, $\mathcal{T}_\beta(t_0) - \mathcal{T}_\alpha(t_0) = (\beta - \alpha)\Phi_{t_0}^\alpha \mathbb{W}_{t_0}^\beta$ for the number $t_0 > 0$ given in Theorem 10.7. This expression tends to 0 in norm as $\beta \to \alpha$ by Lemma 10.5. $\qquad\square$

10.5   Notes

The existence Theorem 10.7 is based on a perturbation result due to Weiss [221] and Staffans [196]; see also Schnaubelt [187] for corresponding work on nonautonomous problems.

The results on output systems (observations) are contained in Weiss [220]; see in particular Theorem 4.5 and Proposition 4.7. The facts on input systems (control) can be found in Weiss [219], in particular in Theorem 3.9.

These authors work in the framework of so-called regular systems in control theory. For the reader's convenience, we collected in Section 10.1 the concepts and facts needed. (In fact, [196] and [221] contain much more general results.)

The perturbation theorem was proved by Weiss [221, Section 6-7], in the Hilbert space setting; the above version is due to Staffans [196, Chapter 7], especially Theorems 7.1.2, 7.1.8, 7.5.3(iii), and Remark 7.1.3 of [196].[1]

The results of this chapter were obtained in collaboration with R. Schnaubelt; see [20]. Parabolic delay equations with highest-order derivatives in the delay terms were first considered in the Hilbert space setting for special delay operators in Di Blasio, Kunisch, and Sinestrari [60], where also the solution semigroup was constructed in the space X. Theorems 7.4–7.6 in Prüß [169] imply similar results to ours on almost-everywhere solutions of (DEU) assuming slightly less regularity for the initial data.

Proposition 10.9 improves and extends various facts established in Di Blasio, Kunisch, and Sinestrari [61]. See also Nakagiri and Tanabe [157] for an investigation of the (generalized) eigenfunction spaces of A for special scalar valued kernels. Example 10.10 shows that in general stability and hyperbolicity can be destroyed by small delays. A more detailed analysis of the small delay problem like the one we presented in Chapter 7 remains to be done.

Smallness conditions and robust stability results were already used in [61, Theorem 4.5] and Mastinsek [143, Theorem 2] for special classes of η and B. See also [20, Proposition 5.1], where a more detailed analysis is done.

We compare the result of Theorem 10.13 with Prüss' monograph [169]. Assume that A is hyperbolic. Then for every $f \in L^2(\mathbb{R}, Y)$ there is a unique $\binom{u}{v} \in L^2(\mathbb{R}, X) \cap C(\mathbb{R}, X)$ such that

$$\begin{pmatrix} u(t) \\ v(t) \end{pmatrix} = \mathcal{T}(t-s)\begin{pmatrix} u(s) \\ v(s) \end{pmatrix} + \int_s^t \mathcal{T}(t-\tau)\begin{pmatrix} f(\tau) \\ 0 \end{pmatrix} d\tau, \quad t \geq s;$$

see, for example, Chicone and Latushkin [32, Theorem 4.33]. Proposition 10.8 further shows that $u \in L^2_{loc}(\mathbb{R}, X_1) \cap W^{1,2}_{loc}(\mathbb{R}, X) \cap L^2(\mathbb{R}, Y)$ and

$$u'(t) = Bu(t) + \Phi u_t + f(t)$$

---

[1]The feedthrough operator D in [196] equals 0 in our setting.

for a.e. $t \in \mathbb{R}$. This property is called "admissibility of $L^2(\mathbb{R}, Y)$." Admissibility of $L^2(\mathbb{R}, X)$ was obtained in [169, Proposition 12.1] (with slightly less local regularity) also assuming that $H(is) : X \to X_1$ is bounded for $s \in \mathbb{R}$ and that X is a Hilbert space. Conversely, admissibility of $BUC(\mathbb{R}, X)$ implies boundedness of $H(is) : X \to X_1$ for $s \in \mathbb{R}$ by [169, Proposition 11.5]. It is not clear whether admissibility of $L^2(\mathbb{R}, X)$ or $L^2(\mathbb{R}, Y)$ in turn implies the hyperbolicity of $\mathcal{A}$ on $\mathcal{X}$. (Roughly speaking, admissibility and hyperbolicity are equivalent for nonretarded problems; see Chicone and Latushkin [32]. For retarded equations with $\eta \in BV([-1, 0], \mathcal{L}(X))$, see Gühring, Räbiger, and Schnaubelt [94].)

The results of Theorem 10.14 have no counterpart in [169].

The application of operator-valued Fourier multipliers to stability and hyperbolicity questions for this type of equation has not been done yet and might be the subject of a future research project.

Appendix

Vector-Valued Functions

Throughout this book, we used permanently the notions of Banach space valued L^p-spaces and Sobolev spaces. Therefore, we collect in this appendix the definitions and basic properties of these spaces.

A.1   The Bochner Integral

We briefly recall some basic notions of Bochner integrals. For further references on this topic, we refer to Dunford and Schwartz [65, Section III.2], and to Hille and Phillips [110, Section III.3.7], and to the overview in Arendt et.al. [6, Section A.1.1]. Here we follow Diestel and Uhl [63].

Let X be a Banach space and (Ω, Σ, μ) a finite measure space. A function $f : \Omega \to X$ is called *simple* if there exist $x_1, \ldots, x_n \in X$, and $E_1, \ldots, E_n \in \Sigma$ such that $f = \sum_{i=1}^{n} x_i \chi_{E_i}$, where χ_{E_i} denotes the characteristic function of E_i. We say that a function $f : \Omega \to X$ is *measurable* if there is a sequence of simple function (f_n) with $\lim_{n \to \infty} \|f - f_n\| = 0$ μ-almost everywhere.[1]

Definition A.1. For a measurable set $E \in \Sigma$, we define the integral of a simple function $f = \sum_{i=1}^{n} x_i \chi_{E_i}$ as

$$\int_E f \, d\mu := \sum_{i=1}^{n} x_i \, \mu_i(E \cap E_i).$$

A measurable function $f : \Omega \to X$ is called *Bochner-integrable* if there exists a sequence of simple functions (f_n) such that

$$\lim_{n \to \infty} \int_\Omega \|f - f_n\| = 0.$$

---

[1] In the literature, this property is usually called *strong measurability*.

In this case, $\int_E f \, d\mu$ is defined for every $E \in \Sigma$ by

$$\int_E f \, d\mu := \lim_{n \to \infty} \int_E f_n \, d\mu.$$

The following characterization is very useful.

Theorem A.2. *A measurable function $f : \Omega \to X$ is Bochner-integrable if and only if $\int_\Omega \|f\| \, d\mu < \infty$.*

Most of the properties of the Lebesgue integral hold also for the Bochner integral, e.g., Lebesgue's Dominated Convergence Theorem and Fubini's Theorem.

Definition A.3. Let $1 \leq p < \infty$. We define the *Lebesgue-Bochner space* $L^p(\Omega, X)$ as the set of all equivalence classes of measurable functions $f : \Omega \to X$ such that

$$\|f\|_p := \left(\int_\Omega \|f\|^p d\mu \right)^{\frac{1}{p}} < +\infty.$$

Theorem A.4. *The Lebesgue-Bochner space $(L^p(\Omega, X), \|\cdot\|_p)$ is a Banach space. Moreover, if X is a Hilbert space and $p = 2$, then $(L^2(\Omega, X), \|\cdot\|_2)$ is a Hilbert space.*

The following lemma will be very useful (see Neidhart [160, Theorem 4.2] and Schnaubelt [185, Lemma 2.1]).

Lemma A.5. *Let X be a Banach space and let I be an interval in $\mathbb{R}$. Assume that $u : [c, d] \to L^p(I, X)$, $1 \leq p < \infty$, is Bochner-integrable and that $u(t) = v(t, \cdot)$ for a.e. $t \in [c, d]$, where $v : [c, d] \times I \to X$ is measurable. Then $v(\cdot, s)$ is Bochner-integrable and*

$$\left(\int_c^d u(t) \, dt \right)(s) = \int_c^d v(t, s) \, dt \qquad \text{for a.e. } s \in I.$$

Now we turn our attention to L^p-valued Bochner spaces. The following argument is due to U. Schlotterbeck.

Let (Ω_1, μ_1) and (Ω_2, μ_2) be σ-finite measure spaces, $1 \leq p < \infty$, and for $f \in \mathcal{L}^p(\Omega_1 \times \Omega_2, \mu_1 \otimes \mu_2)$ and $x \in \Omega_1$ let $\Phi_f(x)$ be the function $y \to f(x, y)$ on Ω_2.

Then $\Phi_f(x) \in \mathcal{L}^p(\Omega_2, \mu_2)$ for almost every $x \in \Omega_1$, and $\Phi(f) : x \to \Phi_f(x)$ is in $\mathcal{L}^p(\Omega_1, L^p(\Omega_2))$.

If $f_1 = f_2$ $(\mu_1 \otimes \mu_2)$ a.e., then $\Phi(f_1) = \Phi(f_2)$ (μ_1) a.e., hence Φ defines in a natural way a linear map $\hat{\Phi}$ from $L^p(\Omega_1 \times \Omega_2, \mu_1 \otimes \mu_2)$ into $L^p(\Omega_1, L^p(\Omega_2))$. The next result plays an important role in the stability theory of positive delay semigroups. Though it seems to be folklore, no proof of it was found.

Theorem A.6. *The mapping $\hat{\Phi}$ is an isometric lattice isomorphism from $L^p(\Omega_1 \times \Omega_2, \mu_1 \otimes \mu_2)$ into $L^p(\Omega_1, L^p(\Omega_2))$.*

Proof. It is a consequence of Fubini's Theorem that $\hat{\Phi}$ is an isometry and that $\hat{\Phi}$ is positive. Since the range of $\hat{\Phi}$ contains the functions $\chi_A g$ for any μ_1-measurable A of finite measure and any $g \in L^p(\Omega_2, \mu_2)$ and since the linear hull of these functions is dense in $L^p(\Omega_1, L^p(\Omega_2))$ (by definition), $\hat{\Phi}$ must be surjective. The fact that $\hat{\Phi}$ is a lattice isomorphism is a consequence of the following lemma. $\qquad\Box$

Lemma A.7. *Let α be a positive isometry from a Banach lattice E into a Banach lattice F, and suppose that the norm of F is strictly monotone on the positive cone, i.e., $\|y_1 + y_2\| > \|y_1\|$ whenever $y_1, y_2 > 0$. Then α is a lattice isomorphism.*

Proof. It suffices to show that $\alpha(|z|) = |\alpha(z)|$ for all $0 \neq z \in E$.

Since α is positive,

$$\alpha(|z|) \geq |\alpha(z)|,$$

and since α is an isometry,

$$\|\alpha(|z|)\| = \| \, |z| \, \| = \|z\| = \|\alpha(z)\| = \| \, |\alpha(z)| \, \|.$$

If we suppose $\alpha(|z|) > |\alpha(z)|$, then

$$\alpha(|z|) = |\alpha(z)| + y \qquad \text{for some } y > 0,$$

hence

$$\|\alpha(|z|)\| > \| \, |\alpha(z)| \, \|$$

by the strict monotonicity of the norm of F, which is a contradiction. $\qquad\Box$

A.2   Vector-Valued Sobolev Spaces

We collect here some basic results about vector-valued Sobolev spaces. For further references, we refer to Amann [2, Section III.1.2], Barbu [10, Section I.2.2], and Yosida [226, Section V.5].

Definition A.8. Let X be a Banach space, $J \subset \mathbb{R}$ a relatively compact interval, and $1 \leq p < \infty$. We define the set $W^{1,p}(J, X)$ as follows:

$$W^{1,p}(J, X) := \{u \in L^p(J, X) \ : \ \text{there exists } v \in L^p(J, X)$$

$$\text{such that } u(s) = u(a) + \int_a^s v(r) \, dr, \ a \in J\}.$$

For $u \in W^{1,p}(J, X)$, we define the *weak derivative*

$$\frac{d}{ds} u := v.$$

Finally, we define the Sobolev norm of a function $u \in W^{1,p}(J, X)$ as

$$\|u\|_{1,p} := \|u\|_p + \left\| \frac{d}{ds} u \right\|_p.$$

The space $(W^{1,p}(J, X), \| \cdot \|_{1,p})$ is called *(vector-valued) Sobolev space.*

Theorem A.9. *The Sobolev space $(W^{1,p}(J, X), \| \cdot \|_{1,p})$ is a Banach space. Moreover, if X is a Hilbert space and $p = 2$, then it is a Hilbert space.*

Bibliography

[1] R. A. Adams, *Sobolev Spaces*, New York, NY: Academic Press, 1975.

[2] H. Amann, *Linear and Quasilinear Parabolic Problems*, Boston, MA: Birkhäuser Verlag, 1995.

[3] ———, "Operator Valued Fourier Multipliers, Vector Valued Besov Spaces, and Applications," *Math. Nachr.* 186 (1997): 5–56.

[4] W. Arendt, "Gaussian Estimates and Interpolation of the Spectrum in L^p," *Diff. Int. Eqs.* 7 (1994): 1153–1168.

[5] W. Arendt and C. J. K. Batty, "Asymptotically Almost Periodic Solutions of Inhomogeneous Cauchy Problems on the Half-Line," *Bull. London Math. Soc.* 31 (1999): 291–304.

[6] W. Arendt, C. J. K. Batty, M. Hieber, and F. Neubrander, *Vector-Valued Laplace Transforms and Cauchy Problems*, vol. 96, Monographs in Mathematics, Basel: Birkhäuser Verlag, 2001.

[7] W. Arendt and A. Rhandi, "Perturbation of Positive Semigroups," *Arch. Math.* 56 (1991): 107–119.

[8] O. Arino and I. Győri, "Necessary and Sufficient Condition for Oscillation of a Neutral Differential System with Several Delays," *J. Differential Equations* 81 (1989): 98–105.

[9] F. V. Atkinson, H. Langer, R. Mennicken, and A. A. Shkalikov, "The Essential Spectrum of Some Matrix Operators," *Math. Nachr.* 167 (1994): 5–20.

[10] V. Barbu, *Nonlinear Semigroups and Differential Equations in Banach Spaces*, Leyden: Noordhoff International Publishing, 1976.

[11] A. Bátkai, "Second Order Cauchy Problems with Damping and Delay," Ph.D. diss., Tübingen, 2000.

[12] ——, "Hyperbolicity of Linear Partial Differential Equations with Delay," *Integral Eqs. Operator Th.* 44 (2002): 383–396.

[13] A. Bátkai, P. Binding, A. Dijksma, R. Hryniv, and H. Langer, "Spectral Problems for Operator Matrices," To appear in *Math. Nachr.*, 2004.

[14] A. Bátkai and K.-J. Engel, "Exponential Decay of 2×2 Operator Matrix Semigroups," To appear in *J. Comp. Anal. Appl.* Available from World Wide Web (http://www.fa.uni-tuebingen.de/research/tuebe/1999).

[15] A. Bátkai and B. Farkas, "On the Effect of Small Delays to the Stability of Feedback Systems," *Progress in Nonlinear Differential Equations* 55 (2003): 83–94.

[16] A. Bátkai, E. Fašanga, and R. Shvidkoy, "Hyperbolicity of Delay Equations via Fourier Multipliers," *Acta Sci. Math. (Szeged)* 69 (2003): 131–145.

[17] A. Bátkai and S. Piazzera, "Damped Wave Equations with Delay," *Fields Inst. Comm.* 29 (2001): 51–61.

[18] ——, "Partial Differential Equations with Relatively Bounded Operators in the Delay Term," *Semigroup Forum* 64 (2001): 71–89.

[19] ——, "Semigroups and Linear Partial Differential Equations with Delay," *J. Math. Anal. Appl.* 264 (2001): 1–20.

[20] A. Bátkai and R. Schnaubelt, "Asymptotic Behavior of Parabolic Problems with Delays in the Highest Order Derivatives," *Semigroup Forum* 69 (2004): 369–399.

[21] C. J. K. Batty, "Differentiability and Growth Bounds of Solutions of Delay Equations," *J. Math. Anal. Appl.* 299 (2004): 133–146.

[22] C. J. K. Batty and R. Chill, "Bounded Convolutions and Solutions of Inhomogeneous Cauchy Problems," *Forum Math.* 11 (1999): 253–277.

[23] I. Becker and G. Greiner, "On the Modulus of One-Parameter Semigroups," *Semigroup Forum* 34 (1986): 185–201.

[24] A. Belleni-Morante, *Applied Semigroups and Evolution Equations*, New York, NY: Oxford University Press, 1979.

[25] S. Boulite, L. Maniar, A. Rhandi, and J. Voigt, "The Modulus Semigroup for Linear Delay Equations," *Positivity* 8 (2004): 1–9.

[26] S. Brendle, "On the Asymptotic Behavior of Perturbed Strongly Continuous Semigroups," *Math. Nachr.* 226 (2001): 35–47.

[27] S. Brendle, R. Nagel, and J. Poland, "On the Spectral Mapping Theorem for Perturbed Strongly Continuous Semigroups," *Arch. Math.* 74 (2000): 365–378.

[28] J. A. Burns, T. L. Herdman, and H. W. Stech, "Linear Functional Differential Equations as Semigroups on Product Spaces," *SIAM J. Math. Anal.* 14 (1983): 98–116.

[29] V. Casarino and S. Piazzera, "On the Stability of Asymptotic Properties of Perturbed C_0-Semigroups," *Forum Math.* 13 (2001): 91–107.

[30] G. Chen and R. Triggiani, "Proof of Extensions of Two Conjectures on Structural Damping for Elastic Systems," *Pac. J. Math.* 136 (1989): 15–55.

[31] ———, "Characterization of Domains of Fractional Powers of Certain Operators Arising in Elastic Systems, and Applications," *J. Differential Equations* 88 (1990): 279–293.

[32] C. Chicone and Y. Latushkin, *Evolution Semigroups in Dynamical Systems and Differential Equations*, vol. 70, Mathematical Surveys and Monographs, Providence, RI: American Mathematical Society, 1999.

[33] R. Chill and J. Prüß, "Asymptotic Behavior of Linear Evolutionary Integral Equations," *Integral Eqs. Operator Th.* 39 (2001): 193–213.

[34] S. Clark, Y. Latushkin, S. Montgomery-Smith, and T. Randolph, "Stability Radius and Internal versus External Stability in Banach Spaces: An Evolution Semigroup Approach," *SIAM J. Control. Optim.* 38 (2000): 1757–1793.

[35] P. Clément, O. Diekmann, M. Gyllenberg, H. Heijmans, and H. Thieme, "Perturbation Theory for Dual Semigroups. I. The Sun-Reflexive Case," *Math. Ann.* 277 (1987): 709–725.

[36] P. Clément, H. J. A. M. Heijmans, S. Angenent, C. J. van Duijn, and B. D. Pagter, *One-Parameter Semigroups*, vol. 5, CWI Monographs, Amsterdam: North-Holland, 1987.

[37] P. Clément and J. Prüss, "On Second-Order Differential Equations in Hilbert Spaces," *Boll. Un. Mat. Ital.* 3 (1989): 623–638.

[38] B. D. Coleman and V. J. Mizel, "Norms and Semigroups in the Theory of Fading Memory," *Arch. Rational Mech. Anal.* 23 (1966): 87–123.

[39] J. B. Conway, *A Course in Functional Analysis*, vol. 96, Graduate Texts in Mathematics, New York, NY: Springer-Verlag, 1985.

[40] ———, *Functions of One Complex Variable*, vol. 98, Graduate Texts in Mathematics, New York, NY: Springer-Verlag, 1986.

[41] S. Cox and E. Zuazua, "The Rate at which Energy Decays in a Damped String," *Comm. Partial Diff. Eqs.* 19 (1994): 213–243.

[42] ———, "The Rate at which Energy Decays in a String Damped at One End," *Indiana Univ. Math. J.* 44 (1995): 545–573.

[43] R. F. Curtain and H. J. Zwart, *An Introduction to Infinite Dimensional Linear Systems Theory*, New York, NY: Springer-Verlag, 1995.

[44] R. Datko, "Not All Feedback Stabilized Systems are Robust with Respect to Small Time Delays in Their Feedback," *SIAM J. Control. Optim.* 26 (1988): 697–713.

[45] ———, "Two Questions concerning the Boundary Control of Elastic Systems," *J. Differential Equations* 92 (1991): 27–44.

[46] ———, "Is Boundary Control a Realistic Approach to the Stabilization of Vibrating Elastic Systems?" in *Evolution Equations*, eds. G. Ferreyra, G. R. Goldstein, and F. Neubrander, New York, NY: Marcel Dekker, 1994, pp. 133–140.

[47] R. Datko, J. Lagnese, and M. P. Polis, "An Example on the Effect of Time Delays in Boundary Feedback of Wave Equations," *SIAM J. Control. Optim.* 24 (1986): 152–156.

[48] R. Datko and Y. C. You, "Some Second Order Vibrating Systems Cannot Tolerate Small Time Delays in Their Damping," *J. Optim. Th. Appl.* 70 (1991): 521–537.

[49] R. Dautray and J. L. Lions, *Mathematical Analysis and Numerical Methods for Science and Technology*, vols. *1–5*, New York, NY: Springer-Verlag, 1988.

[50] E. B. Davies, *One-Parameter Semigroups*, London: Academic Press, 1980.

[51] ———, *Spectral Theory and Differential Operators*, Cambridge, UK: Cambridge University Press, 1995.

[52] M. C. Delfour, "State Theory of Linear Hereditary Differential Systems," *J. Math. Anal. Appl.* 60 (1977): 8–35.

[53] ———, "The Largest Class of Hereditary Systems Defining a C_0-Semigroup on the Product Space," *Canadian J. Math.* 32 (1980): 969–978.

[54] M. C. Delfour and S. K. Mitter, "Hereditary Differential Systems with Constant Delays I; General Case," *J. Differential Equations* 12 (1972): 213–235.

[55] ———, "Hereditary Differential Systems with Constant Delays II; A Class of Affine Systems and the Adjoint Problem," *J. Differential Equations* 18 (1975): 18–28.

[56] R. Derndinger, "Über das Spektrum positiver Generatoren," *Math. Z.* 172 (1980): 281–293.

[57] W. Desch and W. Schappacher, "On Relatively Bounded Perturbations of Linear C_0-Semigroups," *Ann. Scuola Norm. Sup. Pisa, Cl. Sci.* 11 (1984): 327–341.

[58] ———, "Some Generation Results for Perturbed Semigroups," in *Semigroup Theory and Applications*, eds. P. Clément, S. Invernizzi, E. Mitidieri, and I. I. Vrabie, vol. 116, Lecture Notes in Pure and Applied Mathematics, New York, NY: Marcel Dekker, 1989, pp. 125–152.

[59] G. Di Blasio, "Differentiability of the Solution Semigroup for Delay Differential Equations," in *Evolution Equations. Festschrift on the Occasion of the 60th Birthday of J. A. Goldstein*, eds. G. Goldstein, R. Nagel, and S. Romanelli, New York, NY: Marcel Dekker, 2003, pp. 147–158.

[60] G. Di Blasio, K. Kunisch, and E. Sinestrari, "L^2-Regularity for Parabolic Partial Integrodifferential Equations with Delay in the Highest-Order Derivatives," *J. Math. Anal. Appl.* 102 (1984): 38–57.

[61] ———, "Stability for Abstract Linear Functional Differential Equations," *Israel J. Math.* 50 (1985): 231–263.

[62] O. Diekmann, S. A. van Gils, S. M. V. Lunel, and H.-O. Walther, *Delay Equations*, vol. 110, Applied Mathematical Sciences, New York, NY: Springer-Verlag, 1995.

[63] J. Diestel and J. J. J. Uhl, *Vector Measure*, Providence, RI: American Mathematical Society, 1977.

[64] G. Dore, "L^p-Regularity for Abstract Differential Equations," in *Functional Analysis and Related Topics*, ed. H. Komatsu, vol. 1540, Lecture Notes in Mathematics, New York, NY: Springer-Verlag, 1993, pp. 25–38.

[65] N. Dunford and J. T. Schwartz, *Linear Operators I. General Theory*, New York, NY: Interscience Publisher, 1958.

[66] K.-J. Engel, "Operator Matrices and Systems of Evolutions Equations," Manuscript. Available from World Wide Web (http://giotto.mathematik.uni-tuebingen.de/people/klen/omsee.ps).

[67] ———, "On Dissipative Wave Equations in Hilbert Space," *J. Math. Anal. Appl.* 184 (1994): 302–316.

[68] ———, "Positivity and Stability for One-Sided Coupled Operator Matrices," *Positivity* 1 (1997): 103–124.

[69] ———, "Matrix Representation of Linear Operators on Product Spaces," in *International Workshop on Operator Theory*, eds. P. Aiena, M. Pavone, and C. Trapani, No. 56 in Rendiconti del Circolo matematico di Palermo (2) Supplemento, Palermo: Circolo matematico di Palermo, 1998, pp. 219–224.

[70] ———, "Spectral Theory and Generator Property of One-Sided Coupled Operator Matrices," *Semigroup Forum* 58 (1999): 267–295.

[71] K.-J. Engel and R. Nagel, "Cauchy Problems for Polynomial Operator Matrices on Abstract Energy Spaces," *Forum Math.* 2 (1990): 89–102.

[72] ———, *One-Parameter Semigroups for Linear Evolution Equations*, vol. 194, Graduate Texts in Mathematics, New York, NY: Springer-Verlag, 2000.

[73] B. Farkas, "Perturbations of Bi-Continuous Semigroups with Applications to Transition Semigroups on $C_b(H)$," *Semigroup Forum* 68 (2004): 87–107.

[74] E. Fašanga and J. Prüß, "Evolution Equations with Dissipation of Memory Type," in *Topics in Nonlinear Analysis*, eds. J. Escher and G. Simonett, Basel: Birkhäuser, 1999, pp. 213–250.

[75] E. Fašanga and J. Prüß, "Asymptotic Behavior of a Semilinear Viscoelastic Beam Model," *Arch. Math.* 77 (2001): 488–497.

[76] H. O. Fattorini, *The Cauchy Problem*, Reading, MA: Addison-Wesley, 1983.

[77] ———, *Second Order Linear Differential Equations in Banach Spaces*, vol. 108, North-Holland Mathematical Studies, Amsterdam: Elsevier, 1985.

[78] A. Favini, L. Pandolfi, and H. Tanabe, "Singular Differential Equations with Delay," *Diff. Int. Eqs.* 12 (1999): 351–371.

[79] A. Fischer and J. M. A. M. van Neerven, "Robust Stability of C_0-Semigroups and an Application to Stability of Delay Equations," *J. Math. Anal. Appl.* 226 (1998): 82–100.

[80] G. Fragnelli, "Delay Equations with Non-Autonomous Past," Ph.D. diss., Tübingen, 2002.

[81] G. Fragnelli and G. Nickel, "Partial Functional Differential Equations with Nonautonomous Past in L^p-Phase Spaces," *Diff. Int. Eqs.* 16 (2003): 327–348.

[82] L. Gearhart, "Spectral Theory for Contraction Semigroups on Hilbert Space," *Trans. Amer. Math. Soc.* 236 (1978): 385–394.

[83] B. R. Gelbaum and J. M. H. Olmsted, *Counterexamples in Analysis*, San Francisco, CA: Holden-Day, 1964.

[84] M. Girardi and L. Weis, "Operator-Valued Fourier Multiplier Theorems on Besov Spaces," *Math. Nachr.* 251 (2003): 34–51.

[85] I. Gohberg, S. Goldberg, and M. A. Kaashoek, *Classes of Linear Operators I.*, Basel: Birkhäuser Verlag, 1990.

[86] S. Goldberg, *Unbounded Linear Operators*, New York, NY: McGraw-Hill, 1966.

[87] J. A. Goldstein, *Semigroups of Operators and Applications*, New York, NY: Oxford University Press, 1985.

[88] J. A. Goldstein and J. T. Sandefur, "Equipartition of Energy for Higher Order Abstract Hyperbolic Equations," *Comm. Partial Diff. Eqs.* 7 (1982): 1217–1251.

[89] ———, "An Abstract d'Alembert Formula," *SIAM J. Math. Anal.* 18 (1987): 842–856.

[90] G. C. Gorain, "Exponential Energy Decay Estimate for the Solutions of Internally Damped Wave Equations in a Bounded Domain," *J. Math. Anal. Appl.* 216 (1997): 510–520.

[91] G. Greiner, "A Typical Perron-Frobenius Theorem with Applications to an Age-Dependent Population Equation," in *Infinite-Dimensional Systems*, eds. F. Kappel and W. Schappacher, vol. 1076, Lecture Notes in Mathematics, New York, NY: Springer-Verlag, 1984, pp. 86–100.

[92] ———, "Some Applications of Fejér's Theorem to One Parameter Semigroups," *Semesterbericht Funktionalanalysis* (WS 1984/85): 33–49.

[93] G. Greiner and R. Nagel, "On the Stability of Strongly Continuous Semigroups of Positive Operators on $L^2(\mu)$," *Ann. Scuola Norm. Sup. Pisa, Cl. Sci.* 10 (1983): 257–262.

[94] G. Gühring, F. Räbiger, and R. Schnaubelt, "A Characteristic Equation for Non-Autonomous Partial Functional Differential Equations," *J. Differential Equations* 181 (2002): 439–462.

[95] I. Győri, F. Hartung, and J. Turi, "Preservation of Stability in Delay Equations under Delay Perturbations," *J. Math. Anal. Appl.* 220 (1998): 290–313.

[96] I. Győri and T. Krisztin, "Oscillation results for Linear Autonomous Partial Delay Differential Equations," *J. Math. Anal. Appl.* 174 (1993): 204–217.

[97] I. Győri and M. Pituk, "Stability Criteria for Linear Delay Differential Equations," *Diff. Int. Eqs.* 10 (1997): 841–852.

[98] I. Győri and J. Turi, "An Oscillation Result for Singular Neutral Equations," *Canad. Math. Bull.* 37 (1994): 54–65.

[99] K. P. Hadeler, "Delay Equations in Biology," in *Functional Differential Equations and Approximation of Fixed Points*, eds. H.-O. Peitgen and H.-O. Walther, vol. 730, Lecture Notes in Mathematics, New York, NY: Springer-Verlag, 1979, pp. 136–156.

[100] J. K. Hale, "Linear Functional-Differential Equations with Constant Coefficients," *Contributions to Differential Equations* 2 (1963): 291–317.

[101] ———, *Functional Differential Equations*, vol. 3, Appl. Math. Sci., New York, NY: Springer-Verlag, 1971.

[102] J. K. Hale and J. Kato, "Phase Space for Retarded Equations with Infinite Delay," *Funkcialaj Ekvac.* 21 (1978): 11–41.

[103] J. K. Hale and S. M. Verduyn Lunel, *Introduction to Functional-Differential Equations*, vol. 99, Applied Mathematical Sciences, New York, NY: Springer-Verlag, 1993.

[104] ———, "Effects of Time Delays on the Dynamics of Feedback Systems," in *EQUADIFF'99, International Conference on Differential Equations, Berlin 1999*, eds. B. Fiedler, K. Grger, and J. Sprekels, Singapore: World Scientific, 2000, pp. 257–266.

[105] ———, "Effects of Small Delays on Stability and Control," in *Operator Theory and Analysis, The M.A. Kaashoek Anniversary Volume*, eds. H. Bart, I. Gohberg, and A. C. M. Ran, vol. 122, Operator Theory: Advances and Applications, Basel: Birkhäuser Verlag, 2001, pp. 275–301.

[106] G. H. Hardy, J. E. Littlewood, and G. Pòlya, *Inequalities*, Cambridge, UK: Cambridge University Press, 1999, reprint of 2nd Edition.

[107] I. W. Herbst, "The Spectrum of Hilbert Space Semigroups," *J. Operator Th.* 10 (1983): 87–94.

[108] M. Hieber, "Operator Valued Fourier Multipliers," in *The Herbert Amann Anniversary Volume*, eds. J. Escher and G. Simonett, Topics in Nonlinear Analysis, Basel: Birkhäuser, 1999, pp. 363–380.

[109] ———, "A Characterization of the Growth Bound of a Semigroup via Fourier Multipliers," in *Evolution Equations and Their Applications in Physical and Life Sciences*, eds. G. Lumer and L. Weis, New York, NY: Marcel Dekker, 2001, pp. 121–124.

[110] E. Hille and R. Phillips, *Functional Analysis and Semi-Groups*, vol. 31, Colloquium Publications, Providence, RI: American Mathematical Society, 1957.

[111] Y. Hino, S. Murakami, and T. Naito, *Functional Differential Equations with Infinite Delay*, vol. 1473, Lecture Notes in Mathematics, New York, NY: Springer-Verlag, 1991.

[112] A. Holderrieth, "Matrix Multiplication Operators Generating One-Parameter Semigroups," *Semigroup Forum* 42 (1991): 155–166.

[113] J. Howland, "On a Theorem of Gearhart," *Integral Eqs. Operator Th.* 7 (1984): 138–142.

[114] F. Huang, "Characteristic Conditions for Exponential Stability of Linear Dynamical Systems in Hilbert Spaces," *Ann. Diff. Eqs.* 1 (1985): 43–55.

[115] S.-Z. Huang, "On Energy Decay Rate of Linear Damped Elastic Systems," *Tübinger Berichte zur Funktionalanalysis* 6 (1996/97): 65–91.

[116] ———, "A Local Version of Gearhart's Theorem," *Semigroup Forum* 58 (1999): 323–335.

[117] M. A. Kaashoek and S. M. V. Lunel, "An Integrability Condition on the Resolvent for Hyperbolicity of the Semigroup," *J. Differential Equations* 112 (1994): 374–406.

[118] F. Kappel, "Semigroups and Delay Equations," in *Semigroups, Theory and Applications-II (Proceedings Trieste 1984)*, eds. H. Brezis, M. Crandall, and F. Kappel, vol. 152, Pitman Research Notes in Mathematics, New York, NY: John Wiley and Sons, 1986, pp. 136–176.

[119] F. Kappel and K. Zhang, "Equivalence of Functional-Differential Equations of Neutral Type and Abstract Cauchy Problems," *Mh. Math.* 101 (1986): 115–133.

[120] T. Kato, *Perturbation Theory for Linear Operators*, vol. 132, Grundlehren der mathematischen Wissenschaften, New York, NY: Springer-Verlag, 1980.

[121] W. Kerscher and R. Nagel, "Asymptotic Behavior of One-Parameter Semigroups of Positive Operators," *Acta Appl. Math.* 2 (1984): 297–309.

[122] V. Komornik, *Exact Controllability and Stabilization. The Multiplier Method*, Chichester: John Wiley and Sons, 1994.

[123] M. Kramar, D. Mugnolo, and R. Nagel, "Semigroups for Initial Boundary Value Problems," in *Evolution Equations: Applications to Physics, Industry, Life Sciences and Economics.*, eds. M. Iannelli and G. Lumer, vol. 55, Progress in Nonlinear Differential Equations, Basel: Birkhäuser, 2003, pp. 275–292.

[124] N. N. Krasovskii, *Stability of Motion. Applications of Lyapunov's Second Method to Differential Systems and Equations with Delay*, Stanford, CA: Stanford University Press, 1963.

[125] M. G. Kreĭn and H. Langer, "On Some Mathematical Principles in the Linear Theory of Damped Oscillations of Continua. I," *Integral Eqs. Operator Th.* 1 (1978): 364–399.

[126] ———, "On Some Mathematical Principles in the Linear Theory of Damped Oscillations of Continua. II," *Integral Eqs. Operator Th.* 1 (1978): 539–566.

[127] S. G. Krein, *Linear Differential Equations in Banach Spaces*, vol. 29, American Mathematical Society Translations, Providence, RI: American Mathematical Society, 1970.

[128] K. Kunisch and M. Mastinšek, "Dual Semigroups and Structural Operators for Abstract Functional Differential Equations," *Diff. Int. Eqs.* 3 (1990): 733–756.

[129] K. Kunisch and W. Schappacher, "Necessary Conditions for Partial Differential Equations with Delay to Generate C_0-Semigroups," *J. Differential Equations* 50 (1983): 49–79.

[130] Y. Latushkin and R. Shvydkoy, "Hyperbolicity of Semigroups and Fourier Multipliers," in *Systems, Approximation, Singular Integral Operators, and Related Topics. Proceedings of the 11th International Workshop on Operator Theory and Applications, IWOTA 2000, Bordeaux*, eds. A. A. Borichev and N. Nikolski, vol. 129, Operator Theory: Advances and Applications, Basel: Birkhäuser, 2001, pp. 341–363.

[131] G. Lebeau, "Equation des Ondes Amorties," in *Algebraic and Geometric Methods in Mathematical Physics*, eds. A. Boutet de Monvel and V. Marchenko, Dodrecht: Kluwer, 1996, pp. 73–109.

[132] G. Lebeau and E. Zuazua, "Decay Rates for the Three-Dimensional Linear System of Thermoelasticity," *Arch. Ration. Mech. Anal.* 148 (1999): 179–231.

[133] J. L. Lions, "Une Remarque sur les Applications du Théorème de Hille-Yosida," *J. Math. Soc. Japan* 9 (1957): 62–70.

[134] J. López-Gómez, "On the Linear Damped Wave Equation," *J. Differential Equations* 134 (1997): 26–45.

[135] A. Lunardi, *Analytic Semigroups and Optimal Regularity in Parabolic Problems*, Basel: Birkhäuser, 1995.

[136] Y. Luo and D. Feng, "On Exponential Stability of C_0-Semigroups," *J. Math. Anal. Appl.* 217 (1998): 624–636.

[137] L. Maniar and A. Rhandi, "Inhomogeneous Retarded Differential Equation in Infinite Dimensional Banach Space via Extrapolation Spaces," *Rend. Circolo. Math. Palermo* 47 (1998): 163–194.

[138] ———, "Non-Autonomous Retarded Wave Equations," *J. Math. Anal. Appl.* 263 (2001): 14–32.

[139] L. Maniar and J. Voigt, "Linear Delay Equations in the L_p-Context," in *Evolution Equations. Festschrift on the Occasion of the 60th Birthday of J. A. Goldstein*, eds. G. Goldstein, R. Nagel, and S. Romanelli, New York, NY: Marcel Dekker, 2003, pp. 319–330.

[140] M. Mastinšek, "Adjoints of Solution Semigroups and Identifiability of Delay Differential Equations in Hilbert Spaces," *Acta Math. Univ. Comenian. (N.S.)* 63 (1994): 193–206.

[141] ———, "Dual Semigroups for Delay Differential Equations with Unbounded Operators Acting on the Delays," *Diff. Int. Eqs.* 7 (1994): 205–216.

[142] ———, "Norm Continuity and Stability for a Functional Differential Equation in Hilbert Space," *J. Math. Anal. Appl.* 269 (2002): 770–783.

[143] ———, "Stability Conditions for Abstract Functional Differential Equations in Hilbert Space," *Semigroup Forum* 63 (2003): 140–150.

[144] J. Milota, "Stability and Saddle-Point Property for a Linear Autonomous Functional Parabolic Equation," *Comment. Math. Univ. Carolinae* 27 (1986): 87–101.

[145] I. Miyadera, "On Perturbation Theory for Semi-Groups of Operators," *Tôhoku Math. J.* 18 (1966): 299–310.

[146] L. Monauni, "On the Abstract Cauchy Problem and the Generation Problem for Semigroups of Bounded Operators," Control Theory Centre Report 90, Warwick (1980).

[147] S. Montgomery-Smith, "Stability and Dichotomy of Positive Semigroups on L_p," *Proc. Amer. Math. Soc.* 124 (1996): 2433–2437.

[148] R. Nagel, ed., *One-Parameter Semigroups of Positive Operators*, vol. 1184, Lecture Notes in Mathematics, Berlin: Springer-Verlag, 1986.

[149] R. Nagel, "Towards a Matrix Theory for Unbounded Operator Matrices," *Math. Z.* 201 (1989): 57–68.

[150] ———, "The Spectrum of Unbounded Operator Matrices with Non-Diagonal Domain," *J. Funct. Anal.* 89 (1990): 291–301.

[151] ———, "Characteristic Equations for the Spectrum of Generators," *Ann. Sc. Norm. Super. Pisa, Cl. Sci.* 24 (1997): 703–717.

[152] ———, "Stability Criteria through Characteristic Equations of Linear Operators," *Rev. Unión Argent.* 41 (1998): 91–98.

[153] R. Nagel and S. Piazzera, "On the Regularity Properties of Perturbed Semigroups," in *International Workshop on Operator Theory*, eds. P. Aiena, M. Pavone, and C. Trapani, No. 56 in Rendiconti del Circolo matematico di Palermo (2) Supplemento, Palermo: Circolo matematico di Palermo, 1998, pp. 99–110.

[154] R. Nagel and J. Poland, "The Critical Spectrum of a Strongly Continuous Semigroup," *Adv. Math.* 152 (2000): 120–133.

[155] S. Nakagiri, "Optimal Control of Linear Retarded Systems in Banach Spaces," *J. Math. Anal. Appl.* 120 (1986): 169–210.

[156] ———, "Structural Properties of Functional Differential Equations in Banach Spaces," *Osaka J. Math.* 25 (1988): 353–398.

[157] S. Nakagiri and H. Tanabe, "Structural Operators and Eigenmanifold Decomposition for Functional-Differential Equations in Hilbert Spaces," *J. Math. Anal. Appl.* 204 (1996): 554–581.

[158] J. M. A. M. v. Neerven, *The Adjoint of a Semigroup of Linear Operators*, vol. 1529, Lecture Notes in Mathematics, Berlin: Springer-Verlag, 1992.

[159] ———, *The Asymptotic Behavior of Semigroups of Linear Operators*, Basel: Birkhäuser, 1996.

[160] H. Neidhardt, "On Abstract Linear Evolution Equations I," *Math. Nachr.* 103 (1981): 283–293.

[161] F. Neubrander, "Well-Posedness of Higher Order Abstract Cauchy Problems," *Trans. Amer. Math. Soc.* 295 (1986): 257–290.

[162] ———, "Integrated Semigroups and Their Application to Complete Second Order Cauchy Problems," *Semigroup Forum* 38 (1989): 233–251.

[163] E. Obrecht, "Phase Space for an n-th Order Differential Equation in Banach Spaces," in *Semigroup Theory and Evolution Equations*, eds. P. Clément, E. Mitidieri, and B. de Pagter, vol. 135, Lecture Notes in Pure and Applied Mathematics, New York, NY: Marcel Dekker, 1991, pp. 391–399.

[164] A. Pazy, *Semigroups of Linear Operators and Applications to Partial Differential Equations*, New York, NY: Springer-Verlag, 1983.

[165] R. Phillips, "Perturbation Theory for Semi-Groups of Linear Operators," *Trans. Amer. Math. Soc.* 74 (1953): 199–221.

[166] S. Piazzera, "Qualitative Properties of Perturbed Semigroups," Ph.D. diss., Tübingen, 1999.

[167] ———, "An Age Dependent Population Equation with Delayed Birth Process," *Mathematical Models in the Applied Sciences* 4 (2004): 427–439.

[168] J. Prüß, "On the Spectrum of C_0-Semigroups," *Trans. Amer. Math. Soc.* 284 (1984): 847–857.

[169] ———, *Evolutionary Integral Equations and Applications*, Basel: Birkhäuser, 1993.

[170] F. Räbiger, A. Rhandi, R. Schnaubelt, and J. Voigt, "Non-Autonomous Miyadera Perturbations," *Diff. Int. Eqs.* 13 (2000): 341–368.

[171] R. Rebarber and S. Townly, "Robustness with Respect to Delays for Exponential Stability of Distributed Parameter Systems," *SIAM J. Control. Optim.* 37 (1998): 230–244.

[172] M. Renardy, "On the Type of Certain C_0-Semigroups," *Comm. Partial Diff. Eqs.* 18 (1993): 1299–1307.

[173] ———, "On the Linear Stability of Hyperbolic PDEs and Viscoelastic Flows," *Z. Angew. Math. Phys.* 45 (1994): 854–865.

[174] A. Rhandi, "Dyson-Phillips Expansion and Unbounded Perturbations of Linear C_0-Semigroups," *J. Comp. Appl. Math.* 44 (1992): 339–349.

[175] ———, "Extrapolation Methods to Solve Non-Autonomous Retarded Partial Differential Equations," *Studia Math.* 126 (1997): 219–233.

[176] W. Ruess, "Asymptotic Behavior of Solutions to Delay Functional-Differential Equations," in *Recent Developments in Evolution Equations (Glasgow, 1994)*, ed. A. McBride, vol. 324, Pitman Research Notes in Mathematics Series, Harlow, UK: Longman Scientific and Technical, 1995, pp. 220–230.

[177] ———, "Existence of Solutions to Partial Functional Differential Equations with Delay," in *Theory and Applications of Nonlinear Operators of Accretive and Monotone Type*, ed. A. G. Kartsatos, vol. 178, Lecture Notes in Pure and Applied Mathematics, New York, NY: Marcel Dekker, 1996, pp. 259–288.

[178] ———, "Existence and Stability of Solutions to Partial Functional Differential Equations with Delay," *Adv. Diff. Eqs.* 4 (1999): 843–876.

[179] W. Ruess and W. Summers, "Compactness in Spaces of Vector-Valued Continuous Functions and Asymptotic Almost Periodicity," *Math. Nachr.* 135 (1988): 7–33.

[180] ———, "Almost Periodicity and Stability for Solutions to Functional-Differential Equations with Infinite Delay," *Diff. Int. Eqs.* 6 (1996): 1225–1252.

[181] Y. Sakawa, "Feedback Control of Second Order Evolution Equations with Damping," *SIAM J. Control. Optim.* 22 (1984): 343–361.

[182] J. Sandefur, "Higher Order Abstract Cauchy Problems," *J. Math. Anal. Appl.* 60 (1977): 728–742.

[183] ———, "Convergence of Solutions of a Second-Order Cauchy Problem," *J. Math. Anal. Appl.* 100 (1984): 470–477.

[184] H. H. Schaefer, *Banach Lattices and Positive Operators*, New York, NY: Springer-Verlag, 1974.

[185] R. Schnaubelt, "Exponential Bounds and Hyperbolicity of Evolution Families," Ph.D. diss., Tübingen, 1996.

[186] ———, "Asymptotically Autonomous Parabolic Evolution Equations," *J. Evol. Eqs.* 1 (2001): 19–37.

[187] ———, "Feedbacks for Non-Autonomous Regular Linear Systems," *SIAM J. Control. Optim.* 41 (2002): 1141–1165.

[188] ———, "Well-Posedness and Asymptotic Behavior of Non-Autonomous Linear Evolution Equations," *Progress in Nonlinear Differential Equations* 50 (2002): 311–338.

[189] ———, "Parabolic Partial Differential Equations with Asymptotically Autonomous Delay," *Trans. Amer. Math. Soc.* 356 (2003): 3517–3543.

[190] S. N. Shimanov, "On a Theory of Linear Differential Equations with After-Effect," *Differencialniye Uravnenija* 1 (1965): 102–116.

[191] M. Shubov, "Asymptotics of the Spectrum and Eigenfunctions for Nonselfadjoint Operator Generated by Radially Nonhomogeneous Wave Equation," *Asymptotical Analysis* 16 (1998): 245–272.

[192] ———, "Spectral Operators Generated by the 3 Dimensional Damped Wave Equation and Application to Control Theory," in *Spectral and Scattering Theory*, ed. A. G. Ramm, New York, NY: Plenum Press, 1998, pp. 177–188.

[193] E. Sinestrari, "A Noncompact Differentiable Semigroup Arising from an Abstract Delay Equation," *C. R. Math. Rep. Acad. Sci. Canada* 6 (1984): 43–48.

[194] ———, "Wave Equation with Memory," *Discrete Cont. Dyn. Syst.* 5 (1999): 881–896.

[195] H. Sohr, "Ein neues Surjektivitätskriterium in Hilbertraum," *Mh. Math.* 91 (1981): 313–337.

[196] O. Staffans, "Well–Posed Linear Systems Part I: General Theory," Manuscript. Available from World Wide Web (http://www.abo.fi/~staffans).

[197] M. Stein, H. Vogt, and J. Voigt, "The Modulus Semigroup for Linear Delay Equations III." *J. Funct. Anal.* 220 (2005): 388–400.

[198] H. Tanabe, *Equations of Evolution*, London: Pitman, 1979.

[199] ———, *Functional Analytical Methods for Partial Differential Equations*, vol. 204, Lecture Notes in Pure and Applied Mathematics, New York, NY: Marcel Dekker, 1997.

[200] L. T. Tébou, "Well-Posedness and Energy Decay Estimates for the Damped Wave Equation with L^r Localizing Coefficient," *Comm. Partial Diff. Eqs.* 23 (1998): 1839–1855.

[201] C. Travis and G. Webb, "Existence, Stability, and Compactness in the α-Norm for Partial Functional Differential Equations," *Trans. Amer. Math. Soc.* 240 (1978): 129–143.

[202] R. Triggiani, "Wave Equations on a Bounded Domain with Boundary Dissipation: An Operator Approach," *J. Math. Anal. Appl.* 137 (1989): 1839–1855.

[203] A. Vagabov, "Expansion in a Fourier Series of a Solution of a Mixed Problem for the Wave Equation with Delay Conditions," *Russian Acad. Sci. Dokl. Math.* 50 (1995): 53–57.

[204] S. M. Verduyn Lunel, "About Completeness of a Class of Unbounded Operators," *J. Differential Equations* 120 (1995): 108–132.

[205] ——, "Spectral Theory for Delay Equations," in *Systems, Approximation, Singular Integral Operators, and Related Topics. Proceedings of the 11th International Workshop on Operator Theory and Applications, IWOTA 2000, Bordeaux*, eds. A. A. Borichev and N. Nikolski, vol. 129, Operator Theory: Advances and Applications, Basel: Birkhäuser, 2001, pp. 465–507.

[206] K. Veselic, "Energy Decay of Damped Systems," *Z. Angew. Math. Phys.* 84 (2004): 856–864.

[207] J. Voigt, "The Modulus Semigroup for Linear Delay Equations II." *Note Mat.*, to appear.

[208] ——, "On the Perturbation Theory for Strongly Continuous Semigroups," *Math. Ann.* 229 (1977): 163–171.

[209] ——, "Stability of the Essential Type of Strongly Continuous Semigroups," *Trans. Steklov Math. Inst.* 203 (1994): 469–477.

[210] ——, "Absorption Semigroups, Feller Property, and Kato Class," *Oper. Theory, Adv. Appl.* 78 (1995): 389–396.

[211] V. Volterra, "Sur la Théorie Mathématique des Phénomènes Héréditaires," *J. Math. Pures Appl.* 7 (1928): 249–298.

[212] G. Webb, "Autonomous Nonlinear Functional Differential Equations and Nonlinear Semigroups," *J. Math. Anal. Appl.* 46 (1974): 1–12.

[213] ——, "Functional Differential Equations and Nonlinear Semigroups in L^p-spaces," *J. Differential Equations* 20 (1976): 71–89.

[214] J. Weidmann, *Linear Operators in Hilbert Spaces*, vol. 68, Graduate Texts in Mathematics, New York, NY: Springer-Verlag, 1980.

[215] L. Weis, "The Stability of Positive Semigroups on L_p-Spaces," *Proc. Amer. Math. Soc.* 123 (1995): 3089–3094.

[216] ——, "A Short Proof for the Stability Theorem for Positive Semigroups on $L_p(\mu)$," *Proc. Amer. Math. Soc.* 126 (1998): 3253–3256.

[217] ——, "Operator-Valued Fourier-Multiplier Theorems and Maximal L^p-Regularity," *Math. Ann.* 319 (2001): 735–758.

[218] L. Weis and P. Kunstmann, "Maximal L_p-Regularity for Parabolic Equations, Fourier Multiplier Theorems and H^∞-Functional Calculus," in *Functional Analytic Methods for Evolution Equations*, eds. M. Ianelli, R. Nagel, and S. Piazzera, vol. 1855, Lecture Notes in Mathematics, New York, NY: Springer-Verlag, 2004, pp. 65–311.

[219] G. Weiss, "Admissibility of Unbounded Control Operators," *SIAM J. Control Optim.* 27 (1989): 527–545.

[220] ———, "Admissible Observation Operators for Linear Semigroups," *Israel J. Math.* 65 (1989): 17–43.

[221] ———, "Regular Linear Systems with Feedback," *Control Signals Systems* 7 (1994): 23–57.

[222] J. Wu, *Theory and Applications of Partial Functional Differential Equations*, vol. 119, Applied Mathematical Sciences, New York, NY: Springer-Verlag, 1996.

[223] A. Wyler, "Über die Stabilität von Dissipativen Wellengleichungen," Ph.D. diss., Zürich, 1992.

[224] T. Xiao and J. Liang, *The Cauchy Problem for Higher Order Abstract Differential Equations*, vol. 1701, Lecture Notes in Mathematics, New York, NY: Springer-Verlag, 1998.

[225] K. Yosida, "An Operator-Theoretical Integration of the Wave Equation," *J. Math. Soc. Japan* 8 (1956): 79–92.

[226] ———, *Functional Analysis*, Berlin: Springer-Verlag, 1965.

[227] Q. Zheng, "Strongly Continuous M, N-Families of Bounded Operators," *Integral Eqs. Operator Th.* 19 (1994): 105–119.

Index